THE ELECTROCHEMISTRY OF OXYGEN

The Electrochemistry of Oxygen

JAMES P. HOARE

Research Laboratories
General Motors Company
Warren, Michigan

INTERSCIENCE PUBLISHERS

A Division of John Wiley and Sons, New York · London · Sydney · Toronto

CHEMISTRY

SBN 470   401524

Library of Congress Catalog Card Number: 68-16055

Printed in the United States of America

Preface

In the past 100 years, a tremendous amount of work has been done in investigations of the processes occurring at oxygen electrodes. The results of a bewildering number of experimental techniques have been presented along with an equally perplexing maze of theories and explanations to account for the observed results. Since renewed interest in the electrochemistry of oxygen in recent times has been intensified by research in fuel cells and storage batteries, it seemed appropriate to attempt an organized critical examination of this highly controversial field in the light of present-day scientific understanding.

The purpose of this monograph is to present the reader with a consensus of what is known about the oxygen electrode.

The first half of the book is concerned with the electrochemistry of the oxygen reaction occurring at inert electrodes, and the last four chapters deal with the uses and applications of oxygen electrodes and the interaction of oxygen with non-noble metals, particularly those metals which are of interest in battery research. Because such subjects are beyond the scope of this monograph, a discussion of the interaction of oxygen with semiconductors and with so-called valve metals, such as aluminum, tantalum, and zirconium, is not included. Wherever possible, a critical discussion of the points of interest is conducted.

It is hoped that this review of the oxygen electrode will not only be a source of information to those interested in the many facets of the electrochemistry of oxygen, but also will stimulate them to carry out further research in this highly controversial field.

The author is deeply indebted to Prof. P. J. Elving of the University of Michigan and to Prof. I. M. Kolthoff of the University of Minnesota for providing the opportunity and encouragement to write this monograph. Equally indebted is he to the management of the General Motors Corporation, particularly to Mr. John M. Campbell, Assistant to the Vice President of Research, and to Dr. Seward E. Beacom, Head of the Electrochemistry Department of General Motors Research Laboratories, for permission to undertake such a time-consuming task.

He wishes to express his sincere gratitude to Dr. Raymond Thacker of the Electrochemistry Department of General Motors Research Laboratories for reading the complete manuscript and for his many valuable suggestions and criticisms.   Since the first five chapters are an expansion and updating of the chapter, "The Oxygen Electrode on Noble Metals," appearing in Volume 6 of *Advances in Electrochemistry and Electrochemical Engineering*, edited by Prof. Paul Delahay, the author wishes to take this opportunity to express his thanks to Prof. Paul Delahay of New York University, Prof. A. N. Frumkin of the Institute of Electrochemistry, Academy of Sciences of the U.S.S.R., Moscow, and Prof. Brian E. Conway of the University of Ottawa for their many helpful comments and suggestions.   Mr. Sigmund Schuldiner of the U.S. Naval Research Laboratory read the manuscript and offered valuable comments which the author is most happy to acknowledge.   Mrs. Jeanne Burbank, also of the U.S. Naval Research Laboratory, contributed most helpful comments to Chapter VII, and Dr. R. R. Witherspoon of the Electrochemistry Department of General Motors Research Laboratories provided able assistance in the preparation of Chapter VIII with many valuable discussions. Prof. I. M. Kolthoff was most obliging by providing sources of material for the development of Chapter VI.   Chapter IX was critically read by Dr. Roger L. Saur and Mr. James D. Thomas of the Electrochemistry Department of General Motors Research Laboratories. Sincere thanks are offered to the following for their helpful comments and suggestions: Dr. J. J. Lander of Delco-Remy Division of General Motors Corporation; Dr. Paul Winter of the Chemistry Department of General Motors Research Laboratories; Mr. D. W. Hardesty, Dr. S. G. Meibuhr, Dr. S. M. Selis, Mr. T. F. Sharpe, Mr. G. F. Wheeler, and Mr. J. P. Wondowski of the Electrochemistry Department of General Motors Research Laboratories.   The authorization from societies and publishers to reproduce the various diagrams of distinguished authors is duly acknowledged.

The author is particularly grateful to Miss G. Sobieska for her expert editing and preparation of the manuscript.   The help of Miss S. K. Giulioli and Miss N. J. Schim is most appreciated.   To Mr. G. H. Tucker, Mr. K. O. Wetter, and their associates of the Publications Section of the Technical Information Department of General Motors Research Laboratories under Mr. A. P. Bohn for their outstanding cooperation and skill in producing the manuscript in its final form and to Mr. J. C. Seifert of the Photographic Section for his excellent photographic skills, he wishes to express his sincere gratitude. The library staff of General Motors Research Laboratories under Mr. R. W. Gibson was most cooperative.

Finally, the author wishes to acknowledge his deep-felt thanks to his wife, Terry, his mother, Mary, and his children, Karen, Patrick, and John for their sympathetic understanding and loyal support during the long hours required for the writing of the manuscript.

JAMES P. HOARE

May, 1968
Warren, Michigan

Contents

chapter I

Introduction

In recent years, through technological advances made in the fields of electronics and solid-state physics, it has been possible to make more precise measurements on a given electrochemical system using novel current interrupter, potentiostatic, potential sweep, and a variety of fast pulse techniques. A study of the literature presents the reader with a bewildering array of experimental results obtained by this wide variety of techniques and with an equivalent collection of theories derived from or developed otherwise to explain these results. Considerable interest in the properties of the oxygen electrode has been generated by fuel cell investigations and bioelectrochemical studies. It is the purpose of this monograph to arrive at a consensus of what is known about the nature of the electrode processes occurring at an oxygen electrode.

Since oxygen is the most abundant of the elements on earth and interacts chemically with all the other elements except the noble gases, many of the processes most important to man's day-to-day existence and to his creature comforts involve reactions with oxygen. It is no wonder, then, that a knowledge of the mechanisms of oxygen reactions is most desirable.

A. FUNDAMENTAL CONCEPTS IN ELECTROCHEMISTRY

Because this monograph is primarily concerned with the electrochemistry of oxygen, one must make a distinction between a chemical and an electrochemical process. A chemical reaction occurs when electrons are exchanged between the reacting chemical species at the reaction site, that is, one reaction partner loses electrons (oxidation) while the other gains electrons (reduction). In an electrochemical process, the oxidation and reduction processes occur at different sites remote from one another, a situation which produces a separation of charge. Physically, this is accomplished by separating by an electrolyte the sites at which oxidation occurs from those at which reduction occurs. Charge is collected at the reaction sites by electronic conductors called electrodes; at the anode, oxidation takes place, and at the cathode, reduction. Current is transported through the electrolyte by ions, and charge is transferred from a particle in solution across the electrical double layer to the electrode or from a particle chemisorbed on the conducting surface to the electrode.

At all phase boundaries a double layer of charge builds up because the activities of an ion common to both phases is different in each phase. The resulting free energy difference causes ions to drift from the more concentrated phase to the less, which in turn produces a separation of charge.   Finally a steady state is reached where the driving force of the concentration gradient across the phase boundary is equal to the electrostatic forces of the separation of charge.   In one phase on one side of the boundary there exists a layer of charged particles, and in the other phase on the opposite side of the boundary is a layer of oppositely charged particles.   This structure of parallel planes of opposite charge existing at the phase boundary is called the electrical double layer.   Its existence gives rise to contact potentials at a solid–solid boundary, junction potentials at a liquid–liquid boundary, and electrode potentials at a solid–liquid boundary.

In the field of electrochemistry, most attention is directed toward the solid–liquid phase boundary, and when reference is made to the electrical double layer it is to the double layer at the solid–liquid boundary that is meant.   Since the dielectric constant of the metal is approximately infinite, all the charge exists on the electrode surface, but on the solution side a distribution of charge extends from the phase boundary into the bulk of solution.   Much of the charge distribution appears in the region about 4 or 5 Å from the metal surface across which a linear drop in potential occurs.   This region is called the Helmholtz double layer region.   The rest of the charge appears in the diffuse double layer region where the potential falls off exponentially to the value in the bulk of solution.   In colloid chemistry, the diffuse part of the double layer corresponds to the zeta potential.   Grahame (17) visualized the Helmholtz double layer as consisting of two subsections, the outer plane which corresponds to the closest approach of a particle to the electrode surface without being specifically adsorbed and the inner plane which corresponds to the plane constructed through the centers of the specifically adsorbed particles.

During the course of an electrochemical process, charge is transferred across the Helmholtz double layer, and the potential across this part of the double layer influences the rate of the electrochemical reaction.   For dilute solutions, the potential drop across the diffuse layer may be considerable, and a correction to the measured total electrode potential must be made.   If, however, a supporting electrolyte is used, the contribution to the total electrode potential by the potential across the diffuse double layer is reduced to a negligible quantity.

It must be kept in mind that the particles transporting charge in the bulk of the electrolyte may not necessarily be the same particles from which electrons are transferred to the electrode. Electrolytic or ionic conduction is always accompanied by a chemical change at a phase boundary in contrast to electronic conduction which is only accompanied by a physical change, such as the appearance of heat, light, or a magnetic field. The sum of the anodic and cathodic reactions gives the overall electrochemical reaction which is identical to the chemical equation written for the corresponding chemical reaction.

Whenever a separation of charge exists between two points, a difference of potential will be set up between these points since it requires work to transport a unit charge from one point to the other. The potential difference between the anode and cathode is known as the electromotive force, emf, of the cell. When the anode and cathode are connected by an electronic conductor, such as a copper wire, current will flow through this external circuit, and useful energy may be taken from the cell. Thus, chemical energy is converted to electrical energy in an electrochemical cell, and if more than one cell is connected in series or parallel, the system is known as a battery. The reverse process of the electrochemical cell is electrolysis in which electrical energy is transformed into chemical energy. An example of these processes is found in the fuel cell in which hydrogen is burned electrochemically in a cell composed of a hydrogen anode and an oxygen cathode in alkaline or acid electrolytes to produce electrical energy. In the reverse process, electrolysis of water, electrical energy from an external source may be applied to the cell to produce hydrogen at the cathode and oxygen at the anode from the decomposition of water in the acid or alkaline electrolyte. When current is drawn from the cell, potential losses produced by the internal resistance of the cell and by the existence of activation energy barriers along the electrochemical reaction path are incurred; and now the potential difference across the cell is called the terminal voltage, which is always less than the emf.

If the cell potential is studied as a function of the various parameters of the system, little useful information about the electrode reactions can be obtained. This is true because the cell potential is the algebraic sum of the individual electrode potentials. When a change is made in one of the experimental parameters, it is not known whether the observed changes in the cell potential are produced by changes in the oxidation, reduction, or both electrode potentials. Therefore, it is important to make single electrode potential measurements.

The potential difference of interest in electrochemical measurements is the Galvani potential difference, φ, which is defined as the potential difference between a point in the bulk of phase A and a point in the bulk of phase B. However, φ is not a measurable quantity. In fact, as pointed out by Guggenheim (18), it cannot be calculated theoretically because the potential is a function of the free energy, G, which is related to the chemical potential, μ, which in turn is a function of the activity of a single ion which is not defined. On the other hand, the Volta potential, ψ, which is defined as the potential difference between a point just outside of phase A and a point just outside of phase B, is a measurable quantity. Such measurements may be obtained with an electrometer. The Galvani potential is related to the Volta potential by the surface potential, χ, defined as the potential difference between a point on the surface of phase A and a point on the surface of phase B, according to the equation

$$\varphi = \psi + \chi \qquad (1.1)$$

Unfortunately, χ is also an unmeasurable quantity; and hence φ cannot be determined, which means that an absolute electrode potential cannot be obtained.

What can be measured is the difference of Galvani potential differences, $\Delta\varphi$, which is the quantity measured by a voltmeter between the anode and cathode of an electrochemical cell. If one wishes to measure changes in the potential of a single electrode, say, the cathode, a third electrode called a reference electrode must be employed. Since the potential of the reference electrode is the reference point for the experimental measurements and as a result must have a stable value, it is desirable to choose a highly reversible electrode system. In this case the reference potential will not deviate from the stable value because of interactions of the reference system with the experimental environment. The primary reference standard is the normal hydrogen electrode (NHE) which is defined as the potential of a platinum wire in a solution of hydrogen ions at unit activity over which hydrogen gas at unit fugacity is bubbled. This potential is arbitrarily taken as zero volt at all temperatures. The equilibrium open-circuit potential of all other systems is measured with respect to NHE; and if the activities or fugacities of the reacting species in the electrode reaction are unity at 25°C, the potential is the standard potential, E_0. The open-circuit rest potential, E, is related to the activities of the reacting species, a_i, by the Nernst relationship,

$$E = E_0 - \frac{RT}{nF} \ln \prod_i a_i$$

where F is the Faraday and n is the number of electrons exchanged. Usually the NHE is not a convenient reference electrode system for most practical cases, so a stable secondary reference electrode system which does not involve gas bubbling is employed. Some common secondary reference electrode systems are: Hg/HgO, 0.098 V; Hg/Hg$_2$SO$_4$, 0.615 V; saturated calomel (SCE), 0.2445 V; and Ag/AgCl, 0.222 V vs. NHE.

The E_0 value is an equilibrium value and is related to the standard free energy, $\Delta G°$, of the electrode reaction, $\Delta G° = nFE_0$. Under these conditions, the rates of the forward and reverse reactions are equal. Since the rate of an electrochemical reaction is identical to the current density, i, the rate at equilibrium is designated as i_0 and is known as the exchange current density, $i_0 = \overrightarrow{i} = \overleftarrow{i}$.

A word should be said about sign conventions at this point. According to the Gibbs-Stockholm convention, the standard electrode potential, E_0, of a given electrode reaction is given a positive sign if the electrode is the positive terminal of a cell in which the counterelectrode is the normal hydrogen electrode. In the overall cell reaction, the spontaneous electrode process taking place at such an electrode is a reduction process. Since the cell potential for the spontaneous cell reaction is positive, the potential of this electrode reaction, written as a reduction process (electrons on the left-hand side), is given a positive sign. For example, the Ag/Ag$^+$ couple is more noble than hydrogen so that the potential of the reaction written as Ag$^+$ + e → Ag is positive. If the electrode reaction were written as an oxidation process, the potential would be negative as used by Latimer (27) but opposite in sign to that used by de Bethune (8). In general, electrode reactions will be written as reduction processes, and the Gibbs-Stockholm convention will be used throughout this work unless noted otherwise. Van Rysselberghe (38) has presented the latest conventions and definitions of electrochemical terms as adopted by CITCE.

In certain cases the material of the electrode enters into the electrochemical reaction such as in the Cu/Cu^{2+} system whose reaction is

$$Cu^{2+} + 2e \rightleftharpoons Cu \qquad (1.2)$$

In other cases, the electrode material is inert to the electrochemical system. Such an inert electrode acts only as the catalytic surface on which the electrochemical reaction takes place and as a source or sink of electrons. An example is the hydrogen electrode in which the Pt wire does not enter into the overall reaction,

$$2H^+ + 2e \rightleftharpoons H_2 \qquad (1.3)$$

The value of i_0 is a measure of the catalytic activity of the surface of an inert electrode for a given electrode reaction.

A third type of electrode in common usage is one in which the metal of the electrode is in equilibrium with an ion in solution other than the metal ion.   This is accomplished by covering the metal surface with a slightly soluble salt of the metal and placing the electrode in a solution containing a soluble salt of the common anion.   Examples are the Ag/AgCl, Cl$^-$ and the Hg/HgO, OH$^-$ electrodes whose reactions are

$$AgCl + e \rightleftharpoons Ag + Cl^- \tag{1.4}$$

or

$$HgO + H_2O + 2e \rightleftharpoons Hg + 2OH^- \tag{1.5}$$

When current is drawn from a cell, the electrode reaction proceeds in the forward direction, i.e., in the direction in which the change in free energy, ΔG, is negative.   Because activation energies are involved in electrode reactions, energy must be used in traversing the energy barrier.   Consequently, the electrode potential is shifted from its value when current is not flowing.   The magnitude of this potential shift is called the electrode or electrochemical polarization.   Electrochemical polarization may arise from kinetic considerations (activation polarization) or from mass transfer effects (concentration polarization).   Under polarographic conditions, concentration polarization is demanded since it is required that the electrode process be diffusion controlled.   In this case, the concentration of the test species in solution is related to the limiting current.   In most cases, concentration polarization may be minimized to a negligible influence by adequate stirring of the electrolyte so that electrode kinetic and activation polarization studies may be made.   Vetter (39) recognizes two types of activation polarization; first, when the rate-controlling step involves the transfer of an electron (electron-transfer polarization) and, second, when a chemical step preceding or following the electron transfer step is rate determining (chemical polarization).   In general, electrode polarization is a measure of the activation energy accompanying the processes involved in the overall electrochemical reaction.

If the potential of the reference electrode is the thermodynamic reversible potential of the system to be investigated, the activation polarization is known as the overvoltage, η, i.e.,

$$\eta = E - E_{rev} \tag{1.6}$$

It may be seen that η is a measure of how far the system is removed from equilibrium by the passage of a given current density and is uniquely defined only for a given value of i.

Experimentally, the overvoltage is determined by measuring the potential difference with a high input impedance voltmeter between the test electrode and an identical electrode as a reference electrode if a reversible system such as H_2 on Pt is being studied. Otherwise, some other reversible secondary reference electrode, such as SCE, may be used if the potential difference between it and the potential of the thermodynamic reversible test electrode is known. A high input impedance voltmeter is required so that current will not be drawn between the test and reference electrodes producing a measurable polarization of the reference electrode.

It can be shown that the current density is related to the overvoltage by the following expression:

$$i = i_0 \left(e^{-A\eta} - e^{B\eta}\right) \tag{1.7}$$

where A and B are constants. At large positive or negative values of the overvoltage, one of the exponential terms drops out, and η is a logarithmic function of i,

$$\eta = a + b \log i \tag{1.8}$$

Equation 1.8 is known as Tafel's equation, and the slope of an $\eta - \log i$ plot is known as the Tafel slope, b. Under certain conditions, b may take on characteristic values for various mechanisms. As an example, it is often found that for a mechanism in which an electron transfer step is rate controlling, b will have a value close to 0.12. However, it is dangerous to base a proposed mechanism solely on the b value without taking into consideration additional independent data.

If an electrode reaction were truly reversible, an activation energy barrier would not exist, and as a result, η would go to zero. When the Tafel slope is extrapolated to a point on the $\log i$ axis where $\eta = 0$, the value of i determined at the intercept is i_0. This procedure for finding i_0 assumes, of course, that the electrode reaction at very low current densities is the same as that at high current densities. When this assumption is not valid, the value of i_0 so determined may have questionable significance.

At low current densities where $\eta < 1$, the values of the two exponential terms in Eq. 1.7 have comparable magnitudes, and one may not be neglected with respect to the other. This is interpreted to mean that the back reaction contributes significantly to the overall current density,

$$i = \overrightarrow{i} - \overleftarrow{i} \tag{1.9}$$

Under these conditions, the exponentials may be expanded in series form;

and if only the first two terms of the series are retained, it is found that η is a linear function of i; $i = -i_0(A+B)\eta$ or

$$\eta = Ki \qquad (1.10)$$

The slope of the linear plot of η as a function of i gives a value of K which is a measure of the rate of the overall electrode reaction.

The constants, A and B in Eq. 1.7, are equal to $\alpha F/RT$ and $(1-\alpha)F/RT$ for a one-electron process, where α is the transfer coefficient sometimes referred to as the symmetry factor. αE represents that fraction of the electrode potential, E, which drives the electrode reaction in the forward direction and $(1-\alpha)E$, that in the reverse direction. If the potential energy barrier is symmetrical, α would have a value of $\frac{1}{2}$, and experimentally, most electrode reactions are found to have values of α close to 0.5. In passing, one may mention one other kinetic parameter, the stoichiometric number, ν, which is defined as the number of times the rate-determining step must take place for one complete occurrence of the electrode process.

Since a dimer of the electron is not known, electron transfer steps involving the exchange of only one electron are considered possible. Consequently, any electrode mechanism involving steps in which more than one electron is transferred simultaneously is rejected.

All solid-phase surfaces are physically and possibly chemically inhomogeneous, and as a result, the active centers or active sites on a metal surface are different energetically. It is possible that on the same metal surface an oxidation partial reaction may take place at one site while a reduction partial reaction may take place at another. Since the metal surface is a good electronic conductor, a separation of charge cannot be maintained between the anodic and cathodic sites, and consequently this surface is an equipotential surface. On such an equipotential surface, the anodic and cathodic reactions must occur at the same potential which will take on a value between the calculated open-circuit values, E, of the individual reactions. In other words, the potential of each partial reaction is polarized or shifted from the open-circuit values to this common potential known as a mixed potential. If the metal is in contact with an electrolyte, electrons will flow in the metal and ions will migrate in the solution between the anodic and cathodic sites. The current flowing between the anodic and cathodic sites is called the local cell or corrosion current, and this electrochemical system is referred to as a local cell. Corrosion processes proceed by local cell mechanisms. If the metal is removed from contact with the electrolyte and maintained in a dry condition, the circuit is broken and the corrosion process cannot take place. Such electrodes at which more than one electrode process takes place are called polyelectrodes.

B. TOPICS COVERED

Although oxygen interacts with all metal surfaces, the noble metals make the closest approximation to an inert electrode material for the construction of an oxygen electrode. With this in mind, the first chapters of this monograph are devoted to a survey of what is known about the oxygen reaction,

$$O_2 + 4H^+ + 4e \rightleftarrows 2H_2O \qquad (1.11)$$

on the noble metals. Most work has been carried out on Pt, but a large amount of work has been done with Au and Pd electrodes. Studies of the other noble metals are much less abundant. Following the initial chapters which are concerned with the nature of the oxygen electrode, a chapter on the electroanalytical chemistry of oxygen is presented. This involves a survey of the electrochemical methods of detecting and determining oxygen. A discussion of the electroanalytical techniques themselves is not included because adequate coverage of such topics may be found in numerous monographs. A chapter is included concerning the nature of metal–metal oxide electrode systems important in various commercial battery systems. A chapter devoted to a discussion of the oxygen diffusion electrode and applications to the fuel cell follows. The role of oxygen in corrosion mechanisms is the subject of the last chapter.

Very little will be said about the interaction of oxygen with the surfaces of the so-called valve metals, Ta, Nb, Al, Zr, Hf, W, Bi, and Sb, and the formation of barrier oxide layers, since this subject is well covered by Young's monograph (44). The reader is referred to past reviews of the oxygen electrode in which the subject is treated from the viewpoint of a redox system (39), adsorption kinetics (4), and mixed potentials (19).

C. LITERATURE

For those interested in an understanding of the fundamentals of electrochemistry, three standard textbooks (24,31,35) are recommended. Vetter's book (39) gives an excellent comprehensive coverage of the entire field of electrochemical kinetics, and the recent book by Delahay (10) presents the reader with a thorough, critical analysis of the current understanding of electrode kinetics. A new book on electrode kinetics by Conway (7) is now available. The only monograph on the double layer is that written by Delahay (10), although several reviews (1,11a, 14,15,17,30,34) may be mentioned. Two collections of authoritative periodic reviews (3,11) of various topics in electrochemistry and several collections of symposia (5,12,21,36,42,45) are available. For compendia of electrochemical data, the reader is referred to the following references (6,8,27,33).

Ives and Janz (20) have written an excellent book on the kinds and uses of reference electrodes, and Delahay (9) has written a clear account of the standard experimental techniques used in electrochemical investigations. A remarkably comphrehensive treatment of the dropping mercury electrode and its application to electroanalytical systems may be found in the two-volume work of Kolthoff and Lingane (23). The wide variety of experimental techniques applied in the field of electroanalytical chemistry are adequately covered by Lingane (29). The classical work of Levich (28) is now available in an English translation. A two-volume survey of the advances in polarographic studies edited by Zuman and Kolthoff (46) covers the field up to about the year 1960.

A most complete treatment of the bewildering array of corrosion systems is clearly presented in a book written by Evans (13), and an authoritative account of mixed potentials may be obtained from the classic paper of Wagner and Traud (41).

In recent years a number of books have been written on fuel cells, but the best general treatment in English is to be found in William's book (41a). There exist four collections of papers (2,16,43) taken from symposia on fuel cells sponsored by the American Chemical Society. Four other important books are by Mitchell (32), Justi and Winsel (22), Vielstich (40), and Marechal (30a).

Finally, definitions of electrochemical terms may be found in papers of Lange and co-workers (25,26), of Guggenheim (18), of Rüetschi and Delahay (37), and of Van Rysselberghe (38).

References

1. Agar, J. N., and J. E. B. Randles, *Ann. Rept. Chem. Soc.*, **51**, 103 (1954).
2. Baker, B. S. *Hydrocarbon Fuel Cell Technology*, Academic Press, New York, 1965.
3. Bockris, J. O'M., Ed., *Modern Aspect of Electrochemistry*, Butterworths, London, Vol. I, 1954; Vol. II, 1959; Vol. III, 1964; Vol. IV, 1966.
4. Breiter, M. W., in *Advances in Electrochemistry and Electrochemical Engineering*, Vol. 1, P. Delahay, Ed., Interscience, New York, 1961, p. 123.
5. *Can. J. Chem.* **37**, 120–323 (1959).
6. Conway, B. E., *Electrochemical Data*, Elsevier, Amsterdam, 1952.
7. Conway, B. E., *Theory and Principles of Electrode Processes*, Ronald Press, New York, 1965.

8. de Bethune, A. J., and N. A. S. Loud, *Standard Aqueous Electrode Potentials and Temperature Coefficients*, Clifford A. Hampel, Skokie, Ill., 1964.
9. Delahay, P., *New Instrumental Methods in Electrochemistry*, Interscience, New York, 1954.
10. Delahay, P., *Double Layer and Electrode Kinetics*, Interscience, New York, 1965.
11. Delahay, P., and C. W. Tobias, Eds., *Advances in Electrochemistry and Electrochemical Engineering*, Interscience, New York, Vol. 1, 1961; Vol. 2, 1962; Vol. 3, 1963; Vol. 4, 1966; Vol. 5, 1967; Vol. 6, 1967.
11a. Devanthan, M. A. V., and B. V. K. S. R. A. Tilak, *Chem. Rev.*, **65**, 635 (1965).
12. *Discussions Faraday Soc.*, **1**, 1–338 (1947).
13. Evans, U. R., *The Corrosion and Oxidation of Metals*, St. Martin's Press, New York, 1960.
14. Frumkin, A. N., *Z. Elektrochem.*, **59**, 807 (1955).
15. Frumkin, A. N., *J. Electrochem. Soc.*, **107**, 461 (1960).
16. Gould, R. F., *Fuel Cell Systems*, Am. Chem. Soc., Applied Pub., Washington, D. C., 1965.
17. Grahame, D. C., *Chem. Rev.*, **41**, 441 (1947).
18. Guggenheim, E. A., *J. Phys. Chem.*, **33**, 842 (1929); **34**, 1758 (1930).
19. Hoare, J. P., in *Advances in Electrochemistry and Electrochemical Engineering*, Vol. 6, P. Delahay, Ed., Interscience, New York, 1967, p. 201.
20. Ives, D. J. G., and G. J. Janz, *Reference Electrodes*, Academic Press, New York, 1961.
21. *J. Chim. Phys.*, **49**, C3–C218 (1952).
22. Justi, E. W., and A. W. Winsel, *Kalte Verbrennung*, F. Steiner, Weisbaden, 1962.
23. Kolthoff, I. M., and J. J. Lingane, *Polarography*, Vols. 1 and 2, 2nd ed., Interscience, New York, 1952.
24. Kortüm, G., and J. O'M. Bockris, *Textbook of Electrochemistry*, 2 vol., Elsevier, Amsterdam, 1951.
25. Lange, E., and H. Göhr, *Z. Elektrochem*, **63**, 74 (1959).
26. Lange, E., and P. Van Rysselberghe, *J. Electrochem. Soc.*, **105**, 420 (1958).
27. Latimer, W. M., *Oxidation Potentials*, 2nd ed., Prentice-Hall, Englewood Cliffs, N. J., 1952.
28. Levich, V. G., *Physicochemical Hydrodynamics*, Prentice-Hall, Englewood Cliffs, N. J., 1962.
29. Lingane, J.J., *Electroanalytical Chemistry*, 2nd ed., Interscience, New York, 1958.
30. Macdonald, J. R., and C. A. Barlow in *Proceedings of First Australian Conference on Electrochemistry*, A. Friend and F. Gutmann, Eds., Pergamon Press, Oxford, 1964, p. 199.
30a. Marechal, A., *Les Piles à Combustile*, Éditions Technip, Paris, 1965.
31. Milazzo, G., *Electrochemistry*, English translation by P. J. Mill, Elsevier, Amsterdam, 1963.
32. Mitchell, W., *Fuel Cells*, Academic Press, New York, 1963.
33. Parsons, R., *Handbook of Electrochemical Data*, Butterworths, London, 1959.
34. Parsons, R., in *Advances in Electrochemistry and Electrochemical Engineering*, Vol. I, P. Delahay, Ed., Interscience, New York, 1961, p. 1.
35. Potter, E. C., *Electrochemistry*, Macmillan, New York, 1956.
36. *Reports of 4th Soviet Conference on Electrochemistry*, 1956; 3 vol., English translation by Consultants Bureau, New York, 1958.
37. Rüetschi, P., and P. Delahay, *J. Chem. Phys.*, **23**, 697 (1955).

38. Van Rysselberghe, P., *Electrochim. Acta*, **5**, 28 (1961); **8**, 543 (1963); *J. Electroanal. Chem.*, **2**, 265 (1961).
39. Vetter, K. J., *Elektrochemische Kinetik*, Springer, Berlin, 1961, English translation by Scripta Technica, Academic Press, New York, 1966.
40. Vielstich, W., *Brennstoffelemente*, Verlag Chemie, Weinheim, 1965.
41. Wagner, C., and W. Traud, *Z. Elektrochem.*, **44**, 391 (1938).
41a. Williams, K. R., *An Introduction to Fuel Cells*, Elsevier, New York, 1966.
42. Yeager, E., Ed., *Transactions of Symposium on Electrode Processes*, Wiley, New York, 1966.
43. Young, G. J., *Fuel Cells*, Reinhold, New York, Vol. I, 1960; Vol. II, 1963.
44. Young, L., *Anodic Oxide Films*, Academic Press, New York, 1961.
45. *Z. Elektrochem.*, **59**, 591–822 (1955).
46. Zuman, P., and I. M. Kolthoff, *Progress in Polarography*, Interscience, New York, 1962.

chapter **II**

The Oxygen Electrode at Rest

A. THEORETICAL CALCULATED VALUES

Before one can investigate a given electrochemical system intelligently, one must have some understanding of the behavior of the system in a standard state. The standard state for a given electrode couple is the condition that the activities of all participating species in the electrochemical reaction be at unit activity at a temperature of 25°C. The equilibrium potential of this system under these conditions is the standard electrode potential, E_0, and is equal to the emf of a cell composed of the given electrode couple and a standard (normal) hydrogen electrode (NHE) since the E_0 for the NHE is arbitrarily taken as zero volt at all temperatures.

The connecting link between electrochemical and thermodynamic concepts is the relationship

$$\Delta G = -nFE \tag{2.1}$$

where ΔG is the free energy change of the electrode reaction; n, the number of electrons transferred; E, the electrode potential; and F, the Faraday or the charge on one mole of univalent ions (96,484 coulombs). From the mass action law, it may be shown that

$$E = E_0 - \frac{RT}{nF} \ln \prod_i a_i^{\nu_i} \tag{2.2}$$

where R is the gas constant; T, the absolute temperature; a_i, the activities of the species participating in the electrode reaction; and ν_i, the stoichiometric coefficient. Equation 2.2 is known as the Nernst relationship.

The electrode reaction for the oxygen electrode is

$$O_2 + 4H^+ + 4e \rightleftharpoons 2H_2O \qquad E_0 = 1.229 \text{ V} \tag{2.3}$$

in acid solutions, and

$$O_2 + 2H_2O + 4e \rightleftharpoons 4OH^- \qquad E_0 = 0.401 \text{ V} \tag{2.4}$$

in alkaline solutions. If the oxygen couple, either Eq. 2.3 or 2.4, is measured against a hydrogen electrode,

$$H^+ + e \rightleftharpoons \tfrac{1}{2}H_2 \qquad E_0 = 0.0 \text{ V} \tag{2.5}$$

13

in acid solutions, or

$$H_2O + e \rightleftarrows \tfrac{1}{2}H_2 + OH^- \qquad E_0 = -0.828 \text{ V} \qquad (2.6)$$

in alkaline solutions, it is seen that the overall cell reaction requiring the transfer of four electrons is

$$O_2 + 2H_2 \rightleftarrows 2H_2O \qquad (2.7)$$

The measured emf of a cell composed of a hydrogen and an oxygen electrode, the so-called Grove cell (99), is independent of pH according to Eq. 2.7, this was confirmed experimentally by the investigations of Smale (224).

Electrical contact is made in the case of a gas electrode couple (e.g., Eq. 2.3) by an inert electrode around which the gas is bubbled and which is immersed in the given electrolyte. According to thermodynamic principles, the inert electrode acts only as a source or sink of the electrons and the catalytic surface on which the electrode reaction takes place. In earlier times, it was thought that the noble metals were excellent inert materials from which an oxygen electrode could be made.

At first, a value of 1.08 V* was generally accepted (48,224,256) as the emf of the oxygen–hydrogen cell, but subsequent work by Bose (25), Wilsmore (256), and Czepinski (49) indicated that this potential should be closer to a value of 1.10 V. In all cases, the potentials were obtained from the direct measurement of the potential of an oxygen electrode against a hydrogen electrode in the same solution.

Lewis (159) determined a value of 1.217 V for the hydrogen–oxygen cell from the measured potentials of the Ag/Ag_2O, OH^-, and the Ag/Ag^+ couples corrected for differences in the concentration of Ag^+ and OH^- ions by the Nernst relationship. From the measured potential of the cell, Hg/HgO, $NaOH$, H_2/Pt, and the heat of dissociation of HgO, Brönsted (36) calculated a value of 1.238 V. Using the thermal data obtained for the heat of dissociation of water vapor at high temperatures, Nernst and von Wartenberg (186) arrived at a value of 1.225 V. Later Lewis and Randall (160) recorded a value of 1.227 V determined from the free energy of formation of H_2O calculated from specific heat data.

Finally, the best recorded values (156) for the standard heat and entropy of formation of liquid water are $-68,317$ cal/mole and -39.0 cal deg^{-1} mole^{-1}, respectively, which give a value of $-56,900$ cal/mole for the standard free energy of formation of liquid water. According to Eq. 2.1, the standard potential of the hydrogen–oxygen cell, Eq. 2.7,

---

* Unless stated otherwise, all potentials are recorded against the NHE and all species adsorbed on the electrode surface are considered to be hydrated.

is 1.229 V.   It has been defined that the free energy of formation of the elements and H^+ ion in the standard state (unit activities) is zero. Consequently, the standard potential of O_2/H_2O couple, Eq. 2.3, in acid solutions is 1.229 V.   If a value of $-37,595$ cal/mole (156) is taken for the standard free energy of formation of OH^- ion in solution, the standard potential of the O_2/H_2O couple, Eq. 2.4, in alkaline solution is 0.401 V.   It is seen that a Pt/H_2 electrode in an alkaline solution of unit OH^- ion activity is -0.828 V.   These potential values are the present-day accepted reversible potentials for the O_2/H_2O couple (55,156).

B. OBSERVED EXPERIMENTAL VALUES

The first person to use an oxygen electrode was Grove (99), who fabricated it from a piece of platinized Pt gauze immersed in an acid solution. Oxygen was bubbled over the gauze.   The potential of the oxygen electrode was measured against a hydrogen electrode in the same solution made by bubbling H_2 over a platinized Pt gauze.   Not only did Grove show that current could be drawn from this combination of electrodes, but also, that if a battery were applied to two Pt electrodes in an acid electrolyte, oxygen would be evolved at the anode, while hydrogen would be evolved at the cathode.   The electrode process was reversible.   It is considered that this is the beginning of fuel cell research.

Early values obtained for the open-circuit potential of the Grove cell range from about 1 V to a high value of about 1.12 V.   Without special preparation, the potential of a Pt/O_2 electrode is observed to be about 1.06 V even in modern times (116).   Such large discrepancies between the experimental and the theoretical values for the reversible oxygen potential may be traced to the fact that the noble metals are not truly inert to saturated electrolytes.   As early as 1898, Mond, Ramsey, and Shields (178) concluded from measuring the heat evolved during the adsorption of O_2 on Pt black that a layer of $Pt(OH)_2$ was formed on the Pt surface.

Even if the electrode surface were inert, there is the problem that activation energy barriers exist along the electron reaction path.   To overcome these barriers, either the potential hump may be lowered by using a more active catalyst, or the fraction of reactant molecules with the energy required to surmount the activation peak may be increased by raising the temperature of the system.   Lewis (159) recognized early the important role of the electrode surface as a catalyst for the electrode reaction when he pointed out that to reach the reversible oxygen potential a more powerful catalyzer than Pt black was needed.   On the other hand, Haber and co-workers (105–107) used a cell composed of a thin film of noble metal deposited on both sides of a glass plate. The plate was sealed in a glass tube with O_2 on one side and H_2 on the other, and the cell was heated to temperatures ranging between 340° and 575°C. By replacing the glass with porcelain, temperatures between 800° and 1105 °C were investigated. With glass electrolytes a cell potential of 1.167 V at 340°C was found and 1.165 V at 475°C on Au, but porcelain electrolyte cells gave lower values. These potentials are within 70 mV of the theoretical potential of the Grove cell.

Since it was realized that the noble metals are not inert to oxygen-saturated electrolytes, electrochemists have occupied themselves with a variety of investigations leading to various explanations for the nature of the electrode phenomena responsible for the experimentally observed irreversibility and poor reproducibility of the oxygen electrode on the noble metals. Some of these explanations will now be considered. What follows will pertain chiefly to the Pt/O_2 system. The other noble metals will be considered in later sections.

1. The Oxide Theory

After the experiments of Mond, Ramsay, and Shields (178), it was only natural that the first attempts to explain the observed potentials of oxygen electrodes revolved about the so-called "oxide theory." According to this theory, the electrode is covered with an oxide film, and the open-circuit or rest potential is determined by the oxide/oxygen couple instead of the metal/oxygen couple. Usually the test electrode was anodized to a given potential, after which the circuit was opened and a constant cathodic current was applied to the system. The potential of the various metal–metal oxide couples was correlated with the potentials corresponding to the plateaus in the potential–time discharge curves. The standard potentials of such metal–metal oxide systems were either calculated from the heats and free energies of formation of chemically prepared oxides or were measured experimentally on electrodes prepared either by pasting oxides on metal screens or by forming slurries of the suspended oxides in the given electrolyte.

The oxide theory was described by Lorenz and Hauser (165,167), who were the first to use the term. Haber (104) concluded that the low potentials of the Grove cell, as compared with the potentials of the high temperature glass cell, were caused by the oxidized surface of the Pt electrode in the Grove cell. According to Haber, only with glass or porcelain electrolytes was it possible to obtain a good oxygen electrode.

a. The Oxides of Pt. In a series of articles, Wöhler (259,263,266–268) describes methods for preparing a number of the oxides of Pt. Apparently the only stable oxide of Pt is PtO_2. If $PtCl_4$ is boiled with NaOH (259), a white precipitate is formed when acetic acid is added. As this mixture is boiled, the precipitate turns yellow and has the composition $PtO_2 \cdot 3H_2O$, which is soluble in alkalis and gives complex aquosalts. The trihydrate loses a molecule of water over sulfuric acid to give the brown compound $PtO_2 \cdot 2H_2O$, which is also soluble in alkalis. A black monohydrate, $PtO_2 \cdot H_2O$, is obtained by heating $PtO_2 \cdot 2H_2O$ at 100°C, and this compound is very insoluble in all solvents, including HCl and aqua regia. By heating $PtO_2 \cdot H_2O$ further, anhydrous black PtO_2 may be formed, although some PtO_2 is decomposed (263) in the process. Lorenz and Spielmann (169) report the making of $PtO_2 \cdot 4H_2O$ by boiling H_2PtCl_6 in NaOH, filtering the solution, cooling, and adding acetic acid. The yellow tetrahydrate, which is soluble in H_2SO_4, is precipitated out. The $PtO_2 \cdot 4H_2O$ is converted to $PtO_2 \cdot 3H_2O$ by drying. A commercial product known as Adams' catalyst is brown and is used for catalytic hydrogenation processes (245). It is made by fusing H_2PtCl_6 with $NaNO_3$ (1,248) and is probably $PtO_2 \cdot 2H_2O$. In the catalytic process the oxide is reduced, and the true catalyst is a high surface area Pt black.

Another oxide, PtO or $Pt(OH)_2$ in the hydrated form, is oxidized in air to PtO_2, so it must be prepared in an inert CO_2 atmosphere (169,259, 263). The $Pt(OH)_2$ is obtained from the action of hot alkali hydroxide on $PtCl_2$ or K_2PtCl_4 (169), or by the reduction of K_2PtCl_6 with SO_2 followed by the treatment with hot KOH (266). Platinous oxide is a gray powder obtained by heating $Pt(OH)_2$ in a CO_2 atmosphere, but even then some of the PtO decomposes. This highly unstable oxide (222) disproportionates to Pt and PtO_2 (226,266).

$$2PtO \rightleftharpoons Pt + PtO_2 \qquad (2.8)$$

Moore and Pauling (179) made an oxide of Pt by fusing $PtCl_2$ with KNO_3 and studied the structure of the resulting compound by x-ray diffraction techniques. The lines they obtained were broad and diffuse, and a structure could not be determined. Since some of the line series resembled those obtained for PdO, they suggested that the compound was PtO. Galloni and Busch (89) are in disagreement with Moore and Pauling because lines obtained from $PtO \cdot 2H_2O$ made by Wöhler's method (266) were not the same and because fusion with nitrates oxidizes Pt to $PtO_2 \cdot 2H_2O$ (1,248). It appears that the diffraction pattern for PtO is similar to that for pure Pt with enlarged lattice parameters (89).

It is reported (18,263,268) that a hydrated sesquioxide, Pt_2O_3, is obtained by the action of NaOH on the complex sulfate of trivalent Pt obtained by dissolving freshly prepared $PtO_2 \cdot 4H_2O$ in H_2SO_4 followed by reduction with oxalic acid. It is a dark brown color, and its properties lie between those of PtO and PtO_2 (ref. 222, p. 1606).

When a solution of $PtO_2 \cdot 3H_2O$ in KOH was electrolyzed at 3.5 V, layers of a yellow-green precipitate flaked off the anode (267). The precipitate was filtered off, and from chemical analysis, the composition $3PtO_3 \cdot K_2O$ was assigned to it. After treating the $3PtO_3 \cdot K_2O$ with dilute acetic acid, the reddish brown solid, PtO_3, was obtained, which is very unstable since a suspension in water continuously evolves oxygen.

Rädlein (193) concluded from measurements of the Volta potential (90) on Pt by a condenser method that three oxides of Pt exist. Only in a vacuum is PtO stable, whereas PtO_2 is stable in air. Under all conditions, it was found that PtO_3 is unstable.

Under a set of poorly defined conditions, Galloni and Roffo (90) report that an oxide formed on a Pt wire heated "red hot" to purify a radon stream by the catalytic removal of O_2 and H_2 has a composition of Pt_3O_4 as determined from x-ray diffraction studies. Reproducible results were not obtained. Waser and McClanahan (251) suggested that the oxide on the heated wire was PtO_2 because they made x-ray diffraction measurements on Pt_3O_4 obtained by fusing Na_2PtCl_6 with Na_2CO_3 according to a method of Jorgensen (142) and the diffraction lines did not agree in the two cases. Wöhler (259) has pointed out that the Jorgensen fusion method produces a mixture of PtO_2 and PtO, and Galloni and Busch (89) note that the lines for Pt_3O_4 obtained by the Jorgensen method are similar to the PtO_2 lines (2,74). Although Busch (40) has reported from x-ray studies that Pt_3O_4 may be obtained by heating Pt wires in pure oxygen at 550°C for 45 days and although Nagel and Dietz (184) calculate a standard potential for a Pt/Pt_3O_4 electrode, it seems that the existence of Pt_3O_4 must be held tentatively. The potential–pH diagram for the Pt/H_2O system is available (240a).

b. The Metal–Metal Oxide Potentials. Both Lorenz and Spielamn (166,168–170,226) and Grube (100) measured the potentials of the metal–metal oxides electrodes constructed from the chemically prepared oxides. A summary of the results obtained by these investigators is presented in Table 2.1. A definite potential was not obtained for a $Pt/Pt(OH)_2$ electrode (100) which is to be expected (226) since $Pt(OH)_2$ is very unstable as mentioned above.

c. Application of Oxide Theory to the Pt Electrode. When the open-circuit potential of a preanodized Pt electrode was followed as a function

TABLE 2.1

Standard Potentials for Various Platinum Oxide Couples
in Acid Solutions[a] as Reported in the Literature

Electrode couple	Lorenz and Spielmann (169,170,226)	Grube (100)	Nagel and Dietz (184)	Latimer (156) and de Bethune (55)	Hoare (116)
Pt/Pt-O					0.88 V
Pt/PtO		0.9 V			
Pt/Pt(OH)$_2$			0.98 V	0.98 V	
Pt/PtO·2H$_2$O	0.95 V	1.04 V			
Pt/PtO$_2$·2H$_2$O	0.96 V				
Pt/PtO$_2$·3H$_2$O	0.98 V				
Pt/PtO$_2$·4H$_2$O		1.06 V			
Pt/PtO$_3$		1.5 V			
Pt/PtO$_4$		>1.6 V			
Pt/Pt$_3$O$_4$			1.11 V		
Pt(OH)$_2$/PtO$_2$			1.1 V	1.1 V	

[a] The potential values in alkaline solutions may be obtained by subtracting 0.83 V from the corresponding value in acid solutions.

of time, plateaus corresponding to any of these measured oxide potentials could not be found. However, many of the early investigators (77, 147,185,209,228) subscribe to some form of the oxide theory. Foerster (77) observed that the potential of a Pt electrode anodized to 1.5 V fell under open-circuit conditions without hesitation through the potential values of 1.23, 1.1, and 1.06 V. From his investigations, he concluded that the various potentials observed on open circuit were the result of different combinations or ratios of an ill-defined oxide, PtO_x [approximating PtO or Pt(OH)$_2$], and PtO_2. Above 1.23 V the potentials may be due to higher oxides such as PtO_3 or PtO_4 as suggested by Grube (100), but Foerster considered these to be very unstable, decomposing to mixtures of PtO_x and PtO_2. It was observed by Klemenc (147) that the rest potential varied with the partial pressure of oxygen, P_{O_2}, but the Pt/O$_2$ potential was produced by PtO_2 in equilibrium with the P_{O_2} above the solution. At high P_{O_2} up to about 100 psi, Tamman and Runge (228) found large hysteresis effects with changes in P_{O_2}, yet they came to the same conclusions as Foerster about the potential-determining species on the electrode at high potentials. Náray-Szabó (185) used oxygen diffusion electrodes made from platinized carbno in 2N H$_2$SO$_4$ solution and suggested that the potential was caused by the presence of PtO on the

Pt black surface. Recently Watanabe and Devanathan (252) have assigned various oxide potentials to certain rest potentials.

Attempts to use the oxide theory to account for the rest potentials observed on any noble metal–oxygen electrode have failed. This was demonstrated by the work of Bain (8) in which he studied the rest potential on electrodes formed by plating thin films of Pt, Pd, Au, Rh, Os, Ir, and Ru on glass plates. He made the oxides by heating the electrodes in a bunsen flame and recorded the potentials in $1N$ H_2SO_4, $1N$ NaOH, and $1N$ KCl solutions. Although the initial potentials were different, they approached a common value with time. He reasoned that if the potential alone were determined by a metal–metal oxide reaction or by a metal oxide in equilibrium with a P_{O_2} equal to the dissociation pressure of the oxide, the rest potentials would be widely differing. Since this was not observed, the oxide theory cannot be held.

2. The Peroxide Theory

Brislee (35) measured the open-circuit potential of the Grove cell which rose to a value of 1.082 V in 24 hr and to 1.120 V in 14 days. Afterward, the potential fell to a constant value of 1.054 V in 30 days, and peroxide was detected in the solution. When H_2O_2 was added to the system, the potential fell to 0.98 V but rose again to 1.054 V in 28 hr. If the primary product of the Grove cell were H_2O_2 instead of H_2O, Lewis (159) suggested that the rest potential could be determined by the peroxide present in solution. Tilley and Ralston (234) concluded that the potential of the oxygen electrode depended on the H_2O_2 concentration as well as on OH^- ion concentration. Since it was found that the potential of the Pt/O_2 electrode depended on stirring rate, Tarter and Wellman (230) decided that the potential was due to the concentration of H_2O_2 present in the solution and not to oxides on the electrode surface.

It should be noted that in recent investigations (116,123) peroxides were not detected at Pt/O_2 electrodes under open-circuit conditions. Possibly, impurities were present in the early work, creating local cell phenomena which produced the H_2O_2. Consequently, it seems unlikely that the rest potential of a Pt/O_2 electrode is determined by the presence of peroxide alone. As pointed out by Bain (8), few investigators adhere to the peroxide theory. In fact, Tarter and Walker (229) later interpreted the stirring effects in terms of adsorption potentials. Initially, ions are adsorbed from solution on the electrode surface; and when oxygen is added to the system, it too is adsorbed, modifying the ion adsorption layer and causing the observed drift in potential. Since the effects of specifically adsorbed ions on electron processes are a highly complex matter indeed, a discussion of these phenomena is deferred to a later section.

3. Constant Current Charging Curves

Although neither the oxide theory, not the peroxide theory, nor the adsorption potential theory could explain adequately the experimental facts, the results of the early work clearly showed that the noble metals interact with their environment and thin films are formed on the noble metal surface. One of the major problems in the understanding of the nature of the oxygen electrode is the determination of the nature of the thin films adsorbed on the noble metal surface.

Convincing evidence for the presence of adsorbed oxygen films on Pt electrodes has been obtained from constant current charging curve techniques (58) in which the electrode system is anodized or cathodized at a constant current. The change in potential with time is plotted, and plateaus appearing in these curves correspond to the formation or reduction of a potential-determining species adsorbed on the electrode surface. From a determination of the number of coulombs passed during the plateau interval, some knowledge of the nature of the adsorbed film may be obtained. If the experimental conditions are chosen so that the system is diffusion controlled, the mathematical description of the system obeys Sand's equation (203), and the technique is called chronopotentiometry (161). This technique will be discussed in Chapter VI.

A number of investigators (3,26,32,41,62,67,69,71,73,79,86–88,94, 95,103,110,150,153,157,162,163,182,187,188,190,204,205,212,215,223, 242,276) have studied the formation and reduction of adsorbed films on bright or platinized Pt in O_2, N_2, H_2, or He saturated acid or alkaline solutions by recording anodic and cathodic charging curves. Most information about the nature of the adsorbed film itself is obtained in solutions saturated with an inert gas, such as N_2 or He, since then the system is not complicated by the ionization of adsorbed hydrogen or the reduction of adsorbed oxygen. A generalization of the results of these investigators may be described by the idealized charging curves in Fig. 2.1.

Consider a Pt electrode in N_2 saturated acid solution and consider the system to be cathodized to a point where H_2 is evolved. Then the circuit is broken, and a constant anodic current is applied to the electrode. A potential–time curve similar to the one shown on the left-hand side of Fig. 2.1 is recorded. An induction period relating to the ionization of

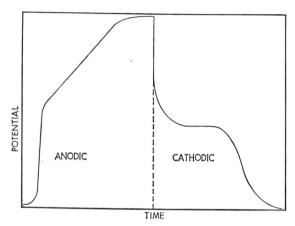

Fig. 2.1. Sketch of charging curves obtained on bright Pt electrodes in N_2 stirred solutions. Anodic charging at constant current is plotted on the left of the broken line; cathodic, to the right.

the adsorbed hydrogen appears in the initial part of the curve at potentials more negative than 0.4 V. Once the hydrogen layer has been stripped from the electrode surface, the potential rises abruptly from about 0.4 V (the limit of hydrogen adsorption) to about 0.8 V where oxygen is first adsorbed. In this potential range between 0.4 and 0.8 V, it is considered (86) that neither hydrogen nor oxygen is adsorbed. This region is known as the double layer region (83) because, according to Frumkin and co-workers (87), most of the current goes into the charging of the double layer.

Starting at a potential of about 0.8 V, a linear increase in potential with time is observed which may be called the oxygen adsorption region, for in this region above 0.8 V a layer of adsorbed oxygen is formed. The curve becomes nearly flat above 1.6 V where the evolution of oxygen gas begins.

After a potential is reached where oxygen is evolved, the current is reversed, and a constant cathodic current is applied to the electrode. The recorded charging curve is similar to that shown in the right-hand side of Fig. 2.1. An initial rapid fall of potential is followed by a gradual fall leading to a fairly flat plateau at a potential somewhat below the potential (0.65 and 0.7 V) where oxygen is first adsorbed (0.8 V). This indicates that an activation energy is required to remove the adsorbed oxygen layer. When this layer has been stripped from the electrode surface, the potential drifts to a value where hydrogen gas is evolved.

An important point to be noted is the observation that, in general, the number of coulombs corresponding to the formation of the adsorbed oxygen layer during the anodic charging process is larger than the number required to remove it during the cathodic stripping process.

a. The Thickness of the Adsorbed Oxygen Film. Bowden (26) was the first to pay attention to the control of impurities in studies of the oxygen electrode and to demonstrate that if a roughness factor is considered the number of coulombs determined from the length of the oxygen adsorption region corresponds to a monolayer of adsorbed oxygen atoms. In addition, it was pointed out by Butler and co-workers (41,190) that one atom of adsorbed oxygen is associated with one surface atom of Pt, and since the amount of adsorbed oxygen may be determined from the charging curve, a measure of the true surface area may be made.   Using such assumptions, Schuldiner and Roe (212) recently developed a method for arriving at the true electrode area from an oscillographic analysis of charging curves obtained with constant high current pulses.   Such an oscillograph is presented in Fig. 2.2.

The linear region between 0.89 and 1.77 V in Fig. 2.2 is the oxygen atom adsorption region, and the product of the transition time of this region and the applied current gives the number of coulombs consumed, Q_T. Since contributions to the total number of coulombs arises from both the charging of the double layer in this potential range (0.89–1.77 V) and the formation of adsorbed oxygen atoms, a correction for the

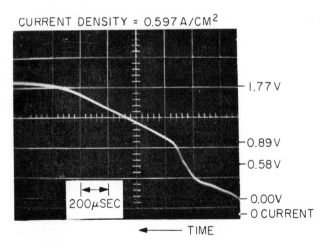

Fig. 2.2. A typical oscillographic trace of a charging curve taken on a Pt electrode (212). Read time scale from right to left. (By permission of The Electrochemical Society.)

double layer charging must be made. This correction may take the form $Q_{dl} = C\Delta E$, where C is the average value of the double layer capacity over the potential range, $\Delta E = 1.77 - 0.89$ V. The corrected value, $Q_c = Q_T - Q_{dl}$, is divided by the charge associated with one oxygen atom ($2 \times 1.6 \times 10^{-19}$ coulombs) to arrive at the number of adsorbed oxygen atoms. Finally, the ratio of the number of adsorbed atoms to the number of surface atoms, 1.31×10^{15} Pt atoms/cm² (212), yields the roughness factor, and hence, the true area if a one-to-one correspondence between oxygen and Pt atoms on the surface is assumed.

b. Number of Arrests in Charging Curve. Most investigators (e.g., 41,62,73,110,153,190,212,215,242) have found one adsorption region in the anodic and one reduction plateau in the cathodic charging curve. However, El Wakkad and Emara (67) and Lee et al. (157) reported curves in which more than one arrest appears in the anodic branch but only one reduction step is observed in the cathodic branch. El Wakkad and Emara obtained these curves using very low current densities ($\sim$1 μA/cm²) requiring 1 hr to complete one curve. It is noted that these arrests in the anodic curve were not very prominent and virtually disappeared when the curves were obtained oscillographically with a higher current density ($\sim$1 mA/cm²). Lee et al. anodized the Pt electrode with MnO_4^- ion or Ce^{4+} ion, and the arrests in this case are even less sharp than those found by El Wakkad. In both cases (67,157) the first arrest is correlated with the appearance of a film of PtO at about 0.82 V, and the second with a film of PtO_2 at about 1.05 V.

c. The Ratio Q_a/Q_c. Those investigators (41,62,73,153,242) who used slow charging techniques (low current densities) found that the number of coulombs required to form a monolayer of adsorbed oxygen, Q_a, is more than that required to reduce it, Q_c. On the other hand, those who used fast charging techniques (high current densities) (190,212,215) and those who used rapid potential sweep techniques (21,254) found that $Q_a = Q_c$. Evidently, in the anodic charging process, side reactions such as the evolution of oxygen must be taking place. Yet, as pointed out by Vetter (242), nearly all investigators report that the values of both Q_a and Q_c are independent of the current density. This has led Vetter to the conclusion that the adsorbed oxygen layer is reduced only to H_2O_2. Since this reduction process requires only two electrons and the deposition of the oxygen layer requires four, this explanation would agree with the observation $Q_a/Q_c = 2$ (41,62,73,153,242). It is known that Pt is a good peroxide-decomposing catalyst (e.g., 16,92,126). The peroxide should be decomposed according to the equation $H_2O_2 \rightarrow H_2O + \frac{1}{2}O_2$ to form oxygen which in turn can be reduced so that effectively a four-electron process is obtained. Therefore, the peroxide explanation would seem to be a less favorable one in the presence of such an active catalyst as Pt. In fact, recent work by Frumkin and co-workers (85) using a rotating ring electrode in conjunction with triangular voltage pulses has shown that the reduction of adsorbed oxygen does not lead to the formation of H_2O_2.

Feldberg, Enke, and Bricker (73) found that initially $Q_a/Q_c = 2$ on a Pt electrode in $0.8M$ $HClO_4$, but after about 20 cycles of anodic polarization followed by cathodic polarization between 0.5 and 1.4 V, this ratio approached unity. It was observed that the higher the current density used, the fewer were the number of cycles required to reduce Q_a/Q_c to 1. Q_c was constant for all conditions, however, and independent of the current density. Such experimental facts led these authors to prefer a mechanism involving a slow, one-electron transfer step to a "half-oxidized" state, such as $Pt(OH)_x$, followed by a fast, one-electron transfer step to the oxidized state, such as $(PtO)_x$, at a Pt anode. Probable values for x are 1 or 2. With the reduction of such an oxidized Pt electrode, the fast step occurs first followed by the slow step. Rapid cycling of the electrode produces a situation in which the surface is always covered by a layer of $Pt(OH)_x$. Only the fast step conversion of $Pt(OH)_x$ to $Pt(O)_x$ can follow the rapid alteration of current. Such a viewpoint could explain the fact that $Q_a/Q_c = 2$ for one cycle of constant current polarization because cathodic reduction removes only half the anodically deposited oxygen. It also explains the fact that $Q_a/Q_c = 1$ with repeated cycles of polarization because only the rapidly exchanged oxygen need be deposited during anodic charging. One may find it difficult, however, to accept the proposition that the cathodic removal of an adsorbed hydroxyl radical, $Pt(OH)_x$, should be any more difficult in acid solutions than the conversion of an adsorbed oxygen atom, $Pt(O)_x$, to $Pt(OH)_x$.

Somewhere in between the convictions of Vetter and Berndt and Feldberg, Enke, and Bricker rest those of Dietz and Göhr (62), who also found $Q_a/Q_c = 2$. They believe that a surface oxide about two or three layers thick is formed with the anodic polarization of Pt in N_2 stirred $1N$ H_2SO_4 solution, but with cathodic polarization only half of the adsorbed film is reduced to peroxide. The remaining part is reduced only after the potential has been lowered to the point where the adsorption of hydrogen has begun. The reduction of this oxide produces a loosely packed layer of Pt atoms which only slowly rearrange to the normal lattice and which Dietz and Göhr suggest may be the reason for the increased activity exhibited by preanodized Pt indicator electrodes used by the electroanalytical chemists in certain redox sustems. This notion will be discussed in more detail in Chapter VI. Why part of the adsorbed film should be reduced in preference to all of it is not answered by this explanation.

According to Laitinen and Enke (153), $Q_a > Q_c$ because O_2 is evolved along with the formation of the adsorbed oxygen layer during the anodic charging of Pt in $HClO_4$ solution. Such an explanation does not explain why $Q_a/Q_c = 1$ with repeated charging and discharging of the electrode (73) or with the fast charging techniques.

d. Effect of pH. In general, the shape of the charging curves is the same in both acid and alkaline solutions (67,71,95,182,188,190,223,242). Vetter and Berndt (242) have shown that the potentials at corresponding points on the charging curves taken in solutions of various pH are shifted by RT/F mV (59.2 mV at 25°C) per unit of pH, as shown in Figs. 2.3 and 2.4. Although the H_2- and O_2-adsorption regions are similar in acid and alkaline solutions, the slope of the curve in the double layer region is much more gradual in alkaline than in acid solutions. Hickling (110) finds the value of the corresponding capacity to be about twice as much in alkaline as in acid solutions, and as he points out, it is doubtful if this section of the curve can be attributed only to the charging of the double layer. It may be possible that the adsorption of hydroxyl ions contrib-

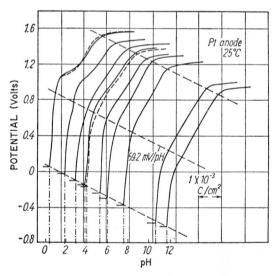

Fig. 2.3. The anodic charging curves on Pt showing the dependence on pH at $i = 5 \ \mu A/cm^2$ and 25°C (242). The broken curve has a slope of 59.2 mV per unit pH. (By permission of Bunsengesellschaft.)

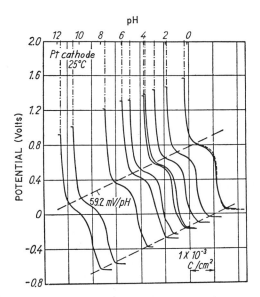

Fig. 2.4. The cathodic charging curves on Pt showing the dependence on pH at $i =$ -5 $\mu A/cm^2$ and 25°C. (242). (By permission of Bunsengesellschaft.)

utes to the charge measured in the double layer region. Hydrogen ionization processes may also contribute charge since Slygin and Frumkin (223) found that the hydrogen adsorption region extended into the oxygen adsorption region in KOH from which they concluded that hydrogen is more tightly bound to the Pt surface in alkaline than in acid solutions. About three times as much oxygen is taken up by Pt in H_2SO_4 than in NaOH solutions, as reported by Obrucheva (188).

4. Potential Sweep Techniques

Another experimental technique from which information about the nature of the adsorbed oxygen layer on Pt may be obtained is the triangular wave potential sweep method. With this method the electrode is polarized between two given potentials by a potential source which is varied at a constant rate of change with an electronic triangular wave function generator. Some early electronic designs were given by Randles (194). Both Randles (194) and Sevcik (218) independently were the first to give the theory behind this method. Delahay and Perkins (59) discussed this method using single and repetitive triangular sweeps for dropping mercury electrodes, and Elving and co-workers (144) described the recording of such sweeps oscillographically.

This early work was concerned with the study of electrode processes controlled by mass transfer steps and is summarized by Vogel (246). As the electrode reaction proceeds in a solution containing an excess of inert electrolyte, the concentration of reactants at the electrode surface decreases and diffusion processes set in. During the potential sweep, the thickness of the diffusion layer increases (lowering the diffusion current), but the concentration gradient across the diffusion layer also increases (increasing the diffusion current). The opposing effects of these two processes on the diffusion current produce a peak in the current. The visible result is the replacement of steps in the polarogram with peaks. Randles (194) has shown that the peak value of the diffusion current may be related to the concentration of the reactant in the bulk of solution.

Will and Knorr (254) first used the output of a triangular sweep generator to control the reference potential of a potentiostat having a fast response time. In this way a potentiostatically controlled voltage sweep is presented to the cell, and E–i curves containing information about electrode processes involving chemisorption may be obtained, provided the experimental conditions are chosen in this case to minimize the effects of mass transfer (e.g., strongly stirred solutions). Under these conditions, a current peak corresponds to the existence of a Faradaic process, and so the potential range over which the given electrode process takes place may be determined.

Typical curves obtained by this method on bright Pt (21,80,220,243, 254) may be seen in Fig. 2.5 for acid solutions and in Fig. 2.6 for alkaline solutions. In acid solutions it may be seen that oxygen adsorption

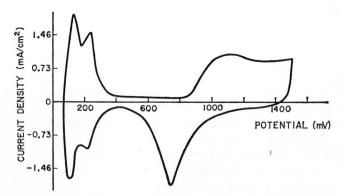

Fig. 2.5. A plot of the potential sweep trace obtained on Pt in 2.3M H$_2$SO$_4$ at 25°C and a sweep rate of 0.5 V/sec (21). (By permission of Pergamon.)

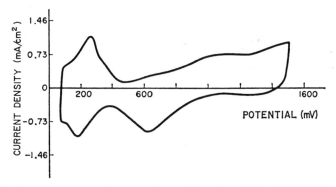

Fig. 2.6. A plot of the potential sweep trace obtained on Pt in 0.1M NaOH at 25°C
and a sweep rate of 0.5 V/sec (21). (By permission of Pergamon.)

begins at about 0.85 V and oxygen evolution above 1.4 V. Below 0.4 V
the hydrogen adsorption peaks are observed, and between 0.4 and 0.85 V
is the low current double layer region. During the cathodic sweep the
oxygen reduction peak is displaced from the oxygen adsorption peak
toward less noble potentials, which indicates that an activation energy
hump must be surmounted before reduction can take place. This overvoltage was also observed in the charging curves of Fig. 2.1. In contrast to oxygen adsorption, it is seen that hydrogen adsorption occurs
reversibly; the adsorption and reduction peaks occur at the same potential.

Böld and Breiter (21) and Burshtein and co-workers (220) found that
for fast sweeps $Q_a = Q_c$. The value of Q is determined by integration
under current peaks since the sweep rate is constant and known. A
correction to the charge due to double layer charging currents must be
made and is usually obtained from the low current double layer region.
For slow sweep speeds, Böld and Breiter observed that $Q_a > Q_c$ because,
as they suggest, oxygen enters the Pt metal on the anodic scan. Burshtein recorded that the value found for Q_c is larger for a single sweep
than for repetitive sweeps.

It has been observed (21,80,254) that the adsorbed oxygen layer builds
up with anodic polarization to a point (about a monolayer thick) and
then no further. Such a behavior would be expected if the layer of adsorbed oxygen were electronically conducting.

Bagotskii and co-workers (172a,233a) found two peaks in the reduction sweep which was first observed by Will and Knorr (254) and later
discounted as an experimental artifact by Böld and Breiter (21). They
(172a) found that if the electrode is swept beyond 1.4 V the first peak at

0.85 V becomes so large that the small peak at 0.65 V is covered up. Because the potential was usually swept beyond 1.4 V [as Böld and Breiter (21) did] this is the reason why two peaks rarely are observed. Bagotskii (172a,233a) interprets his results in terms of at least two forms of adsorbed oxygen on Pt electrodes. Similar to the findings of Feldberg et al. (73), Bagotskii found that with slow sweeps $Q_a = 2Q_c$, but with repeated sweeps, Q_a became equal to Q_c, which corresponds to a monolayer of adsorbed oxygen. He also found, in agreement with other investigators (21,80), that the oxygen adsorbed during the first sweep modified the adsorption process in the second sweep, and it required several cycles of triangular sweeps to come to the steady-state curve.

In alkaline solutions (Fig. 2.6), the hydrogen and oxygen adsorption regions overlap in agreement with the observations of Slygin and Frumkin (223). Also in agreement with Obrucheva's findings (188), the oxygen reduction peak is larger in the acid case than in the alkaline case. In passing, it is interesting to note that only one hydrogen adsorption peak appears in NaOH solutions, whereas two appear in acid solutions.

There is no doubt that the triangulr wave potential sweep method is an excellent technique for studying quickly and qualitatively the potential regions of a given electrochemical system in which various components of the system are adsorbed. As Conway (45,46) cautions for quantitative experiments, complicating Faradaic and non-Faradaic processes may be present and may contribute current to the recorded current. Because steady-state conditions are never achieved, non-Faradaic processes are always present, and the corrections for the double layer charging currents obtained in the double layer region may not be the same in other potential regions. An even more serious complication arises from currents contributed by the Faradaic processes of charging the electrode with a film of oxide or some other adsorbed intermediate. These currents, which correspond to an adsorption pseudo-capacitance (47) and may be much larger than the double layer charging currents, depend on the sweep rate and confound the true $E–i$ behavior of the system.

It is seen that the results with hydrogen and oxygen adsorption agree very well with those obtained by other methods, so for such relatively simple systems, the triangular sweep methods give valid results. In complicated systems involving more than one component in competition for an adsorption site or those involving many intermediates (e.g., hydrocarbon oxidation systems), Conway's comments are certainly valid.

He shows that if rigorous attention is paid to the double layer and adsorption pseudocapacitance effects the mathematics becomes rather unwieldy and the interpretation complex.

5. Optical Methods

The optical method known as ellipsometry (197) is interesting because it is a direct method of examining films on the reflecting electrode surface during the course of the electrochemical experiment with elliptically polarized light.   The experimental setup for an ellipsometer is sketched in Fig. 2.7.   This method is based on the changes produced on the polarization state of the beam of light reflected from the electrode surface by the growth or adsorption of surface films.   Such changes in the polarization of the light beam reflect changes in density at the electrode surface caused by the presence of the film.   With each change of the electrochemical parameters of the system, a change in the setting of the optical analyzer crystal in the reflected beam path must be made.   Since these adjustments require time, the ellipsometer cannot follow rapid changes in the film properties.   The potential of the system is held at a given value with a potentiostat.   An excellent account of the advantages, problems, and disadvantages associated with ellipsometric measurements has been presented clearly by Kruger (151a).

Reddy, Genshaw, and Bockris (196) have used this technique to study the adsorbed oxygen films on Pt in acid solution.   It was found that films adsorbed from oxygen saturated solutions cannot be detected, but with anodic polarization, the first measurable changes suddenly appear at about 0.98 V, after which the readings increase with potential.   These readings may be calibrated in terms of the thickness of the film, and a

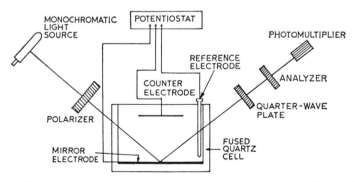

Fig. 2.7. Experimental arrangement of ellipsometer and cell (196).   (By permission of Elsevier.)

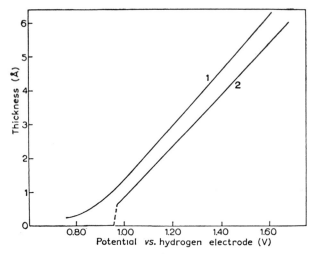

Fig. 2.8. The dependence of film thickness on potential obtained (*1*) by typical coulometric techniques and (*2*) by ellipsometry (196). (By permission of Elsevier.)

plot of such data is given in Fig. 2.8. The agreement with coulometric data obtained from charging curves is noted. Even the reduction curves are similar (not shown in Fig. 2.8) since a large hysteresis is observed. After the system has been anodized, the current was reversed, but no change in the ellipsometer readings was obtained until a potential less noble than one volt was reached, after which the readings fell rapidly with cathodic potential to zero (91).

The index of refraction, n_f, of the film is a complex quantity (196), $n_f = n_f° + iK$, where the imaginary part, iK, is a function of the adsorption coefficient, K, which is a measure of the conductivity of the film at optical frequencies. The results indicate that the conductivity of the adsorbed oxygen film on Pt is of the same order as that of a metal at optical frequencies. Bockris and co-workers conclude that a highly conducting film of oxide rather than a chemisorbed film of oxygen atoms is formed. Under ordinary conditions, this thin film is less than 10 Å thick and increases linearly with potential.

6. The Mixed Potential Theory

As pointed out by Giner (95), the charging curves do not show plateaus corresponding to the oxide potentials obtained by Spielmann, Lorenz, and Grube, and not only are there difficulties in preparing these oxides, but also the chemically prepared oxides most likely do not have the same stoichiometry as the electrochemically formed oxides.   Therefore the rest potential is not determined by oxides alone.   Since the Nernst equation is not obeyed [open-circuit potential decay curves obtained after anodic polarization (77,95,116) pass smoothly through 1.23 V], the rest potential must be a mixed potential (249), and the electrode system is a polyelectrode (154,155,241).

It was recognized early by Richards (198) that the rest potential was not only influenced by the adsorbed oxygen but also by the molecular oxygen dissolved in solution. Although the open-circuit potential was well below 1.23 V, he observed that it was very constant (within a few millivolts) for days in oxygen saturated solutions.

The first to propose a mixed potential mechanism for the rest potential behavior of Pt in O_2 saturated acid solutions was Hoar (115).   He explains the deviation of the rest potential from 1.23 V by the presence of an oxide film with cracks in it or at least permeable to the solution.   As a result, a local cell is set up such that O_2 reduction takes place on the surface of the film.   The process is oxygen depolarized, and Pt oxide is formed on the metal surface.   Because of local cell action, a potential drop occurs at the electrode surface, and the observed potential is depressed below the reversible value of 1.23 V.   To maintain a potential of 1.23 V in an O_2 saturated $0.1N$ H_2SO_4 solution, a countercurrent must be applied externally to the Pt electrode.   The applied current necessary to maintain a potential of 1.23 V decreased with time, rapidly at first, but finally reached a fairly steady value of about 2×10^{-8} A/cm². Hoar interprets the fall in current as a healing of the cracks or a plugging of the pores in the oxide film, and he suggested that if the film were complete and impermeable to the solution, the calculated reversible oxygen potential could be observed.   Other estimations of this local cell current are 2×10^{-7} A/cm² (242) and 3.2×10^{-7} A/cm² (125).

Giner (95) has suggested that the mixed potential is composed of either the O_2/H_2O reaction, Eq. 2.3, or the O_2/H_2O_2 reaction,

$$O_2 + 2H^+ + 2e \rightleftharpoons H_2O_2 \qquad E_0 = 0.682 \text{ V} \qquad (2.9)$$

and a reaction involving bare Pt sites and those covered with adsorbed oxygen.   He could not find any evidence for the presence of a definite oxide, and since H_2O_2 was found in the reduction of oxygen, he considered that the O—O bond was not broken.   Therefore, he reasoned, the form of adsorbed oxygen important in determining the mixed potential is chemisorbed molecular oxygen, Pt-O_2, and the counterreaction is the O_2/H_2O_2 reaction, Eq. 2.9.   This is in disagreement with the evidence for the dissociative adsorption (coverage of the surface with adsorbed oxygen is a function of the square root of the partial pressure of oxygen) of O_2 on Pt (51,195) and the observation that H_2O_2 is not detected under open-circuit conditions at Pt electrodes in O_2 saturated solutions (116, 212).

Lewartowicz (158) suggests that the Pt potentials in the presence of redox systems are mixed potentials with two reactions occurring at different points on the electrode.

The dependency of the rest potential at Pt electrodes in O_2 saturated $2N$ H_2SO_4 solution on the partial pressure of oxygen, P_{O_2}, was determined (116) over a range of values from 0.2 to 1 atm. In this range, the potential changed by about 15 mV per unit of log P_{O_2} which indicated that a four-electron oxygen reaction is potential-determining. Careful attention was paid to the control of impurities in this work because, as pointed out by Bowden (26) and Frumkin (223) and convincingly demonstrated by Bockris and co-workers (5,19), reproducible and reliable results cannot be obtained without rigorous control of impurities. The fact may be mentioned that references to such serious sources of impurities as picene, rubber stoppers, cells open to the atmosphere, Tygon paint, and silicone grease appear even in the modern literature.

Since the observed steady-state rest potential of a Pt electrode was 1.06 V (116) instead of 1.23 V, the potential was concluded to be a mixed potential involving the O_2/H_2O reaction, Eq. 2.3, and a Pt/Pt-O reaction, such as

$$\text{Pt-O} + 2\text{H}^+ + 2e \rightleftharpoons \text{Pt} + \text{H}_2\text{O} \qquad E_0 = 0.88 \text{ V} \qquad (2.10)$$

where Pt-O represents a layer of adsorbed oxygen atoms on the Pt surface instead of a layer of oxide, PtO. A schematic diagram of this mixed potential system is given in Fig. 2.9. In this case, Eq. 2.10 (curve B), is strongly polarized by the local cell current such that at the mixed potential, $E = E_m$ (the point where curve B intersects curve A), Eq. 2.10 is in a limiting current condition. Only Eq. 2.3 (curve A) depends on P_{O_2}, and it is seen that the mixed potential, E_m, varies with P_{O_2} in the same way as the reversible potential of Eq. 2.3, E_A, alone (A_1, A_2, and A_3 are local cell polarization curves for decreasing values of P_{O_2}). The rest potential is controlled by the O_2/H_2O reaction, Eq. 2.3, thus accounting for the 15 mV dependency of the rest potential on log P_{O_2}.

The driving force of the local cell is such that oxygen is reduced by Eq. 2.3 on the electronically conducting Pt-O sites, and Pt-O is formed by the reverse of Eq. 2.10 on the bare Pt sites. Without further consideration, the surface would become completely covered with Pt-O by

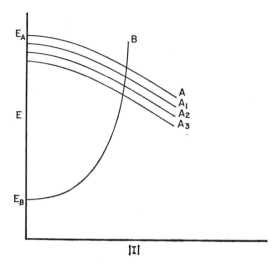

Fig. 2.9. A sketch of the local cell polarization curves for the half-reactions which determine the mixed potential observed in the Pt/O₂ system. The A curves are those for the O_2/H_2O reaction for decreasing values of P_{O_2}; the B curve, for the Pt/ Pt-O reaction. The mixed potential, E_m, is the potential at the point of intersection of the curves; the E_0 value is the potential at the point of intersection of a given curve, and the potential or E axis and I is the absolute value of the local cell current.

the local cell action, the mixed potential mechanism would cease, and the reversible oxygen potential would be observed. This, of course, is not found experimentally, which means that a complete film of Pt-O is unstable in acid solutions. Some Pt-O disappears either by chemical decomposition or by dissolving in the Pt metal. At any rate, a certain number of bare Pt sites are always present, and a steady-state situation is reached. Nesterova and Frumkin (187) found that the amount of oxygen absorbed by Pt black increased with time until a point was reached after which no more oxygen was adsorbed. From cathodic stripping measurements, this maximum quantity of oxygen was less than a complete monolayer.

It has been shown (e.g., 21,71,95,190,196,212) that the potential is a function of the coverage of the electrode surface with oxygen. Therefore, variations in the observed values of the rest potential such as found by Devanathan and co-workers (244,252) may be interpreted in terms of the varying degrees of coverage of the electrode surface with oxygen at the steady-state condition produced by various electrode pretreatments.

If the system is stirred with purified N_2 to remove all dissolved O_2, the potential should no longer be determined by a mixed potential mechanism but solely by the Pt/Pt-O reaction. It was found (116) that a plateau in the rest potential decay-time curve at about 0.88 V was present in N_2-stirred $2N$ H_2SO_4 (pH~0), and this value is preferred as the standard potential of the Pt/Pt-O couple. With time, however, the potential drifted to less noble values (~0.7 V) in purified-N_2-saturated solutions.

Very careful measurements of the rest potential as a function of P_{O_2} in the range 0.0001–0.01 atm were made by Schuldiner and Roe (213) in a gas-tight system. In a purified He atmosphere they observed a plateau at about 0.88 V, but over extended periods of time, the potential decayed to much less noble potentials (from 0.2 to 0.4 V). In a later report, Warner and Schuldiner (250) obtained rest potentials of about 0.7 V in He-stirred solutions. Palladium reference electrodes which were used earlier (213) but eliminated from the system later (250) were held responsible for depolarizing the oxygen electrode to 0.4 V or lower by hydrogen dissolved in the Pd. In addition, Schuldiner and co-workers (213,250) found potential–log P_{O_2} curves with a slope of 60 mV in the low P_{O_2} region, From these observations, they reject the polyelectrode mechanism involving the Pt/Pt-O couple and propose that the system is a single electrode with a one-electron potential-determining reaction involving adsorbed peroxyl radicals, such as

$$O_2 + H^+ + e \rightleftarrows HO_2 \qquad (2.11)$$

even though peroxide was not detected under open-circuit conditions.

The rest potential observations in He-saturated solutions of Schuldiner and Roe may be interpreted in terms of the mixed potential by the possibility that the Pt-O layer is unstable in the acid solution in the absence of dissolved O_2. As the Pt-O layer is decomposed, the activity of the Pt-O becomes less than unity, and according to the Nernst relationship, the potential drifts to less noble values. The 60 mV slope of the potential–log P_{O_2} curve may be explained by the local cell polarization curves as shown in Fig. 2.10. The notation is the same as in Fig. 2.9. In the very low P_{O_2} regions, the polarization of the O_2/H_2O reaction increases as P_{O_2} decreases because of the low concentration of dissolved oxygen available to the O_2/H_2O electrode process. Consequently, the A curves are nearly at a limiting current when they cross the B curve (Pt/Pt-O reaction). As seen in Fig. 2.10, the variation of the mixed potential, E_m, is much larger than the variation of the standard potential, E_a, with P_{O_2}, thus accounting for the 60 mV slope.

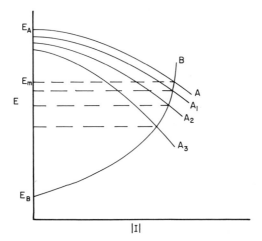

Fig. 2.10. A sketch of local cell polarization curves for the Pt/O_2 system for very low values of P_{O_2}. The notation is the same as given in Fig. 2.9.

In support of this contention, the Bockris group (60) studied the rest potential over a large range of P_{O_2} (from about 10^{-3} to 1 atm) and found that the large dependency of potential at low P_{O_2} (about 60 mV/unit log P_{O_2}) decreased continuously with increasing P_{O_2} to a low value (about 15 mV/unit P_{O_2}) above 0.2 atm.

In the most recent work reported by Schuldiner and co-workers (215b), rest potential studies were carried out in He-saturated acid solutions in a gas-tight system (215a) where the O_2 concentration is controlled to one part in 10^9. Under these conditions, the open-circuit potential is 0.46 V, but the presence of dermasorbed hydrogen or oxygen can shift the rest potential to less or more noble potentials, respectively. Consequently, elaborate preparation of the solutions and electrodes over extended periods of time (up to 2 or 3 months) is required (215b) to arrive at the 0.46 V value. Variations in the reported value of the rest potential of a Pt electrode in acid solutions saturated with an inert gas are without doubt caused by the presence of dermasorbed hydrogen or oxygen.

Bockris and co-workers (51,52,195) have made extensive studies of the rest potential of Pt/O_2 electrodes in acid solution as a function of electrode pretreatment, P_{O_2}, and pH. They have interpreted their results in terms of dipole- or χ-potentials since a dissociative adsorption of oxygen was observed. Earlier, Bowden (27) had suggested that surface forces deformed the adsorbed oxygen atoms, producing a layer of dipoles which controlled the electrode potential.   It is difficult to explain the fact that the rest potential shifts by about 60 mV per pH unit (116, 213) with this mechanism, since the potential should depend only on the number of oxygen dipoles adsorbed on the surface.   The Bockris group found that the potential of an oxygen-free Pt electrode rose to a nearly steady potential of about 0.98 V in a half hour when stirred with O_2.   By cathodic stripping, oxygen coverage called θ_{max} was determined at this time, and θ_{max} on Pt was found to be about 0.22.

During this relatively short time period, oxygen is probably adsorbed by dissociative adsorption until $\theta = 0.22$, and the potential is probably a dipole potential.   However, after this partial layer of Pt-O is formed, the mixed potential mechanism (Eqs. 2.3 and 2.10) sets in.   After about 24 hr, the potential comes to a true steady-state value of 1.06 V (116), and the coverage is closer to a monolayer (187).   Although the initial adsorption of oxygen (first 10 min) involves dipole potentials, the later steady-state condition of the Pt/O_2 electrode is determined by a mixed potential mechanism.

Other mixed potential systems involving peroxide couples are considered less favorable.   The combination of the O_2/H_2O_2 and Pt/Pt-O reactions cannot occur because the observed rest potential is more noble than the E_0 for either Eq. 2.9 or 2.10, and the mixed potential must fall between the two E_0 values (249).   Similarly, a combination of Eq. 2.3 and the H_2O_2/H_2O reaction,

$$H_2O_2 + 2H^+ + 2e \rightleftarrows 2H_2O \qquad E_0 = 1.77 \text{ V} \qquad (2.12)$$

is ruled out because the observed rest potential is less noble than the potential of either of these half-reactions.   A combination of the O_2/H_2O and the O_2/H_2O_2 reactions would set up a local cell in which H_2O_2 would be consumed and not produced.   Calculations show that a mixed potential value of one volt would require a steady-state concentration of about $10^{-5}M$ H_2O_2 in solution.   It is difficult to visualize what the source of this H_2O_2 would be, and consequently, this mechanism is rejected.   Besides, H_2O_2 is not detected under these circumstances (116, 213).   Other combinations are discounted because they require a source of H_2O_2 or are independent of P_{O_2}.

C. CONSENSUS ON THE REST POTENTIAL OF THE Pt/O_2 ELECTRODE

Since the reversible oxygen potential was not observed on noble metal electrodes, it became apparent that the noble metals were not truly inert and did interact with oxygen. It was only natural to ascribe the observed potentials to the influence of oxides present on the metal surface. This approach did not prove fruitful, since the rest potentials of the noble metals are very similar which would not be expected if the potentials were determined by a metal–metal oxide couple or by the dissociation partial pressure of oxygen in equilibrium with the oxide. A reasonable attack seems to be from the concept of mixed potential. Because the mixed potential depends on the relative area of the anodic and cathodic sites which, in turn, depend on the electrode pretreatment, such a viewpoint can explain not only the failure to attain the reversible potential, but also the strong dependence of the potential on the history of electrode preparation.

A nagging problem associated with a basic understanding of the oxygen electrode is the nature of the thin films of adsorbed oxygen produced on noble metal surfaces and an unambiguous determination of the coverage of the surface as a function of electrochemical parameters, such as potential.

1. The PtO/Pt-O Controversy

From all of these investigations described so far, several points may be established. Platinum surfaces in contact with an oxygen saturated medium will adsorb oxygen in the form of an electronically conducting film about a monolayer thick. With anodic polarization, oxygen begins to adsorb at a potential of about 0.8 V, but prolonged anodization does not cause the film to grow beyond a monolayer thick. Oxygen is evolved only after the monolayer of adsorbed oxygen is formed.

The form of the adsorbed oxygen on a Pt surface is still not determined unequivocally. A group of investigators (e.g., 3,80,110,153,182,184, 205) seem to feel that at potentials below about one volt the adsorbed oxygen is in the form of PtO, and above this value, in the form of a mixture of PtO and PtO_2. To confirm this, Anson and Lingane (3) oxidized Pt electrodes for various times under various conditions. Afterwards they chemically stripped the "oxides" formed in a solution containing $0.2M$ HCl + $0.1M$ NaCl. From spectrophotometric analysis for $PtCl^{2-}$ and $PtCl_6^{2-}$, they concluded that two oxides are present, PtO and PtO_2, in a ratio of 6 to 1. It is important to note that only part of the film could be stripped. A curious observation was the fact that under stronger oxidizing conditions more oxide was formed, but the ratio, PtO/PtO_2, remained constant at a value of 6. On general principles, one would have predicted that the concentration of PtO_2 would increase at the expense of the PtO.

Breiter and Weininger (33) obtained open-circuit decay and potential sweep curves on Pt electrodes in solutions of $0.2M$ HCl $+ 0.1M$ NaCl. They concluded that Anson and Lingane's results could be interpreted in terms of a mixed potential mechanism in which the local cathodic current arises from the reduction of the platinum–oxygen layer,

$$\text{Pt-O} + 2\text{H}^+ + 2e \rightarrow \text{Pt} + \text{H}_2\text{O} \qquad I_c \qquad (2.13)$$

and the local anodic current from the reactions of Pt with Cl^- ions,

$$\text{Pt} + 4\text{Cl}^- \rightarrow \text{PtCl}_4{}^{2-} + 2e \qquad I_a{}' \qquad (2.14)$$

and

$$\text{Pt} + 6\text{Cl}^- \rightarrow \text{PtCl}_6{}^{2-} + 4e \qquad I_a{}'' \qquad (2.15)$$

In this case, the Pt in solution comes from Pt metal, not Pt oxide. Also, the local cell currents (not current densities) must be equal since no external current flows,

$$I_c = I_a{}' + I_a{}'' \qquad (2.16)$$

If $I_a{}'/I_a{}''$ is a constant, one would always have a constant ratio of Pt^{2+} to Pt^{4+}. Therefore, although oxides may be on the surface, Anson and Lingane's chemical analysis is not sufficient proof of their existence. This same criticism applies to the recent work of Every and Grimsley (72).

Mayell and Langer (182) anodized electrodes made of bright Pt, platinum black, and polytetrafluoroethylene bonded Pt black to various potentials and deduced the nature of the adsorbed oxygen layers from cathodic stripping curves. Contrary to what has been reported elsewhere, they found that oxygen may be adsorbed in a variety of forms, $Pt(OH)_2$, $Pt(O)_2$, PtO, "tight PtO," $Pt(O)_{2.5}$, and $Pt(O)_4$ (parentheses indicate Pt/O ratios rather than compounds). Various potentials were associated with these oxide forms. However, these workers used what is called a shielded electrode (10) in which a Pt disk electrode is suspended horizontally in an open ended glass tube. Probably, oxygen which was evolved during the anodic forming processes was trapped in the stagnant solution layers next to the electrode and contributed to the reduction current during the cathodic stripping process. Although a shielded electrode may be useful in studying diffusion controlled processes, such as a polarographic determination of oxygen dissolved in solution (see Chapter VI), it is difficult to see how this electrode system could be justified in electrochemical kinetic studies. Most other workers stirred the solution around the electrode virogorously with purified N_2 so that only the adsorbed oxygen would contribute to the reduction current.

Other workers (e.g., 21,26,41,69,86,95,116,188,195,215,242) favor the concept that the oxygen exists on a Pt surface as adsorbed oxygen atoms, Pt-O. On oxygen-free surfaces, Bockris and co-workers (51,195) observed that a square root relationship exists between the partial pressure of oxygen and the amount of oxygen adsorbed on the Pt surface as determined from cathodic stripping techniques. This indicates that oxygen is adsorbed as atoms by a dissociative adsorption process.

The heat capacity and entropy changes produced by the adsorption of a monolayer of oxygen on platinum black were measured recently (76), and an equation was derived to fit the experimental results. It could not be decided whether the adsorbed oxygen was present in the form of atoms or molecules, since the equation could be made to fit the data for either form of adsorbed oxygen. Therefore, the definitive experiment is yet to be done to determine conclusively in what form the adsorbed oxygen on Pt exists.

In spite of this fact, many of the observed phenomena associated with oxide formation may be explained by the concept of dissolved oxygen in the surface layers of the Pt metal. The term "dermasorbed oxygen" has been applied to this sorption form by Schuldiner and Warner (215, 250). Terms such as "dermasorbed oxygen," "Pt-O alloy," or simply "dissolved oxygen" are preferred to the term "phase oxide" since this term implies a certain stoichiometry which is not believed to be present. The oxygen occluded in the interior of the metal is considered to be dissolved in the Pt metal over a continuous range of solid solutions. Hence, "alloy" is a more apt term.

There is good evidence in the literature (110,143,172,188) that oxygen may be dissolved in massive Pt. Kalish and Burshtein (143) found that the work function increased when oxygen was adsorbed on a Pt surface because of the presence of the negative dipoles of the adsorbed oxygen atoms. Similar observations were obtained by Giner and Lange (97). After about 11 hr, however, Kalish and Burshtein found that the work function decreased. They concluded that adsorbed oxygen must diffuse into the Pt metal. To confirm this contention, they allowed oxygen to be adsorbed on Pt at various temperatures and then measured the quantity of oxygen on the surface with cathodic stripping in a N_2 stirred solution. A monolayer of adsorbed oxygen was found. When the circuit was opened and the electrode remained at rest for a given time, more adsorbed oxygen could be measured from a second cathodic charging curve which could be explained by the oxygen diffusing from the interior of the Pt metal. Up to about three layers of adsorbed oxygen could be obtained by repeating this procedure. Apparently the Pt becomes saturated at about the equivalent of three or four monolayers of absorbed oxygen. Only a monlayer exists on the surface, and the rest is dissolved inside the body of the metal. It has been reported (69) that the adsorbed oxygen on a Pt surface becomes more tightly bound with time.

By using rapid (high current density) and slow (low current density) charging pulses, Schuldiner and Warner (215) were able to distinguish between oxygen adsorbed on the surface and that dissolved in the interior of the Pt. With rapid charging, only the surface oxygen is measured, whereas with slow charging both kinds are measured. They found that although oxygen dissolves in the Pt metal relatively quickly it is removed very slowly, and if enough time is allowed, oxygen equivalent to multilayers may be sorbed.

Laitinen and Enke (153) suspected that oxygen was dissolving in the Pt because the anodic current varied with time at constant potential, but they suggested that the oxygen diffused to the grain boundaries where it entered the bulk of the metal. From photomicrographs taken on gold anodes in $HClO_4$ solution, Laitinen and Chao (152) concluded that the adsorbed oxygen was concentrated at the grain boundaries. It was shown (177) that the data from photomicrographs are not sufficient to determine where the oxygen is located on the surface of Pt because one may illuminate the object to give almost any relative shading of the surface. Mohilner and co-workers (177) found that an anodized surface looks the same as an unanodized one, and besides, even if all of the oxygen were concentrated at a grain boundary, it still would be too thin to see. They favor the viewpoint that the oxygen is distributed randomly over the surface of the electrode. This seems to be a reasonable point of view, since the heat of adsorption of oxygen on metals is high (23,238) and since the oxygen film on some metals is immobile (ref. 238, p. 215). Davtyan (53,54), however, suggests that O atoms may migrate between active and inactive sites on platinized Pt.

This concept of dissolved oxygen in Pt may explain some of the inconsistencies mentioned before. With regard to El Wakkad and Emara's observations, the irregularities in the anodic charging curve may be due to changes in the rates of film formation, oxygen evolution, and oxygen dissolution in the Pt produced by interactions between these processes at the very low charging rates. It is significant that the cathodic charging curves exhibit only one plateau corresponding to one type of adsorbed oxygen. Most likely, the series of unusual oxides found in various potential regions by Mayell and Langer (182) and by Watanabe and Devanathan (252) can be explained in terms of combinations of oxygen adsorbed on the surface and oxygen dissolved in the metal leading to various steady mixed potential situations produced by variations in the electrode pretreatment. The "tight PtO" of Mayell and Langer may very well be correlated with dermasorbed oxygen.

A more likely explanation for Feldberg, Enke, and Bricker's findings may also rest on the concept of oxygen dissolution. In the anodic charging, $Q_a > Q_c$ because the current can produce evolved molecular oxygen (probably almost negligible since the oxygen overvoltage is so high) and oxygen which diffuses into the metal interior, in addition to the oxygen adsorbed on the surface. The cathodic stripping measures only the surface oxygen because the dissolved oxygen is difficult to remove. With repeated cycles of anodic and cathodic charging, the Pt interior becomes saturated with dissolved oxygen and Q_a approaches Q_c. This was observed (73). In support of these ideas, a half-oxidized state was not found with Au (73). It is generally agreed (12) that Au does not dissolve gases, so the complication of oxygen dissolution is not met. A similar explanation may be made for the observations of French and Kuwana (80) and of Breiter (29) that a reproducible triangular potential sweep curve can be obtained only after at least 10 complete cycles of polarization have been taken. It was further observed that nearly 50% of the change in shape occurred between the first and second cycles. Accordingly, it required over 10 anodic potential scans to saturate the platinum metal with dissolved oxygen. Breiter suggests that the preliminary anodic sweeps are needed to remove impurities from the electrode surface. Although it is agreed that such a process may account for some of the change in the shape of the curves, it is maintained that the major change is produced by the dissolved oxygen. Once the Pt is saturated with dissolved oxygen, a surface of constant catalytic activity is obtained, and a reproducible potential scan pattern is observed.

The peroxide explanation of Vetter and Berndt (242) may be less satisfactory than the dissolved oxygen idea, since, as pointed out by Laitinen and Enke (153) and by Frumkin and co-workers (85), peroxide is not observed when cathodic stripping is carried out in N_2-saturated solutions. The presence of peroxide is detected only during the cathodic reduction of molecular oxygen in O_2-saturated solutions. These observations lend support to the contention that the oxygen adsorbed on a Pt surface exists in the atomic form.

Anson and Lingane (3) observed that all of the oxygen adsorbed on the Pt electrode could not be stripped by the chloride chemical stripping solution. In all probability this unreactive oxygen was dissolved in the

Pt metal and could not participate in the local cell mechanism of Eqs. 2.13–2.15.

The ellipsometric observations made by the Bockris group (91,196) may be interpreted in terms of oxygen dissolved in the Pt metal rather than the appearance of a definite oxide. Oxygen adsorbed on a Pt surface from O_2-saturated acid solution as in certain rest potential studies (116) does not produce measurable readings in ellipsometric studies (91, 196). Possibly, the threshold potential, 0.98 V, where measurable readings first appear corresponds to the minimum energy required to dissolve O_2 in the Pt surface layers. Schuldiner and co-workers (215b) concluded that dermasorbed oxygen is unstable below 1 V. Since the ellipsometer responds to changes in density, the dissolved oxygen modifies the density of the surface layers, and a change in the polarization of the light beam is obtained. As the electrode is anodized to higher potentials, more oxygen is dissolved. Because this oxygen is tightly bound (69,187,215), a large hystersis in the ellipsometric readings is observed when the current is reversed and the electrode is cathodized. Consequently, the ellipsometric data do not prove that a definite oxide is present or absent on Pt anodes below 1.5 V.

2. The Normal O_2 Electrode

As pointed out before (115,116), the reversible potential of 1.23 V would be observed if a complete monolayer of electronically conducting Pt-O could be formed on the Pt surface. Under these conditions the electrode would be truly inert to oxygen saturated H_2SO_4 or $HClO_4$ solutions. Such a situation was first observed by Bockris and Huq (19), who were able to maintain a potential of 1.24 ± 0.03 V (15 independent experiments) for about 1 hr in O_2 saturated H_2SO_4 solutions, after which the potential decayed to less noble values. This potential was the O_2/H_2O potential because the proper variation of potential with changes in P_{O_2} and pH was observed. They stated that this was the first time that the 1.23 V potential had been observed because no one previously had controlled the concentration of impurities rigorously enough. Since the i_0 for the oxygen reaction is so low (10^{-10} to 10^{-9} A/cm²), many impurity reactions could become potential-determining at potentials below 1.23 V.

Although it is agreed that control of impurities is one essential requisite for observing the reversible O_2 potential, the other essential requisite is the presence of a complete layer of Pt-O. Bockris and Huq prepared their Pt electrodes in pure O_2 at 500°C for 2 hr just before plunging them into prepurified-O_2-saturated acid solution. This preparation produced the complete layer of electronically conducting film of Pt-O. Frumkin and co-workers (70) showed that prolonged heating of Pt in oxygen not only enabled more O_2 to be sorbed by Pt, but also produced more firmly bound oxygen. Since the complete layer of adsorbed oxygen is unstable in acid solutions (116,118), bare platinum sites were exposed, the mixed potential mechanism set in, and the potential fell to less noble values.

A similar explanation may be applied to the results of Watanabe and Devanathan (252), who observed exactly the same phenomena as Bockris and Huq. They (252) found that the 1.23 V potential could be observed only if the Pt were preanodized in solution followed by the raising of the electrodes into the pure O_2 atmosphere above the solution for 48 hr. Such a procedure produced the unstable complete layer of Pt-O, and the reversible potential was maintained for about 1 hr, after which the mixed potential mechanism sets in with the resulting decrease in potential. Bianchi and co-workers (16) also observed this.

When bright Pt is in contact with concentrated HNO_3 for 72 hr or more (118,127), a thin, invisible, protective film is formed on the electrode surface. From double layer capacity measurements (124), it is assumed that the film is a complete, electronically conducting layer of adsorbed oxygen not much thicker than a monolayer. When such an electrode is plunged into oxygen-saturated $2N$ H_2SO_4 solution (118), a potential of 1.225 ± 0.010 V is observed for periods of time well over 24 hr. Similar observations have been made by Wrotenbery, Hurd, and Snavely (272). With time, the potential decays to less noble values of potential, indicating that eventually this much more stable film of adsorbed oxygen becomes impaired and a mixed potential mechanism is set up. The O_2/H_2O reaction is potential-determining at the HNO_3-treated Pt electrode because the potential–log P_{O_2} curve has a slope of 15 mV (118). Also, the correct dependency of the potential on pH is observed.

Other than HNO_3, passivating agents such as 30% H_2O_2, concentrated $HClO_4$, fuming H_2SO_4, and HF did not produce a complete, protective film on Pt (133). Of the noble metals, only Pt and Rh when treated with HNO_3 (118) exhibited the reversible oxygen potential in O_2-saturated sulfuric acid solution.

The complete layer on HNO_3-treated Pt electrodes may also be impaired by anodic or cathodic polarization (118,127). If the complete layer of Pt-O could be stabilized in acid solutions, a stable reversible oxygen electrode (normal oxygen electrode) could be made. However, even after the passive film had been removed with cathodic stripping or by reduction in H_2, a rest potential more noble that 1.06 V by about 0.1 V was observed when the Pt electrode was again placed in O_2-saturated acid solution. It was found (127) that the surface of such electrodes was a better catalyst for the reduction of O_2 and the catalytic decomposition of H_2O_2. Evidently the HNO_3 treatment modified the Pt surface in addition to producing the complete, electronically conducting film of Pt-O.

X-ray diffraction patterns were obtained from HNO_3 treated and untreated bright electrodes, and it was found (133) that the diffraction lines obtained from the treated sample were slightly shifted from those obtained from the untreated sample. From vacuum fusion analyses (133), it was observed that the treated samples contained more oxygen than the untreated. It was concluded (124,127,133) that the HNO_3 treatment enabled the Pt to dissolve much more oxygen than mere anodization could. Since the x-ray lines were shifted, it appears that this oxygen exists in the Pt as an alloy of Pt and O atoms. Probably the alloy is many layers deep because the x-ray pattern was obtained. Such HNO_3-treated electrodes are referred to as Pt-O-alloy electrodes (124,127,133), and the surface of these electrodes has different catalytic properties for various electrode reactions than a pure Pt surface.

3. Rest Potential in Alkaline Solution

Nearly all the studies of the rest potential have been carried out in acid solutions. Since the data obtained from charging curves and potential sweep traces indicate that the nature of the Pt–O_2 system is essentially the same in either medium, it is reasonable to assume that explanations for the behavior of the rest potential in alkaline solutions may be similar to those suggested for acid solutions.

If the rest potential is truly an equilibrium potential, then only the initial and final states of the system are important according to thermodynamic principles. As a result, adsorption of anions would not play a role. However, the rest potential is not a simple equilibrium electrode potential but a steady-state polyelectrode potential. From the shape of the double layer region in the charging curves (110) and the potential sweep curves, Fig. 2.6, OH^- ion adsorption currents most likely contribute to the additional current obtained in the double layer region over that observed in the acid case, Fig. 2.5. Consequently, the potential-determining mechanisms are similar in the acid and alkaline cases only if one assumes that the adsorption of OH^- ions does not interfere with the electrode processes of the local cell.

D. THE REST POTENTIAL ON OTHER NOBLE METALS

1. Gold

Although a highly unstable brown oxide, AuO, may exist (ref. 222, p. 178), the most important oxide of gold is the reddish brown Au_2O_3, the preparation of which is described by Roseveare and Buehrer (201). Several investigators (39,93,192) have attempted to prepare Au_2O, but the black precipitate obtained proved to be a mixture of Au_2O_3 and colloidal gold. It is not possible to prepare an Au/Au_2O electrode (138), but there is no doubt that an Au/Au_2O_3 electrode may be made.

When a gold wire electrode is used to make contact with a slurry of Au_2O_3 (140,141,232) or a Pt wire with a slurry of fine Au crystals and Au_2O_3 (93) in H_2SO_4 solutions, a potential of about 1.36 V is obtained. Hickling (111) concluded that 1.36 V is the E_0 for the Au/Au_2O_3 reaction,

$$Au_2O_3 + 6H^+ + 6e \rightleftharpoons 2Au + 3H_2O \qquad E_0 = 1.36 \text{ V} \qquad (2.17)$$

Initially, the rest potential of an oxide-free Au electrode (previously treated with H_2) in oxygen-saturated acid solution is between 0.7 and 0.8 V (11,57,117) which drifts slowly with time until a steady value of about 0.98 V (117) is reached (after about 48 hr). When a Au surface is anodized above 1.4 V, a film of Au_2O_3 may be formed (e.g.,11,102,255). Although the open-circuit potential of a previously anodized Au electrode drifts toward 1.36 V with time, it passes through this value to less noble potentials (11,117,152). This has been explained (117) as a depolarizing effect of H_2O_2 generated at the cathode during the anodic polarization of the gold electrode. When the soluble H_2O_2 diffused to the Au electrode, the rest potential was depressed by a mixed potential mechanism (130a). Brummer and Makrides (37,38) found that the higher the potential at which the Au electrode was anodized the further the rest potential deviated from 1.36 V toward less noble values after the polarizing circuit had been broken. Since they used a Au counterelectrode and since Au is a poor peroxide-decomposing catalyst (126,130a), it is likely the concentration of H_2O_2 in solution increased with the forming potential, causing the rest potential to be depressed further. If the H_2O_2 was destroyed by stirring the system with H_2, a potential of 1.36 V was observed after the system had been saturated with O_2 again. During the H_2-reduction process, some of the Au_2O_3 had been reduced to Au which made a more intimate contact between the oxide and the metal, a situation similar to that of the slurry system.

The Au_2O_3 forms a flaky, poorly adherent oxide film which continues to grow with anodic polarization (11) in contrast to the monolayer formation of Pt-O on Pt.   From these observations, it is reasonable to assume that these oxide films are better ionic than electronic conductors. Some evidence for this conclusion is obtained from double layer capacity measurements (9,123,207,208) in which it is found that the capacity decreases as oxide is formed on the surface.   This behavior is opposite to the effect found on Pt (124,164a,212).   Gold oxide is unstable in acid solutions and disappears with time (66,111,117,162).   This phenomenon is accompanied by a decrease in the potential (117) to a value of about 0.98 V (after about a week), and the appearance of the electrode surface changes from reddish brown to black.   Some gold may go into solution, probably as a complex (141; ref. 222, p. 177), but most of it has been reduced to form a Au black electrode which exhibits the same rest potential as the bright Au electrode.

With anodization below about 1 A/cm², Au does not build up visible films in alkaline solutions (4,173,181,221,232,) and apparently, only a monolayer of Au_2O_3 is adsorbed.   Auric oxide appears to be more soluble in alkaline than in acid solutions, since the solutions become yellowish with anodic polarization (232).   The rest potential of the Au/Au_2O_3 electrode in KOH solutions (232) was about 1.1 V.   Possibly the deviation from 1.36 V may be accounted for by the presence of a relatively large amount of gold in solution in the alkaline solutions.

a. Charging Curves.   Many workers (4,13,38,66,111,152,162,173,181, 192,221,242) have recorded constant current charging curves to obtain evidence for the nature of the films formed on Au anodes.   Only El Wakkad and El Din (66) report the presence of Au_2O and AuO on the electrode surface from the appearance of small irregularities in the anodic charging curve obtained at low current densities on a gold-plated platinum foil.   It is significant that only one reduction plateau appeared in the cathodic stripping curve in agreement with the other investigators. The general opinion is that only one oxide is formed, and its formation begins above a value of about 1.35 V.

The majority of observers (73,111,152,190) reported that $Q_a = Q_c$ which agrees with the viewpoint that oxygen is not dissolved in Au as it is in Pt.   Although Laitinen and Chao (152) offered evidence from photomicrographs that oxygen could diffuse along grain boundaries into the interior of the Au, Mohilner and co-workers (177) showed that such evidence was not adequate.   Only Vetter and Berndt (242) found $Q_a >$ Q_c, but this was true only at low current densities.   At higher charging rates, Q_a approached Q_c.   Possibly at the low charging rates, some current is used in forming O_2 at "bare" Au sites where the time required for the film of Au_2O_3 to flake off and expose the "bare" Au sites was available, or some Au_2O_3 was dissolved during the relatively long charging time so that this quantity of Au_2O_3 could not contribute to the cathodic discharge current.   At the high current densities, these side processes do not occur, and $Q_a = Q_c$.   It is highly unlikely that the oxide is reduced to H_2O_2 as they (242) suggest, because if this were the case there would not be a variation of Q_a/Q_c with charging rate.

b. Potential Sweep Traces.   Potential sweep experiments have been carried out on Au electrodes (30,102,255).   The results show that only one oxide, Au_2O_3, is formed and does not appear until a potential more noble than 1.35 V is reached.   During the cathodic sweep, the reduction of the oxide appears at less noble potentials than the formation which indicates that the reduction of the Au_2O_3 layer occurs with a rather large overvoltage as seen in Fig. 2.11.   From these data, $Q_a/Q_c \sim 1$ in agreement with the results obtained from charging curves.

c. Problem of Adsorbed Oxygen on Gold.   There is some controversy over the question of whether a gold surface may adsorb oxygen below a potential of 1.3 V.   In the dry state, the experimental evidence (ref. 238, p. 175) supports the conclusion that gold does not adsorb oxygen.   Data obtained from constant current charging curves and from potential sweep traces do not offer any indication of the presence of adsorbed oxygen on Au below 1.3 V.

After a gold wire has been heated in air at 900°C and placed in O_2-saturated acid solution, Deborin and Ershler (57) found that a rest potential of about 0.9 V was observed and the cathodic stripping curve exhibited a step at about 0.8 V.   Clark and co-workers (44) studied the

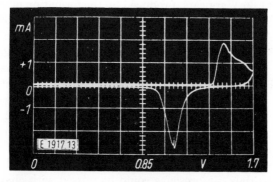

Fig. 2.11. Potential sweep curve on a bright Au electrode in N_2 stirred $8N$ H_2SO_4 solution (255). (By permission of Bunsengesellschaft.)

charging curves obtained on various samples of gold wires preheated in air, and concluded from the data that Deborin and Ershler had used impure gold, since the step in the reduction curve could be explained by the presence of metallic impurities such as iron in the gold surface. Since Clark (44) reported the current density but not the potential at which they anodized the gold, it is not possible to comment on the results of the cathodic reduction curves obtained on preanodized electrodes because it is not known whether or not the potential became more anodic than 1.3 V.

In any case, it was found (117) that gold electrodes which had been heated in air or reduced by H_2 in solution displayed the same potential behavior when placed in O_2-saturated acid solution. Initially the rest potential was between 0.7 and 0.8 V, but with time came to a steady value of 0.98 V. Even a Au electrode which had been anodized well above 1.35 V eventually with time arrived at a steady rest potential of 0.98 V. Apparently, if Au does not adsorb oxygen in the dry state, it does from an O_2-saturated solution, as indicated by the steady rest potential of 0.98 V. Young (275) agrees with this viewpoint.

Additional evidence which favors the presence of adsorbed oxygen, Au-O, on Au below 1 V is obtained from double layer studies. As oxygen is adsorbed from solution, the double layer capacity decreases (123), and when it is removed with H_2 stirring, the capacity increases again. Other investigators (9,207,208) have found that the double layer capacity decreases in the potential range 0.8 to 1.0 V which has been attributed (123) to the presence of Au-O.

Steady-state potentiostatic polarization curves which were obtained on Au anodes in oxygen saturated acid solution (129) show a limiting current region between potentials of 0.9 and 1.36 V before a passivation region above 1.36 V appears, produced by the formation on nonconducting Au_2O_3. This limiting current region is also attributed to the building of a layer of Au-O which is converted to Au_2O_3 at potentials above 1.36 V.

Under open-circuit conditions, the steady-state coverage of the Au surface with adsorbed O atoms in O_2 saturated acid solutions is probably very low. Bockris and co-workers (195) record a value of about 0.03 for θ. It is known that small amounts of substances adsorbed on an electrode surface may change the potential greatly (e.g., 82,98); in fact, the point of zero charge on Pt may be changed from 0.11 to about 1.0 V in going from a H_2 stirred to an O_2 stirred solution (81).

Although such a small amount of adsorbed oxygen could produce the observed potentials, its presence could be missed with charging curves if the charging current was too high. As seen in Fig. 2.12, at the lowest charging current, the curves obtained by Vetter and Berndt (242) show a small kink at a potential value of about 1.0 V in addition to the large one at 1.36 V. At higher charging rates, the low potential kink disappears. Since the potential sweep methods are transient techniques, the observed current always contains a contribution from non-Faradaic processes. At the sweep rates recorded in the literature (30,102,255), the non-Faradaic currents may be so much greater than the Faradaic currents that the small amount of oxygen adsorbed on the surface may not be detected. Even ellipsometry techniques (196,197) cannot detect the presence of adsorbed oxygen atoms on a metal surface. Therefore, none of these methods provides adequate evidence for the absence or presence of adosrbed oxygen atoms on Au electrodes in O_2-saturated solutions in the potential range from 0.8 to 1.3 V. In the absence of other available techniques, it seems that steady-state potential measurements may be the safest means for detecting the presence of adsorbed O atoms on noble metal surfaces.

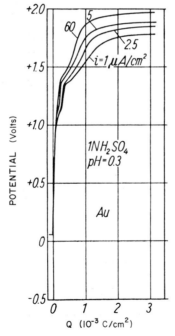

Fig. 2.12. The dependence of the anodic charging curve on the current density (from 1 to 60 μA/cm²) taken on Au electrodes in $1N$ H_2SO_4 solution at 25°C (242). (By permission of Bunsengesellschaft.)

d. Potential-Determining Mechanisms. There seems to be little doubt that the rest potential observed at 1.36 V on Au electrodes in acid solution (130a) is due to a true metal–metal oxide system, the Au/Au_2O_3 electrode, with the potential-determining reaction, Eq. 2.17. In this case, the rest potential varies by 60 mV per unit of pH and is independent of P_{O_2} (117) as required by Eq. 2.17. However, the rest potential at 0.98 V is not so easily explained, because in this case also, the rest potential is independent of P_{O_2} and varies with pH at a rate of 60 mV per unit of pH.

Since true oxides apparently are not present on the Au surface at potentials below 1.3 V, oxide potential mechanisms are rejected to explain the behavior of the rest potential at 0.98 V. One explanation may be found in the concept of mixed potentials. It has been assumed (117) that a Au/Au-O reaction,

$$Au\text{-}O + 2H^+ + 2e \rightleftharpoons Au + H_2O \tag{2.18}$$

exists and that a local cell is set up at the Au surface composed of the O_2/H_2O, Eq. 2.3, and the Au/Au-O, Eq. 2.18, reactions. Because Au is a poor catalyst for the O_2/H_2O reaction (123), Eq. 2.3 is strongly polarized, as seen in Fig. 2.13, and at the mixed potential where the curves cross, the O_2/H_2O reaction is in a limiting current region. Under these conditions, the rest potential is controlled by the Au/Au-O reaction which would account for the experimentally determined dependency of

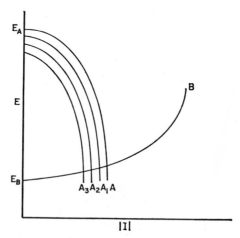

Fig. 2.13. A sketch of local cell polarization curves for the Au/O₂ system. The notation is the same as in Fig. 2.9 with curve B representing the polarization of the Au/Au-O reaction.

the rest potential on pH and P_{O_2}. Variations of the rest potential with time may be traced to changes in the activity of the Au-O layer with the amount of oxygen adsorbed on the Au surface.

The reversible oxygen potential is not observed on Au (117,118,252). Possibly, this is so because a complete, electronically conducting, thin film of adsorbed oxygen cannot be formed on a Au surface, so that consequently the surface is never truly inert to oxygen.

2. Palladium

According to Sidgwick (ref. 222, p. 1558), only one stable oxide of Pd, PdO, exists. What was supposed to be Pd_2O has been shown (264) to be a mixture of PdO and Pd. Although Wöhler and Martin (265) claim to have made Pd_2O_3, which is highly unstable, most likely it exists only in the hydrated form (ref. 222, p. 1573). A very unstable PdO_2 may be made wet which slowly loses O_2 on standing (ref. 222, p. 1575). When Pd is heated in O_2 (260), PdO is formed, which is a very insoluble black powder but which appears as a purple film on the metal surface. This oxide has been identified by Moore and Pauling (179) by x-ray analysis studies. In a manner similar to the formation of PtO_2, PdO may be made by fusing $PdCl_2$ with $NaNO_3$ at 600°C (219). Shriner and Adams (219) report that the black PdO has catalytic properties similar to the brown PtO_2 for several types of reactions. Although Jirsa (139) has reported evidence for the formation of Pd_2O and PdO_3, these "oxides" are probably mixtures of PdO and Pd. A potential–pH diagram for the Pd/H_2O system has been constructed (61b).

a. Charging Curves. The evidence from charging curves (17,42,65, 113,175,242) is that two oxides, PdO and PdO_2, may be formed on Pd electrodes electrochemically. Before O_2 is evolved on a Pd anode, a monolayer of adsorbed oxygen is formed. Butler and Drever (42) concluded that first a layer of PdO was formed which was converted to a monolayer of PdO_2 as O_2 was evolved. On open circuit, the layer of PdO_2 spontaneously decomposed to Pd. The monolayer of oxide is a good electronic conductor, for as noted by Mazitov and co-workers (175), oxygen may be evolved on Pd anodes without an increase in the thickness of the oxide layer. With very strong anodization, Butler and Drever observed that an oxide film several layers deep could be formed on Pd electrodes.

On Pd electrodes in acid solutions, Vetter and Berndt (242) found that $Q_c = 0.5Q_a$, but Hickling and Vrjosek (113) observed that $Q_c = Q_a$. As in the case of Pt, oxygen may be dissolved in the Pd metal (56,225), and under the conditions of slow charging rates used by Vetter and Berndt, oxygen that was dissolved in the metal during the anodic charging was not removed during the cathodic stripping. On the other hand, Hickling and Vrjosek used high current pulses, and enough time was not allowed to dissolve significant amounts of oxygen in the Pd metal during the anodic charging process.

Although the two plateaus are well defined in acid solutions, the two steps become less clear in alkaline solutions.(65) It was found (42) that PdO_2 is more stable in alkaline solutions. Müller and Riefkohl (180) observed only the presence of PdO in Cl^- ion-containing solutions.

b. Potential Sweep Traces. The potential sweep traces taken on a Pd electrode in N_2 stirred acid solution were recorded by Will and Knorr (255). In the hydrogen region, large negative currents are recorded which reflect the remarkable affinity Pd has for hydrogen (84,134). Oxygen first begins to adsorb at 0.75 V, and O_2 is evolved at about 1.5 V. Since the reduction peak is displaced toward more negative values than the formation peak, the reduction of the oxide layer on Pd occurs with a large overvoltage. The determination of the charge under the reduction peak which gives results which correspond to about a monolayer of oxide in agreement with the results obtained from charging curve techniques.

c. Potential-Determining Mechanisms. When an oxide-free (treated with H_2) Pd electrode is placed in an O_2-saturated, peroxide-free acid solution, a steady rest potential of about 0.87 V is observed (120) after the dissolved hydrogen has been removed from the Pd. It was found that the rest potential was independent of pH but dependent on P_{O_2} with a slope of 33 mV for the potential–log P_{O_2} curve. Schuldiner and Roe (214) also observed the E–log P_{O_2} slope of about 30 mV in a certain pH range, but in addition, they found a complicated dependence of the rest potential on pH with a maximum value at a pH of 1.5. They were not able to offer a mechanisms but concluded that a complex mixed potential mechanism must be involved.

Vetter and Berndt (242) have measured the rate of dissolution of Pd in H_2SO_4 solutions as a function of potential. They found that Pd is inert below 0.8 V, becomes active between 0.8 and 1.15 V with a maximum corrosion rate at 0.9 V, is passive between 1.15 and 1.75 V, and is transpassive above 1.8 V. If Pd goes into solution according to the reaction

$$Pd \rightarrow Pd^{2+} + 2e \tag{2.19}$$

Pd^{2+} ions near the electrode could react with dissolved oxygen in solution or with O_2 molecules adsorbed on the surface to produce adsorbed PdO_2 according to the equation

$$Pd^{2+} + O_2 + 2e \rightarrow (PdO_2)_{ads} \tag{2.20}$$

If a local cell composed of Eqs. 2.19 and 2.20 were set up such that the Pd/Pd^{2+} reaction was virtually at a limiting current at the mixed potential (similar to the diagram in Fig. 2.13), the rest potential would be determined by the Pd^{2+}/PdO_2 reaction, Eq. 2.20. Under these conditions, the requirements that the potential be independent of pH but dependent on P_{O2} and the overall potential-determining process be a two-electron reaction (33 mV slope) would be satisfied.

Such a mechanism has been proposed (120). When the oxide-free Pd surface first comes in contact with the O_2-saturated acid solution, oxygen is adsorbed on the metal surface, and a partial layer of PdO is formed. The presence of PdO on the metal surface causes the rest potential to become more noble. Hickling and Vrjosek (113) reported that the standard potential of the Pd/PdO reaction,

$$PdO + 2H^+ + 2e \rightleftharpoons Pd + H_2O \qquad E_0 = 0.85 \text{ V} \qquad (2.21)$$

is 0.85 V. Initially the rest potential depends on pH (60 mV per unit of pH). As the PdO layer builds up, the potential reaches a point where corrosion of the Pd metal becomes important. In this potential region, Pd can go into solution, and Eq. 2.20 can take place. Eventually, the local cell mechanism of Eqs. 2.19 and 2.20 becomes potential-determining. In this case, the rest potential becomes independent of pH as found experimentally.

When a strongly preanodized (above 1.8 V) Pd electrode was placed in O_2-saturated, peroxide-free, $2N$ H_2SO_4 solution (120), a potential of 1.47 V was observed which was steady for several hours before decaying to less noble values. The 1.47 V potential was independent of P_{O2} and of the stirring rate and depended on pH by 60 mV per unit of pH. For the 1.47 V potential, the Pd/PdO_2 reaction,

$$PdO_2 + 4H^+ + 4e \rightleftharpoons Pd + 2H_2O \qquad E_0 = 1.47 \text{ V} \qquad (2.22)$$

is favored (120,128) as the potential-determining reaction. In support of this viewpoint, Butler and Drever (42) concluded that the monolayer of PdO was converted to a monolayer of PdO_2 when the evolution of O_2 was reached but decomposed spontaneously on open circuit. With time, the potential falls from the 1.47 V value to a steady one of 0.87 V, which is attributed (120) to the decomposition of the unstable PdO_2 by a reaction similar to

$$PdO_2 \rightarrow PdO + \tfrac{1}{2}O_2 \qquad (2.23)$$

El Wakkad and El Din (65) favor the reaction

$$PdO_2 + 2H^+ + 2e \rightleftharpoons PdO + H_2O \qquad (2.24)$$

as potential-determining. The reversible O_2 potential has not been observed on Pd (118).

3. Rhodium

According to Sidgwick (ref. 222, p. 1527), compounds with higher oxygen content than that of Rh_2O_3 do not exist. Although Wöhler and co-worlers (269) suggested the existence of RhO and Rh_2O, no evidence for the presence of such oxides was found (206). Such oxides are probably mixtures of Rh_2O_3 and Rh (ref. 222, p. 1516). When Rh is heated above 600°C in air or oxygen, the highly insoluble gray Rh_2O_3 is formed which may decompose when heated above 1150°C. A dioxide, RhO_2, may be made in the hydrated form only. Consequently, Rh_2O_3 is the only stable oxide of Rh. For further information one is referred to the potential–pH diagram for the Rh/H_2O system (239a).

a. Charging Curves. The results obtained from charging curves taken on Rh electrodes in N_2 stirred acid and alkaline solutions (31,42, 164b,164c,238a) show that a layer of adsorbed oxygen is formed on the metal surface before O_2 is evolved. These oxygen films do not grow much beyond a monolayer thick, indicating that these films are good electronic conductors. Supporting evidence for this contention is obtained from double layer capacity measurements (131) since the capacity does not fall when O_2 is adsorbed on Rh. Butler and Drever (42) found that an oxide ("peroxidic oxygen") was formed at potentials where O_2 is evolved. From rest potential studies on Rh (119), a more tightly bound oxygen was obtained with anodic polarization. Although the presence of an oxide, Rh_2O_3, may be assumed (146c,164c), it was concluded that anodic polarization enabled O_2 to be dissolved in Rh in a manner similar to the Pt case to form a Rh-O alloy in the outer layers or skin of the metal surface.

b. Potential Sweep Traces. Oxygen begins to adsorb (~ 0.6 V) at less noble potentials on Rh electrodes (22,146c,238a,255) than on Pt electrodes as seen from potential sweep curves (see Fig. 2.14). This oxygen adsorbed during the anodic sweep is so strongly held to the surface that the oxygen reaction region overlaps the H^+ ion reduction region. In alkaline solutions, Fig. 2.15, the overlap is even more pronounced. A thin skin of Rh-O alloy would be expected to be more difficult to reduce than a layer of adsorbed oxide on the electrode surface. In agreement with this viewpoint, Böld and Breiter (22) found from potential sweep studies that $Q_a > Q_c$ which was attributed to the strong resistance of the adsorbed film to reduction.

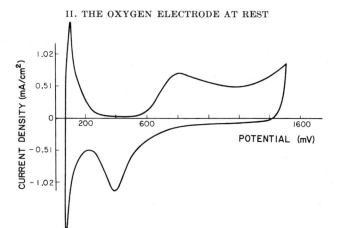

Fig. 2.14. Potential sweep curve on bright Rh in 2.3M H$_2$SO$_4$ with a sweep rate of 0.5 V/sec at 25°C (22).   (By permission of Pergamon.)

c. Potential-Determining Reactions.   It was interpreted from rest potential studies (119) that an oxide-free Rh surface (pretreated with H$_2$) in contact with an O$_2$ saturated acid solution adsorbs oxygen from the solution under open-circuit conditions as a monolayer of adsorbed oxygen atoms.   Within a period of about 24 hr, a steady rest potential was reached at a value of 0.93 V which depended on the P_{O_2} with a slope of the potential–log P_{O_2} curve equal to 15 mV and on the pH with a change of 60 mV per unit of pH.   As in the Pt/O$_2$ case, the system

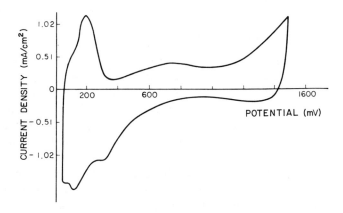

Fig. 2.15. Potential sweep curve on bright Rh in 0.1M NaOH at 25°C with a sweep rate of 0.5 V/sec (22).   (By permission of Pergamon.)

is considered (119) to be a polyelectrode, and the rest potential is a mixed potential. When the system was stirred with N_2, a steady potential of about 810 mV was obtained after a period of about 20 hr.

The local cell responsible for the mixed potential is composed of the O_2/H_2O reaction, Eq. 2.3, and the Rh/Rh-O reaction,

$$\text{Rh-O} + 2\text{H}^+ + 2e \rightleftharpoons \text{Rh} + \text{H}_2\text{O} \qquad E_0 \sim 0.81 \text{ V} \qquad (2.25)$$

In a manner similar to the Pt case, it is reasoned that the Rh/Rh-O reaction is in a virtual limiting current region at the mixed potential point as in Fig. 2.9.   Under these conditions the O_2/H_2O reaction is potential-determining, thus accounting for the experimentally determined dependency of the rest potential on P_{O_2} and pH.   From the interaction of the adsorbed oxygen film on a Rh/Rh-O electrode with dissolved H_2 when the system is stirred with H_2, it was concluded that the oxygen of the Rh-O film is no more tightly bound to the Rh surface than the oxygen of the Pt-O film is to the Pt surface.   When the Rh electrode is anodized, however, the adsorbed oxygen layer produced by anodization is very tightly bound which is indicated by the great difficulty with which it is reduced by dissolved H_2 (119) or by cathodization (see Fig. 2.14).

If a preanodized electrode is placed on O_2-saturated, peroxide-free, $2N$ H_2SO_4 solution (119), a steady rest potential of 0.88 V is observed which is independent of P_{O_2} but dependent on pH (60 mV per pH unit). From galvanostatic polarization studies at very low current densities in N_2 and O_2 stirred acid solutions (119) and from potentiostatic polarization studies (131), this rest potential appears also to be a mixed potential.   The symbol Rh(O)_x will be used to denote the Rh-O alloy system. Then, a reaction such as

$$\text{Rh(O)}_x + 2x\text{H}^+ + 2xe \rightleftharpoons \text{Rh} + x\text{H}_2\text{O} \qquad (2.26)$$

is conceivable, and a local cell consisting of Eq. 2.26 and the O_2/H_2O reaction could be set up.   To explain the independence of P_{O_2}, the Rh/Rh(O)_x reaction must be potential-determining.   As a result, the $O_2/$ H_2O reaction must be at a limiting current at the mixed potential point (see Fig. 2.13) which may be reasonable since evidence in the literature (21,50,130,243) shows that the oxygen layer adsorbed on Pt inhibits the reduction of O_2.

When Rh was heated in air and plunged into O_2 saturated acid solution, a rest potential of 0.875 V was reached in about 3 hr.   The electrochemical properties of this system are the same as those in which the Rh was preanodized.   Apparently, the Rh-O alloy can be made by heating in air also.   It may be concluded that oxygen is bound to Rh in at least two ways: weakly by the adsorption of O_2 dissolved in solution, and strongly by anodization.

The reversible oxygen potential, 1.23 V, was observed on Rh electrodes treated with concentrated nitric acid (118). It is believed that a complete, electronically conducting film of adsorbed oxygen is produced by this treatment, and the electrode surface is truly inert to oxygen.

4. Iridium

The only stable oxide of Ir is IrO_2 which is blue when wet, but black when dry. It may form violet colloidal solutions (78) which have been observed when Ir was anodized in acid solutions by ac polarization (122) techniques. Wöhler and Witzmann (270,271) have studied the $Ir–O_2$ system and concluded that during the thermal decomposition of IrO_2 the reaction went directly to Ir and O_2 without forming Ir_2O_3 ot IrO as intermediates. Bell and co-workers (14a) have also studied the $Ir–IrO_2–O_2$ system. Although IrO does not exist, Ir_2O_3 may be made indirectly (ref. 222, p. 1533) but exists only in the hydrated form since it disproportionates to IrO_2 and Ir when heated. By fusing Ir with sodium peroxide (270), an unstable compound approximating IrO_3 is made which liberates oxygen in H_2SO_4 solutions. The only oxide important in the $Ir–O_2$ electrode system is IrO_2. Also available is the potential–pH diagram for the Ir/H_2O system (239b).

a. Charging Curves. Constant current charging curves have been obtained on Ir electrodes (32,42) in N_2 stirred acid and alkaline solutions. From these data it was concluded that a monolayer of adsorbed oxygen was adsorbed on the surface before the evolution of oxygen took place. The layer of adsorbed oxygen is a good electronic conductor since this film does not grow much beyond a monolayer thick with prolonged anodization and since the double layer capacity does not decrease as oxygen is adsorbed (132). Butler and Drever (42) could not find evidence for the presence of IrO_2 with charging curve techniques, a behavior similar to that found in the Pt/O_2 case for the presence of PtO_2.

b. Potential Sweep Traces. The evidence from potential sweep data obtained on Ir electrodes in N_2 stirred acid solution (22,255) is in general agreement with that found with charging curves. However, it was found (22) that the formation and reduction of the adsorbed oxygen layer occurs reversibly (with very little overvoltage), as seen in Fig.2.16. The reduction currents appear over the same potential region as the oxidation currents. In alkaline solutions, Fig. 2.17, the overvoltage appears larger than in acid solutions. Böld and Breiter (22) observed that a small amount of oxygen is adsorbed even after oxygen is evolved. They suggested that the additional oxygen is dissolved in the surface layers of the metal lattice.   From the appearance of the sweep curves (Figs. 2.16 and 2.17), the oxygen dissolved in Ir is more easily removed than in the Pt case.   In agreement with the data obtained from charging curves, evidence for the presence of IrO_2 is not found from the potential sweep data.   Supporting evidence for these conclusions may be obtained from steady-state polarization curves taken on Ir anodes (121,132).

c. **Potential-Determining Reactions.**   When an oxide-free Ir electrode is placed in an O_2-saturated acid solution, a steady-state rest potential of 1.02 V (121) is obtained relatively rapidly (within 10 hr).   Although the pH dependence of the rest potential is 60 mV per unit of pH, the dependence on the P_{O_2} indicates that the overall potential-determining reaction involves 6–12 electrons, which is unlikely.   After stirring the solution with N_2 for 24 hr, a steady potential of 0.87 V is observed. An explanation is offered in terms of a local cell mechanism.   It is believed (121) that a layer of adsorbed oxygen atoms, Ir-O, is formed on an Ir surface in contact with O_2-saturated acid solutions.   A local cell consisting of the O_2/H_2O reaction, Eq. 2.3, and the Ir/Ir-O reaction,

$$\text{Ir-O} + 2H^+ + 2e \rightleftarrows \text{Ir} + H_2O \qquad E_0 = 0.87 \text{ V} \qquad (2.27)$$

is set up at the Ir electrode.   According to the local cell diagram presented in Fig. 2.18, the mixed potential is determined by both half-

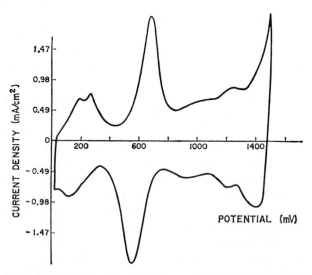

Fig. 2.16. Potential sweep curve on bright Ir in $2.3M$ H_2SO_4 at 25°C with a sweep rate of 0.5 V/sec (22).   (By permission of Pergamon.)

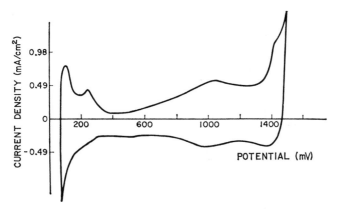

Fig. 2.17. Potential sweep curve on bright Ir in $0.1M$ NaOH at 25°C with a sweep
rate of 0.5 V/sec (22). (By permission of Pergamon.)

reactions since neither polarization curve is in a limiting current region
where the curves cross. As the P_{O_2} is lowered, the variation in the mixed
potential, E_m, with P_{O_2} is smaller than the similar variation in E_0 for the
O_2/H_2O reaction. This would account for the observed small dependence of the rest potential on P_{O_2} which was observed instead of the 15
mV dependency expected if the O_2/H_2O reaction were potential-deter
mining.

As noted by Böld and Breiter (22), there was some increase in the
amount of adsorbed oxygen on Ir after oxygen evolution had commenced.
A possible explanation may be the conversion of the layer of Ir-O to one
of IrO_2 by anodic polarization. A strongly preanodized Ir electrode was
plunged into an O_2-saturated, H_2O_2-free, $2N$ H_2SO_4 solution (121), and
after a period of about 24 hr, a steady rest potential of 0.94 V was observed. This rest potential was independent of P_{O_2} but dependent on
pH with a coefficient of 60 mV per unit of pH. When an Ir wire preheated in air was plunged into this acid solution, a steady potential of
0.935 V was reached quickly (within 4 hr). These rest potentials are
very close to the estimated standard potential (ref. 156, p. 218) of the
Ir/IrO_2 couple,

$$IrO_2 + 4H^+ + 4e \rightleftharpoons Ir + 2H_2O \qquad E_0 = 0.93 \text{ V} \qquad (2.28)$$

It was tempting to suggest (121) that such an electrode system is a true
metal–metal oxide electrode. It was observed further that after about
48 hr the the potential drifted from the 0.93 V value and was no longer
independent of P_{O_2}. Such behavior was attributed to the instability of

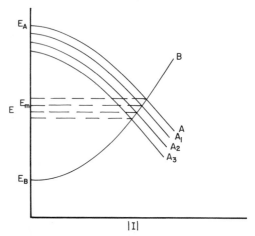

Fig. 2.18. A sketch of local cell polarization curves for the Ir/O₂ system. The notation is the same as in Fig. 2.9 with curve B representing the polarization of the Ir/Ir-O reaction.

IrO_2 in acid solutions, and the system was slowly converted to the Ir/Ir-O surface.

However, steady-state potentiostatic polarization studies (132) give no evidence of the presence of IrO_2. If IrO_2 is formed during anodic polarization, it is unstable and decomposes in the presence of metallic iridium below 1.4 V. Possibly, anodic polarization causes oxygen to be dissolved in the Ir metal which produces a surface on which the reduction of O_2 proceeds with difficulty. The system could be described in terms of a polyelectrode by Eqs. 2.3 and 2.27. In this case the local cell diagram would be similar to that in Fig. 2.13 since the O_2/H_2O reaction would be more strongly polarized for a given current density and the mixed potential would be controlled by the Ir/Ir-O reaction, Eq. 2.27. With time, the oxygen dissolved in the Ir surface layers could escape, and the local cell system described by Fig. 18 would be reestablished. In support of this contention, polarization studies (132) show that the reduction of O_2 is inhibited on a preanodized Ir electrode as compared with the reduced Ir electrode. Consequently, an Ir/IrO_2 electrode is most likely not produced by anodization.

Because a complete layer of electronically conducting adsorbed oxygen is not produced on the Ir surface by the treatment with HNO_3 (118), the reversible oxygen potential was not observed on Ir electrodes.

5. Ruthenium and Osmium

Of the noble metals, only ruthenium and osmium form hexagonal, close-packed crystals; the others form face-centered-cubic crystals. These metals are very inert and are not attacked by acids.

By heating Ru metal to 1000°C in oxygen (261), the dioxide, RuO_2, which can exist as blue crystals and which is stable when heated to a red heat (ref. 222, p. 1475) is formed. At higher temperatures RuO_2 decomposes to the elements. X-ray diffraction studies (171) confirm the contention that RuO_2 is formed when Ru is heated in oxygen. A volatile, poisonous tetroxide, RuO_4, may be obtained indirectly by oxidizing Ru with $KMnO_4$, followed by decomposition of the intermediate product with H_2SO_4 in an inert atmosphere (202). Although RuO_4 forms yellow needle-like crystals, it is rather unstable since it may be decomposed to RuO_2 by the action of sunlight and since water solutions of RuO_4 decompose in a few hours to the lower oxide. A brown form of RuO_4 is known to exist. These two oxides are the only oxides of Ru that exist (ref. 222, p. 1475).

Similarly, only two oxides of osmium, OsO_2 and OsO_4, are known to exist, but unlike Ru the tetroxide is the one formed when Os is heated in air (ref. 222, p. 1504). OsO_4 is an extremely poisonous, volatile, yellowish solid that melts at 40°C and boils at 130°C (247). In air, OsO_2 goes to OsO_4 when heated. OsO_2 may be made by heating Os metal in nitric oxide or osmium tetroxide vapor (262).

a. Electrochemical Data. Determination of the electrochemical behavior of oxygen at Ru and Os electrodes at rest or at Ru and Os anodes or cathodes is, to our knowledge, virtually nonexistent in the published literature.

From thermal data, Latimer (ref. 156, p. 228) records a value of 0.79 V for the standard potential of the Ru/RuO_2 couple,

$$RuO_2 + 4H^+ + 4e \rightleftharpoons Ru + 2H_2O \qquad E_0 = 0.79 \text{ V} \qquad (2.29)$$

Similarly, both Latimer (ref. 156, p. 231) and de Bethune (55) record a value of 0.85 V for the standard potential of the Os/OsO_4 couple,

$$OsO_4 + 8H^+ + 8e \rightleftharpoons Os + 4H_2O \qquad E_0 = 0.85 \text{ V} \qquad (2.30)$$

Potential–pH diagrams constructed by Pourbaix and co-workers are available for the Ru–H_2O (61) and the Os–H_2O (61a) systems.

Bain (8) has recorded the steady-state rest potentials in $1N$ H_2SO_4 solutions for Ru and Os electrodes which have been made by depositing the metal films on glass supports. For Ru electrodes the rest potential was 0.85 V against a hydrogen electrode in the same solution. This potential value was independent of stirring and was virtually the same on bright, black, or oxidized Ru films. Because OsO_4 is so volatile, only films of Os black deposited on a Pt-coated glass support were studied. A steady-state, stirring-independent potential of 0.87 V vs. a Pt/H_2 electrode in the same solution was obtained. It is probable that these electrodes are polyelectrodes and the rest potentials are mixed potentials similar to the other noble metals. In both cases, the rest potential was about 0.1 V more noble in $0.1N$ NaOH solution (Ru, 0.97 V; Os, 0.96 V).

Charging curves were obtained on Os-plated Pt wires (146b) and on Ru-plated Pt wires (146a). In the case of Os, the metal began to dissolve when a potential of 0.88 V was reached and considerable was passed in the double layer regions produced by a Faradaic process. The Ru curves behaved more like Pt.

E. PEROXIDE-STABILIZED SYSTEMS

Whenever an electrochemical process involves the reduction of oxygen, such as a corroding active metal in an aerated solution, it is established (ref. 138, p. 375) that hydrogen peroxide is produced. To remove the influence of metal oxides on oxygen electrode systems, Berl (15) studied the oxygen reaction in KOH solutions on porous carbon which does not form ionic conducting layers. He found that if H_2O_2 was added to the solution a reproducible and reversible oxygen electrode system was obtained. Yeager and co-workers (258) made an extensive study of carbon electrodes in KOH solutions to which H_2O_2 had been added (referred to as peroxide-stabilized KOH solutions) and noted that these systems were very reversible. However, in acid solutions, Hickling and Wilson (114) observed that the O_2 electrode was not reversible on carbon electrodes in H_2O_2-stabilized H_2SO_4 solutions.

The first investigator to observe the presence of H_2O_2 in the reduction of O_2 was Taube (231). Casopis (43) and Heyrovsky (109) showed that the reduction of O_2 on Hg proceeded through the intermediate, H_2O_2, because two polarographic waves were observed. These observations were confirmed by Kolthoff and Miller (149). In alkaline solutions, Koryta (151) found that the O_2/H_2O_2 reaction is reversible on a dropping Hg electrode and measured an E_0 value of 0.68 V against a hydrogen electrode in the same solution. Potential values between 0.66 and 0.69 V (24,146) have been reported. Koryta's value agrees very well with the value of 0.682 V quoted by Latimer (156) and de Bethune (55) for the O_2/H_2O reaction in acid solutions,

$$O_2 + 2H^+ + 2e \rightleftharpoons H_2O_2 \qquad E_0 = 0.682 \text{ V} \qquad (2.31)$$

If one accepts a value of -0.83 V for the potential of the hydrogen electrode in alkaline solutions of pH $= 14$, then a value close to -0.076 V determined from the free energy of formation of peroxide (55,156) for the O_2/H_2O reaction in alkaline solutions,

$$O_2 + H_2O + 2e \rightleftarrows HO_2^- + OH^- \qquad (2.32)$$

is calculated. From rest potential studies of the O_2/H_2O_2 system on carbon and taking the assumption that the activities of OH^- and HO_2^- ions are the same, Yeager and co-workers (274) obtained a value of -0.048 V for the E_0 of Eq. 2.32. Similar studies on Hg electrodes were carried out by Bagotskii and co-workers (6,273) from which a value of -0.045 V was found. Bowen et al. (27a) determined a value of -0.067 V. Apparently, the E_0 value for Eq. 2.32 is shadowed by some doubt.

Since carbon and mercury surfaces are poor peroxide-decomposing catalysts, such electrodes may be expected to produce good O_2/H_2O_2 electrodes. In general, the noble metals are good peroxide-decomposing catalysts, and in these cases the situation is different.

1. Rest Potential on Noble Metals

In oxygen-saturated acid solutions, the rest potential of a Pt electrode becomes less noble as the H_2O_2 concentration in solution increases (20, 92,96,126) until it reaches a value of about 10^{-3} mole/liter after which the rest potential becomes independent of H_2O_2. In addition, the observed rest potential in strong solutions of H_2O_2 at pH $= 0$ is more noble than the accepted reversible value of 0.682 V. The open-circuit potential is also independent of P_{O_2} (20,92,96,126) for solutions above 10^{-3} molar in H_2O_2.

To explain these results, Bockris and Oldfield (20) suggested that the O—O bond of the H_2O_2 was split to give a layer of adsorbed OH radicals on the Pt surface and the potential-determining reaction is the one-electron transfer step,

$$(OH)_{ads} + e \rightarrow (OH^-)_{ads} \qquad (2.33)$$

Using potential energy diagrams, Weiss (253) concluded that the splitting of H_2O_2 into an OH^- ion and an OH radical with the transfer of one electron is the most likely process in alkaline solutions. However, such mechanisms cannot account for the rest potential values between 0.81 and 0.83 V on Pt electrodes in strong H_2O_2 solutions of pH $= 0$ (20,114, 126) since Hickling and Hill (112) estimated a value of 2 V for the E_0 of the OH/OH^- couple.

a. The Mixed Potential Theory. Since the Nernst equation is not obeyed, clearly the peroxide-stabilized system with a Pt electrode must be a polyelectrode and the rest potential a mixed potential. The first to propose a mixed potential mechanism were Roiter and co-workers (200), who coupled the O_2/H_2O_2 reaction, Eq. 2.31, with the following reactions:

$$H_2O_2 \rightarrow H_2O + (O)_{ads} \qquad (2.34)$$

$$(O)_{ads} + 2H^+ + 2e \rightarrow H_2O \qquad (2.35)$$

Many workers (7,92,96,126,257) subscribe to the mixed potential explanation. In solutions containing H_2O_2, concentrations below 10^{-3} mole/liter, the dependency of the rest potential on H_2O_2 concentration and P_{O_2} becomes very complicated (92,96,126), and a simple description of the system is not possible.

In the presence of high concentrations of H_2O_2 (above $10^{-3}M$) where the open-circuit potential is independent of P_{O_2} and H_2O_2 concentration, a local cell mechanism composed of the O_2/H_2O_2 reaction, Eq. 2.31, and the H_2O_2/H_2O reaction,

$$H_2O_2 + 2H^+ + 2e \rightleftarrows 2H_2O \qquad E_0 = 1.77 \text{ V} \qquad (2.36)$$

offers a possible explanation for these experimental observations. Because the E_0 value for Eq. 2.36 is more noble than the E_0 value for Eq. 2.31, the local cell will be set up such that Eq. 2.31 is polarized in the anodic direction while Eq. 2.36 is polarized in the cathodic direction, as shown in Fig. 2.19. The overall reaction is the consumption of peroxide according to the reaction

$$2H_2O_2 \rightarrow 2H_2O + O_2 \qquad (2.37)$$

The A curves represent the local cell polarizarion curves of Eq. 2.31 for increasing concentrations of H_2O_2, and the B curves, those of Eq. 2.36. Because the dependency of the potential on the H_2O_2 concentration is in opposite directions for Eqs. 2.31 and 2.36, the corresponding polarization curves will cross at the same potential, producing the observed effect that the rest potential (mixed potential) is independent of the H_2O_2 concentration.

For large H_2O_2 concentrations, the competition between H_2O_2 and O_2 molecules for surface adsorption sites may be such that even for low values of P_{O_2} the surface is saturated with adsorbed oxygen. Under these conditions, the potential would be independent of P_{O_2} as observed experimentally.

It was found that the rest potential in strong solutions of H_2O_2 (126) was the same on Pt, Au, and Rh electrodes. As a result, it was concluded that metal oxide reactions are not involved in the mixed potential mechanism.

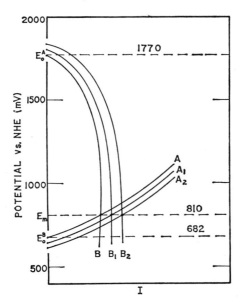

Fig. 2.19. A schematic diagram of the local cell polarization curves for a Pt electrode in O_2-saturated, $0.1M$ H_2O_2-stabilized, $2N$ H_2SO_4 solution (126). The A curves represent the O_2/H_2O_2 reaction, Eq. 2.31; the B curves, the H_2O_2/H_2O reaction, Eq. 2.36. The curves A to A_2 and B to B_2 sketch the change in shape for increasing H_2O_2 concentration. I is the absolute value of the local cell current. (By permission of The Electrochemical Society.)

Recently, Bagotskii and co-workers (233) investigated the rest potential of a reduced Pt probe electrode in a degassed acid or alkaline solution to which a very small amount of O_2 was admitted. They concluded from this work and from triangular sweep studies that the equilibrium between O_2 and H_2O_2, Eq. 2.31, can be established at a reduced Pt electrode in both acid and alkaline solutions provided only small amounts of O_2 are introduced into the system so that the Pt surface does not become covered with adsorbed O_2. Under these conditions a steady potential value of 0.70–0.71 V, which is very close to the reversible O_2/H_2O_2 potential (Eq. 2.31), may be maintained. The presence of adsorbed oxygen on the surface or dissolved oxygen in the metal disrupts the O_2/H_2O_2 equilibrium, and the usual mixed potential is observed (92,126). Possibly a process similar to that described by Bagotskii (233) may account for the recent observations of the rest potential on Pt electrodes in the presence of very small controlled quantities of dissolved oxygen obtained by Schuldiner and co-workers (215a).

2. Catalytic Decomposition of H_2O_2

As early as 1924, Fischer and Krönig (75) came to the conclusion that H_2O_2 in the presence of Pt was catalytically decomposed, and Weiss (253) suggested that in the catalytic decomposition of H_2O_2 the electron transfer was accompanied by a splitting of the H_2O_2 into an OH radical and an OH^- ion in alkaline solutions.   In ^{18}O experiments, Dole and co-workers found (63) that none of the O_2 liberated during the catalytic decomposition of H_2O_2 on Pt came from the water.   So H_2O_2 may be either oxidized or reduced.   Gerischer and Gerischer (92) observed from a study of the polarization of Pt in H_2O_2-containing H_2SO_4 solutions near the rest potential that the anodic and cathodic transfer coefficients, α and β, did not add up to unity and came to the conclusion that the rest potential and catalytic decomposition of H_2O_2 were determined by a local cell mechanism composed of Eqs. 2.31 and 2.36.

An alternative explanation for the catalytic decomposition of peroxide is a mechanism (16,28,100,108,164,174) in which adsorbed oxygen layers or surface oxides play a role.   According to Bianchi et al. (16), the metal surface is oxidized by H_2O_2 to PtO, such as $H_2O_2 + Pt \rightarrow H_2O + PtO$, followed by the interaction of more H_2O_2 with the PtO similar to $PtO + H_2O_2 \rightarrow H_2O + O_2 + Pt$ to form O_2 and bare Pt sites.   Myüller and Nekrasov (183) favor a chemical catalytic decomposition mechanism for the breaking down of H_2O_2 to H_2O.   With such chemical schemes, it may be difficult to account for the experimentally observed dependency of the system on P_{O_2}, pH, and H_2O_2 concentration as found by Giner (96) and Gerischer and Gerischer (92).

From the analysis of the polarization curves obtained for the reduction of oxygen, as discussed in Chapter IV, a chemical decomposition route is favored for the catalytic decomposition of H_2O_2 on a surface on which an adsorbed oxygen layer is present.   After the adsorbed layer is removed by the decomposition process, an electrochemical process sets in on the essentially "bare" metal surface.   Consequently, it is favored that under open-circuit conditions the catalytic decomposition of H_2O_2 may take place according to a chemical process (Bianchi's mechanism) initially because the surface is likely to be covered with adsorbed oxygen in the beginning.   After a period of time, however, when the adsorbed layer is removed, the catalytic decomposition of H_2O_2 takes place by an electrochemical local cell process (Gerischer's mechanism) where the system is in a steady state.   As mentioned in the previous section, metal oxide reactions do not seem to play a part in the steady-state catalytic decomposition of peroxide.

F. OTHER OXYGEN ELECTRODE SYSTEMS

Several metal–metal oxide electrode systems have reasonably stable and reproducible rest potentials. Such systems may make suitable secondary reference electrodes for certain applications. Some of the important metal–metal oxide electrodes which will not be discussed in later chapters will be covered here.

1. The Antimony Electrode

Although many of the applications for the Sb/Sb_2O_3 electrode such as the measurement of pH have been supplanted to a large extent by the glass electrode, there are still special applications, such as determinations of stomach pH where the electrode must be swallowed, in which much interest lies. The first to show that an Sb/Sb_2O_3 electrode could be used as a suitable reference electrode system were Uhl and Kestranek (239).

a. Metal–Metal Oxide Couples of Sb. Three stable oxides of Sb are known: the trioxide Sb_2O_3, the tetroxide Sb_2O_4, and the pentoxide Sb_2O_5. Using pure Sb powder obtained by the electrolysis of the fluoride in a suspension of Sb_2O_3 in perchloric acid solutions under a N_2 atmosphere, Schuhmann (210) determined the standard potential of the Sb/Sb_2O_3 couple,

$$Sb_2O_3 + 6H^+ + 6e \rightleftarrows 2Sb + 3H_2O \qquad E_0 = 0.152 \text{ V} \qquad (2.38)$$

as 0.152 V. In alkaline solutions a value of -0.66 V is recorded for

$$SbO_2^- + 2H_2O + 3e \rightleftarrows Sb + 4OH^- \qquad E_0 = -0.66 \text{ V} \qquad (2.39)$$

The difference between these potential values is the magnitude of the Pt/H_2 electrode in alkaline solution, 0.828 V, as required if the Nernst relationship holds. As one progresses from acid through neutral to alkaline solutions, the form of the antimony in solution progresses through an overlapping series, Sb^{+3}, $SbOH^{+2}$, SbO^+, $SbO \cdot OH$, SbO_2^-, $Sb(OH)_4^-$, which is always in equilibrium with the solid phases Sb and Sb_2O_3 (101; ref. 138, p. 339).

Other couples have been recorded in the literature, but these are not important to the antimony oxide electrode. A summary of the potentials of the Sb–O system is given by Pourbaix and co-workers (191). Latimer (ref. 156, p. 119) records E_0 values of 0.48 V for the Sb_2O_4/Sb_2O_5 couple and 0.692 V for the Sb_2O_3/Sb_2O_5 couple (Pourbaix gives 0.671 V). Ives and Janz (ref. 138, p. 341) estimates E_0 values of 0.330 V for the Sb/Sb_2O_4 couple and 0.861 V for the Sb_2O_3/Sb_2O_4 couples. De Bethune

(55) records a value of 0.581 V for the SbO^+/Sb_2O_5 couple. All of these E_0 values are for the given couples in acid solution.

b. Experimental Observations. In general, it appears that three principle types of the antimony electrode have been studied. The so-called "stick" electrode (64,227), which is made by casting the metal in the form of a bar or rod, is merely placed in solution, and measurements (chiefly pH determinations) are made after a steady state is reached. Kolthoff and Hartong (148) observed that the steady state is reached much more quickly if some Sb_2O_3 is added to the molten Sb before casting into rods. Microelectrodes can be made (227) by sealing the Sb in open ended glass tubes.

The potential of the stick Sb electrode varies from about 0.25 to 0.34 V (148,176,189), and this spread of values probably is caused by the history of the electrode preparation. If properly made (148,189), the potential of the calibrated electrode is stable, and reliable pH values may be obtained. Stirring of the solution shifts the steady-state potential to another steady value, so measurements should be made either without stirring or with constant stirring. It was shown by Tourky and Mousa (236) that oxygen must be present in the system to get a straight-line potential–pH relationship. The effect of the interaction of stirring and the presence or absence of oxygen on the observed potential is well summarized by Ives and Janz (ref. 138, p. 345).

A second form of the Sb electrode is the plated electrode which is usually made by plating a film of Sb on a Pt substrate from a nonaqueous solution of antimony trichloride (34,236,277). The behavior of this type is similar to that of the stick electrode.

Finally, an antimony electrode may be made from powdered antimony or a mixture of powdered Sb and Sb_2O_3 to which electrical contact was made with an Sb-plated Pt wire. Roberts and Fenwick (199) allowed the test solution to flow over the powder held in a vertical tube with cotton wads. Special care was taken in the preparation of the metal and oxide powders, and oxygen was removed from the system with a nitrogen purge. They stress the point that only one crystal form of Sb_2O_3 (cubic) must be present by suppressing the formation of the ortho-rhombic form. Under these conditions, they record a value of 0.144 V for the Sb/Sb_2O_3 electrode. El Wakkad (64) recorded a value of 0.15 V for the powdered Sb/Sb_2O_3 electrode. Stock et al. (227) give a good review of the construction and behavior of the Sb/Sb_2O_3 electrode.

c. Potential-Determining Reactions. Apparently the powdered electrode is a true metal–metal oxide electrode with the potential-determining reaction, Eq. 2.38. As cautioned by Roberts and Fenwick, it is essential to eliminate orthorhombic Sb_2O_3 from the system. In support of this warning, Pourbaix and co-workers (191) record that the standard potential for the Sb/Sb_2O_3 (cubic) is 0.152 V, and for the Sb/Sb_2O_3 (orthorhombic) it is 0.167 V.

Various explanations have been proposed for the fact that the potential of the stick and plated Sb electrodes is about 100 mV more noble than the reversible potential, 0.152 V. El Wakkad (64) suggested that the presence of higher oxides than Sb_2O_3 may account for this discrepancy, but as pointed out by Ives and Janz (ref. 138, p. 342), later anodic polarization studies (68) do not support this contention. An "oxygen overvoltage effect" due to adsorbed oxygen was put forth by Tourky and Mousa (236).

The most satisfying explanation, of course, is obtained from the concept of a polyelectrode system (ref. 138, p. 344). Since the metallic antimony is an active metal and can dissolve in acid or alkaline solutions, a local cell can be set up in the presence of dissolved oxygen in solution. At anodic sites, Sb^{3+} ions go into solution, but at other protected sites covered with oxide formed in the presence of dissolved oxygen, a reduction process may take place. Reduction of oxygen is the likely process occurring on these cathodic sites. Supporting evidence from the observation of Kauko and Knappsberg (145) is the result that the Sb/Sb_2O_3 electrode responds slowly to changes in the P_{O_2} with a coefficient, $dE/d \log P_{O_2}$, equal to 14.5 mV. Evidently the local cell is controlled by the O_2/H_2O reaction in a manner similar to that shown in the diagram in Fig. 2.9. If the rest potential is a mixed potential, any variation in the experimental parameters of the system that would vary the ratio of the number of cathodic to anodic sites would vary the mixed potential because the shape of the local cell polarization curves would change. With this viewpoint an adequate explanation of the observed variations in potential may be made with the history of electrode preparation, with P_{O_2}, and with stirring.

The reason why additions of Sb_2O_3 to the molten Sb electrode (143) shorten the time to reach steady state according to the mixed potential picture is the likelihood that the Sb/Sb_2O_3 reaction, Eq. 2.38 is set up more quickly as the anodic half-reaction of the local cell. Also from this standpoint, the powder electrodes (199) were intimate mixtures of Sb and Sb_2O_3 in a N_2 atmosphere. Under such conditions, the O_2/H_2O reaction could not take place, the local cell could not be set up, the electrode is a simple system, and the rest potential is the equilibrium potential of a true metal–metal oxide electrode, Sb/Sb_2O_3, with an E_0 of 0.152 V.

Excellent summaries of the behavior and applications of the Sb/Sb_2O_3 electrode are available (ref. 138, p. 336; 227).

2. The Arsenic Electrode

Of the three forms of arsenic, yellow, black, and gray, only the gray form is stable under ambient conditions and is metallic (ref. 222, p. 758). Two forms of As_2O_3 exist (216), the octahedral and the monoclinic form. A higher unstable oxide, As_2O_5, is known. The potential–pH diagram for the As–H_2O system is given by Van Muylder and Pourbaix (240). Schuhmann (211) recorded the standard potential of the As/As_2O_3 couple as 0.234 V after taking elaborate precautions in the construction of the arsenic-plated Pt probe in contact with a slurry of As and As_2O_3 powder in N_2-saturated $HClO_4$. In the literature (ref. 156, p. 114;240), the 0.234 V potential is recorded as the best value for the E_0 of the reaction

$$As_2O_3 + 6H^+ + 6e \rightleftharpoons 2As + 3H_2O \qquad E_0 = 0.234 \text{ V} \qquad (2.40)$$

Van Muylder and Pourbaix (240) also record the E_0 values of the As/As_2O_5 couple as 0.429 V and the As_2O_3/As_2O_5 couple as 0.721 V.

The behavior of the As/As_2O_3 electrode is similar to that of the Sb/Sb_2O_3 electrode (ref. 138, p. 355). Apparently, only the powder electrodes in an N_2 atmosphere exhibit the reversible potential, and as noted by Tourky and Mousa (237), the massive electrodes require the presence of oxygen to provide a stable reference potential. The stable steady-state potential is about 0.1 V more noble than the reversible potential. No doubt the behavior of the massive arsenic electrode may be described in terms of a local cell mechanism as in the case of antimony.

3. Other Examples

Schwabe (217) made an extensive study of the bismuth electrode, but as pointed out by Ives and Janz (ref. 138, p. 353), its behavior is disappointing. Since tungsten wires may be sealed in glass tubes, this metal offers certain advantages of convenience in handling, but certain disadvantages encountered in practice (14,135) may lessen its importance. Molybdenum is similar in behavior to tungsten, and some success has been reported (136) in its use for pH determinations. Because tellurium electrodes are insensitive to anions, such as citrate, oxalate, or tartrate, which upset the antimony electrode, interest was generated (137,235) in its use in determining pH.

Because all of these electrodes when used in the stick or massive form in aerated solutions are polyelectrode systems, they must be calibrated before using. Only in the powdered form in an inert atmosphere can the reversible potential be observed.

References

1. Adams, R., and R. L. Shriner, *J. Am. Chem. Soc.*, **45**, 2171 (1923).
2. Altmann, S., and R. H. Busch, *Trans. Faraday Soc.*, **45**, 720 (1949).
3. Anson, F. C., and J. J. Lingane, *J. Am. Chem. Soc.*, **79**, 4901 (1957).
4. Armstrong, G., F. R. Hinsworth, and J. A. V. Butler, *Proc. Roy. Soc. (London)*, **A143**, 89 (1934).
5. Azzam, A. M., J. O'M. Bockris, B. E. Conway, and H. Rosenberg, *Trans. Faraday Soc.*, **46**, 918 (1950).
6. Bagotskii, V. S., and D. L. Motov, *Dokl. Akad. Nauk SSSR*, **71**, 501 (1950).
7. Bagotskii, V. S., and I. E. Yablokova, *Dokl. Akad. Nauk SSSR*, **95**, 1219 (1954).
8. Bain, H. G., *Trans. Am. Electrochem. Soc.*, **78**, 173 (1940).
9. Banta, M. C., and N. Hackerman, *J. Electrochem. Soc.*, **111**, 114 (1964).
10. Bard, A. J., *Anal. Chem.*, **33**, 11 (1961).
11. Barnartt, S., *J. Electrochem. Soc.*, **106**, 722 (1959).
12. Barrer, R. M., *Diffusion in and through Solids*, Cambridge Univ. Press, Cambridge, England, 1941, p. 146.
13. Bauman, F., and I. Shain, *Anal. Chem.*, **29**, 303 (1957).
14. Baylis, J. R., *J. Ind. Eng. Chem.*, **15**, 852 (1923).
14a. Bell, W. E., M. Tagami, and R. E. Inyard, *J. Phys. Chem.*, **70**, 2048 (1966).
15. Berl, W. G., *Trans. Faraday Soc.*, **83**, 253 (1943).
16. Bianchi, G., F. Mazza, and T. Muzzini, *Electrochim. Acta*, **7**, 457 (1962); **10**, 445 (1965); G. Bianchi and G. Caprioglio, *ibid.*, **1**, 18 (1959).
17. Blackburn, T. R., and J. J. Lingane, *J. Electroanal. Chem.*, **5**, 216 (1963).
18. Blondel, M., *Ann. Chim. Phys.*, **6**, 81 (1905).
19. Bockris, J. O'M., and A. K. M. S. Huq, *Proc. Roy. Soc. (London)*, **A237**, 277 (1956).
20. Bockris, J. O'M., and L. F. Oldfield, *Trans. Faraday Soc.*, **51**, 249 (1955).
21. Böld, W., and M. W. Breiter, *Electrochim. Acta*, **5**, 145 (1961).
22. Böld, W., and M. W. Breiter, *Electrochim. Acta*, **5**, 169 (1961).
23. Bond, G. C., *Catalysis by Metals*, Academic Press, New York, 1962, p. 85.
24. Borneman, K., *Z. Anorg. Allgem. Chem.*, **34**, 1 (1903).
25. Bose, E., *Z. Physik. Chem.*, **34**, 730 (1900).
26. Bowden, F. P., *Proc. Roy. Soc. (London)*, **A125**, 446 (1929).
27. Bowden, F. P., *Proc. Roy. Soc. (London)*, **A126**, 107 (1929).
27a. Bowen, R. J., H. B. Urbach, J. H. Harrison, *Nature*, **213**, 592 (1967).
28. Bredig, G., *Z. Physik. Chem., (Leipzig)*, **31**, 258 (1899); **37**, 5 (1901).
29. Breiter, M. W., *J. Electroanal. Chem.*, **8**, 230 (1964); **11**, 157 (1966).
30. Breiter, M. W., *Electrochim. Acta*, **10**, 543 (1965).
31. Breiter, M. W., C. A. Knorr, and R. Meggle, *Z. Elektrochem.*, **59**, 153 (1955).
32. Breiter, M. W., C. A. Knorr, and W. Völkl, *Z. Elektrochem.*, **59**, 681 (1955).
33. Breiter, M. W., and J. L. Weininger, *J. Electrochem. Soc.*, **109**, 1135 (1962).
34. Brinkman, R., and F. J. Buytendiyk, *Biochem. Z.*, **199**, 387 (1928).
35. Brislee, F. J., *Trans. Faraday Soc.*, **1**, 65 (1905).
36. Brönsted, J. N., *Z. Physik. Chem.*, **65**, 84 (1909).
37. Brummer, S. B., *J. Electrochem. Soc.*, **112**, 633 (1965).
38. Brummer, S. B., and A. C. Makrides, *J. Electrochem. Soc.*, **111**, 1122 (1964).
39. Buehrer, T. F., F. S. Wartman, and R. L. Nugent, *J. Am. Chem. Soc.*, **49**, 1271 (1927).

40. Busch, R. H., *Z. Naturforsch*, **5b**, 130 (1950).
41. Butler, J. A. V., and G. Armstrong, *Proc. Roy. Soc. (London)*, **A137**, 604 (1932).
42. Butler, J. A. V., and G. Drever, *Trans. Faraday Soc.*, **32**, 427 (1936).
43. Casopis, C., *Ceskoslav. Lekarn.*, **7**, 242 (1927).
44. Clark, D., T. Dickinson, and W. N. Mair., *Trans. Faraday Soc.*, **55**, 1937 (1959).
45. Conway, B. E., *J. Electroanal. Chem.*, **8**, 486 (1965).
46. Conway, B. E., E. Gileadi, and H. Angerstein, *J. Electrochem. Soc.*, **112**, 341 (1965).
47. Conway, B. E., E. Gileadi, and M. Dzieciuch, *Electrochim. Acta*, **8**, 142 (1963); B. E. Conway and E. Gileadi, *Trans. Faraday Soc.*, **58**, 2493 (1962).
48. Crotogino, F., *Z. Physik. Chem. (Leipzig)*, **25**, 248 (1898).
49. Czepinski, V., *Z. Physik. Chem. (Leipzig)*, **30**, 1 (1902).
50. Damjanovic, A., and J. O'M. Bockris, *Electrochim. Acta*, **11**, 376 (1966).
51. Damjanovic, A., M. L. B. Rao, and M. Genshaw, ASTIA Rept. No. AD 405675, Nov. (1962).
52. Damjanovic, A., M. L. B. Rao, and J. O'M. Bockris, Ext. Abstr. No. 205, Spring Meeting of the Electromechanical Society, Toronto (1964).
53. Davtyan, O. K., *Zh. Fiz. Khim.*, **35**, 2582 (1961).
54. Davtyan, O. K., and E. G. Misyuk, *Zh. Fiz. Khim.*, **36**, 673 (1962).
55. de Bethune, A. J., and N. A. S. Loud, *Standard Aqueous Electrode Potentials and Temperature Coefficients*, Clifford A. Hampel, Skokie, Ill., 1964.
56. De Boer, J. H., and J. D. Fast, *Rec. Trav. Chim.*, **59**, 161 (1940).
57. Deborin, G., and B. V. Ershler, *Acta Physicochim.*, **13**, 347 (1940).
58. Delahay, P., *New Instrumental Methods in Electrochemistry*, Interscience, New York, 1954, p. 22.
59. Delahay, P., and G. Perkins, *J. Phys. Colloid Chem.*, **55**, 586 (1946).
60. Devanathan, M. A. V., and M. L. B. Rao, ASTIA Rept. No. AD 291763, Nov. (1962).
61. De Zoubov, N., and M. Pourbaix, *CEBELCOR Rapp. Tech.*, **58** (1958).
61a. De Zoubov, N., and M. Pourbaix, *CEBELCOR Rapp. Tech.*, **61** (1958).
61b. De Zoubov, N., J. Van Muylder, and M. Pourbaix, *CEBELCOR Rapp. Tech.*, **60** (1958).
62. Dietz, H., and H. Göhr, *Electrochim. Acta*, **8**, 343 (1963); *Z. Physik. Chem. (Leipzig)*, **223**, 113 (1963).
63. Dole, M., F. P. Rudd, G. R. Muchow, and C. Comte, *J. Chem. Phys.*, **20**, 961 (1952).
64. El Wakkad, S. E. S., *J. Chem. Soc.*, **1950**, 2894.
65. El Wakkad, S. E. S., and A. M. S. El Din, *J. Chem. Soc.*, **1954**, 3094.
66. El Wakkad, S. E. S., and A. M. S. El Din, *J. Chem. Soc.*, **1954**, 3098.
67. El Wakkad, S. E. S., and S. H. Emara, *J. Chem. Soc.*, **1952**, 461.
68. El Wakkad, S. E. S., and A. Hickling, *J. Phys. Chem.*, **57**, 203 (1953).
69. Ershler, B. V., *Discussions Faraday Soc.*, **1**, 269 (1947).
70. Ershler, B. V., G. Deborin, and A. N. Frumkin, *Bull. Acad. Sci. URSS*, **1937**, 1065; *Acta Physicochim. URSS*, **8**, 565 (1938).
71. Ershler, B. V., and A. N. Frumkin, *Trans. Faraday Soc.*, **34**, 464 (1939).
72. Every, R. L., and R. L. Grimsley, *J. Electroanal. Chem.*, **9**, 165 (1965).
73. Feldberg, S. W., C. G. Enke, and C. E. Bricker, *J. Electrochem. Soc.*, **110**, 826 (1963).

74. Finch, G. I., C. A. Murison, N. Stuart, and G. P. Thomson, *Proc. Roy. Soc. (London)*, **A141**, 414 (1933).
75. Fischer, F., and W. Krönig, *Z. Anorg. Allgem. Chem.*, **135**, 169 (1924).
76. Fischer, R. A., H. Chon, and J. G. Aston, *J. Phys. Chem.*, **68**, 3240 (1964).
77. Foerster, F., *Z. Physik. Chem. (Leipzig)*, **69**, 236 (1909).
78. Foerster, F., and A. Piguet, *Z. Elektrochem.*, **10**, 714 (1904).
79. Franke, K. F., C. A. Knorr, and M. W. Breiter, *Z. Elektrochem.*, **63**, 226 (1959).
80. French, W. G., and T. Kuwana, *J. Phys. Chem.*, **68**, 1279 (1964).
81. Frumkin, A. N., *Vestn. Mosk. Univ.*, **7**, No. 9, 37 (1952).
82. Frumkin, A. N., *J. Electrochem. Soc.*, **107**, 461 (1960).
83. Frumkin, A. N., *Advances in Electrochemistry and Electrochemical Engineering*, Vol. 3, P. Delahay, Ed., Interscience, New York, 1963, p. 310.
84. Frumkin, A. N., and N. Aladjalova, *Acta Physicochim.*, **19**, 1 (1944).
85. Frumkin, A. N., E. I. Khruschcheva, M. G. Tarasevich, and N. A. Shumilova, *Elektrokhimiya*, **1**, 17 (1965).
86. Frumkin, A. N., and A. Slygin, *Acta Physicochim. URSS*, **5**, 819 (1936); *Compt. Rend. Acad. Sci. URSS*, **2**, 173 (1934).
87. Frumkin, A. N., A. Slygin, and W. Medvedovsky, *Acta Physicochim.*, **4**, 911 (1936).
88. Furguson, A. L., and M. B. Towns, *Trans. Am. Electrochem. Soc.*, **83**, 271, 285 (1943).
89. Galloni, E. E., and R. H. Busch, *J. Chem. Phys.*, **20**, 198 (1952).
90. Galloni, E. E., and A. E. Roffo, *J. Chem. Phys.*, **9**, 875 (1941).
91. Genshaw, M., V. Brusic, and A. Damjanovic, ASTIA Rept. No. AD 616779, Dec. (1964).
92. Gerischer, R., and H. Gerischer, *Z. Physik. Chem. (Frankfurt)*, **6**, 178 (1956).
93. Gerke, R. H., and M. D. Rourke, *J. Am. Chem. Soc.*, **49**, 1855 (1927).
94. Gilman, S., *Electrochim. Acta*, **9**, 1025 (1964); *J. Electroanal. Chem.*, **9**, 276 (1965).
95. Giner, J., *Z. Elektrochem.*, **63**, 386 (1959).
96. Giner, J., *Z. Elektrochem.*, **64**, 491 (1960).
97. Giner, J, .and E. Lange, *Naturwiss*, **40**, 506 (1953).
98. Grahame, D. C., *Chem. Rev.*, **41**, 441 (1947).
99. Grove, W. R., *Phil. Mag.*, **14**, 127 (1839); **21**, 417 (1842); *Proc. Roy. Soc. (London)*, **4**, 463 (1843); **5**, 557 (1845).
100. Grube, G., *Z. Elektrochem.*, **16**, 621 (1910).
101. Grube, G., and F. Schweigardt, *Z. Elektrochem.*, **29**, 257 (1923).
102. Grüneberg, G., *Electrochim. Acta*, **10**, 339 (1965).
103. Güntherschulze, A., and H. Betz, *Z. Elektrochem.*, **44**, 253 (1938).
104. Haber, F., *Z. Anorg. Allgem. Chem.*, **51**, 356 (1906).
105. Haber, F., *Z. Elektrochem.*, **12**, 415 (1906).
106. Haber, F., and F. Fleischmann, *Z. Anorg. Allgem. Chem.*, **51**, 245 (1906).
107. Haber, F., and G. W. A. Forster, *Z. Anorg. Allgem. Chem.*, **51**, 289 (1906).
108. Haber, F., and S. Grinberg, *Z. Anorg. Allgem. Chem.*, **18**, 37 (1898).
109. Heyrovsky, J., *Arkiv. Hem. Ferm.*, **5**, 162 (1931).
110. Hickling, A., *Trans. Faraday Soc.*, **41**, 333 (1945).
111. Hickling, A., *Trans. Faraday Soc.*, **42**, 518 (1946).
112. Hickling, A., and S. Hill, *Trans. Faraday Soc.*, **46**, 557 (1950).
113. Hickling, A., and G. G. Vrjosek, *Trans. Faraday Soc.*, **57**, 123 (1961).

114. Hickling, A., and W. H. Wilson, *J. Electrochem. Soc.*, **98**, 425 (1951).
115. Hoar, T. P., *Proc. Roy. Soc. (London)*, **A142**, 628 (1933).
116. Hoare, J. P., *J. Electrochem. Soc.*, **109**, 858 (1962).
117. Hoare, J. P., *J. Electrochem. Soc.*, **110**, 245 (1963).
118. Hoare, J. P., *J. Electrochem. Soc.*, **110**, 1019 (1963).
119. Hoare, J. P., *J. Electrochem. Soc.*, **111**, 232 (1964).
120. Hoare, J. P., *J. Electrochem. Soc.*, **111**, 610 (1964).
121. Hoare, J. P., *J. Electrochem. Soc.*, **111**, 988 (1964).
122. Hoare, J. P., *Electrochim. Acta*, **9**, 599 (1964).
123. Hoare, J. P., *Electrochim. Acta*, **9**, 1289 (1964).
124. Hoare, J. P., *Nature*, **204**, 71 (1964).
125. Hoare, J. P., *J. Electrochem. Soc.*, **112**, 602 (1965).
126. Hoare, J. P., *J. Electrochem. Soc.*, **112**, 608 (1965).
127. Hoare, J. P., *J. Electrochem. Soc.*, **112**, 849 (1965).
128. Hoare, J. P., *J. Electrochem. Soc.*, **112**, 1129 (1955).
129. Hoare, J. P., *Electrochim. Acta*, **11**, 311 (1966).
130. Hoare, J. P., *J. Electroanal. Chem.*, **12**, 260 (1966).
130a. Hoare, J. P., *Electrochim. Acta*, **11**, 549 (1966).
131. Hoare, J. P., *Electrochim. Acta*, in press.
132. Hoare, J. P., *J. Electroanal. Chem.*, in press.
133. Hoare, J. P., S. G. Meibuhr, and R. Thacker, *J. Electrochem. Soc.*, **113**, 1078 (1966).
134. Hoare, J. P., and S. Schuldiner, *J. Electrochem. Soc.*, **102**, 485 (1955).
135. Holven, A. L., *J. Ind. Eng. Chem.*, **21**, 965 (1929).
136. Issa, I. M., and H. Kahlifa, *Anal. Chim. Acta.*, **10**, 567 (1954).
137. Issa, I. M., H. Kahlifa, and S. A. Awad, *J. Indian Chem. Soc.*, **34**, 275 (1957).
138. Ives, D. J. G., and G. J. Janz, *Reference Electrodes*, Academic Press, New York, 1961, p. 333.
139. Jirsa, F., *Z. Physik, Chem., (Leipzig)*, **113**, 241 (1924).
140. Jirsa, F., and O. Burynek, *Z. Elektrochem.*, **29**, 126 (1923).
141. Jirsa, F., and H. Jelinek, *Z. Elektrochem.*, **30**, 286 (1924).
142. Jörgensen, S. M., *J. Prakt. Chem.*, **16**, 344 (1877).
143. Kalish, T. V., and R. Kh. Burshtein, *Dokl. Akad. Nauk SSSR*, **81**, 1093 (1951); **88**, 863 (1953).
144. Kaufman, D. C., J. W. Loveland, and P. J. Elving, *J. Phys. Chem.*, **63**, 217 (1959).
145. Kauko, Y., and L. Knappsberg, *Z. Elektrochem.*, **45**, 760 (1939).
146. Kern, D. M. H., *J. Am. Chem. Soc.*, **76**, 4208 (1954).
146a. Khomchenko, G. P., T. N. Stayanovskaya, and G. D. Vovchenko, *Zh. Fiz. Khim.*, **38**, 434 (1964).
146b. Khomchenko, G. P., N. G. Ul'ko, and G. D. Vovchenko, *Elektrokhimiya*, **1**, 659 (1965).
146c. Khrushcheva, E. I., N. A. Shumilova, and M. R. Tarasevich, *Elektrokhimiya*, **2**, 277 (1966).
147. Klemenc, A., *Z. Physik. Chem., (Leipzig)*, **185**, 1 (1939).
148. Kolthoff, I. M., and B. D. Hartong, *Rec. Trav. Chim.*, **44**, 113 (1925).
149. Kolthoff, I. M., and C. S. Miller, *J. Am. Chem. Soc.*, **63**, 1013 (1941).
150. Kolthoff, I. M., and N. Tanaka, *Anal. Chem.*, **26**, 632 (1954).
151. Koryta, J., *Chem. Listy*, **46**, 593 (1952).

151a. Kruger, J., *Corrosion*, **22**, 88 (1966).
152. Laitinen, H. A., and M. S. Chao, *J. Electrochem. Soc.*, **108**, 726 (1961).
153. Laitinen, H. A., and C. G. Enke, *J. Electrochem. Soc.*, **107**, 773 (1960).
154. Lange, E., and H. Göhr, *Z. Elektrochem.*, **63**, 74 (1959).
155. Lange, E., and P. Van Rysselberghe, *J. Electrochem. Soc.*, **105**, 420 (1958).
156. Latimer, W. M., *Oxidation Potentials*, 2nd ed., Prentice-Hall, Englewood Cliffs, N. J., 1952, p. 39.
157. Lee, J. K., R. N. Adams, and C. E. Bricker, *Anal. Chim. Acta*, **17**, 321 (1957).
158. Lewartowicz, E., *Compt. Rend.*, **253**, 1260 (1961); *J. Electroanal. Chem.*, **6**, 11 (1963).
159. Lewis, G. N., *J. Am. Chem. Soc.*, **28**, 158 (1906); *Z. Physik. Chem.*, (*Leipzig*), **55**, 534, 544 (1906).
160. Lewis, G. N., and M. Randall, *J. Am. Chem. Soc.*, **36**, 1969 (1914).
161. Lingane, J. J., *Electroanalytical Chemistry*, 2nd ed., Interscience, New York, 1958, p. 617.
162. Lingane, J. J., *J. Electroanal. Chem.*, **1**, 379 (1959).
163. Lingane, J. J., *J. Electroanal. Chem.*, **2**, 296 (1961).
164. Lingane, J. J., and P. J., Lingane, *J. Electroanal. Chem.*, **5**, 411 (1963).
164a. Llopis, J., and F. Colom, *Proc. CITCE*, **8**, 414 (1958).
164b. Llopis, J., and L. Jorge, *Electrochim. Acta*, **9**, 103 (1964).
164c. Llopis, J., and M. Vazques, *Electrochim. Acta*, **9**, 1655 (1964).
165. Lorenz, R., *Z. Elektrochem.*, **14**, 781 (1908).
166. Lorenz, R., *Z. Elektrochem.*, **15**, 661 (1909).
167. Lorenz, R., and H. Hauser, *Z. Anorg. Allgem. Chem.*, **51**, 81 (1906).
168. Lorenz, R., and E. Lauber, *Z. Elektrochem.*, **15**, 205 (1909).
169. Lorenz, R., and P. E. Spielmann, *Z. Elektrochem.*, **15**, 293 (1909).
170. Lorenz, R., and P. E. Spielmann, *Z. Elektrochem.*, **15**, 349 (1909).
171. Lunde, G., *Z. Anorg Allgem. Chem.*, **163**, 345 (1927).
172. Luk'yanycheva, V. I., and V. S. Bagotskii, *Dokl. Akad. Nauk SSSR*, **155**,160 (1964).
172a. Luk'yanycheva, V. I., V. I. Tikhomirova, and V. S. Bagotskii, *Elektrokhimiya*, **1**, 262 (1965).
173. MacDonald, J.J., and B.E. Conway, *Proc. Roy. Soc.* (*London*), **A269**, 419 (1962).
174. MacInnes, D. A., *J. Am. Chem. Soc.*, **36**, 878 (1914).
175. Mazitov, Yu. A., K. I. Rosental', and V. I. Veselovskii, *Zh. Fiz. Khim.*, **38**, 151 (1964); *Dokl. Akad. Nauk SSSR*, **148**, 152 (1963).
176. Mehta, D. N., and S. K. K. Jatkar, *J. Indian Inst. Sci.*, **18A**, 85 (1935).
177. Mohilner, D. M., W. J. Argersinger, and R. N. Adams, *Anal. Chim. Acta*, **27**, 194 (1962).
178. Mond, L., W. Ramsay, and J. Shields, *Z. Physik. Chem.*, (*Frankfurt*), **25**, 657 (1898).
179. Moore, W. J., and L. Pauling, *J. Am. Chem. Soc.*, **63**, 1392 (1941).
180. Müller, F., and A. Riefkohl, *Z. Elektrochem.*, **34**, 744 (1928); **36**, 181 (1930).
181. Müller, W. J., and E. Löw, *Trans. Faraday Soc.*, **28**, 471 (1932); **31**, 1291 (1935).
182. Mayell, J. S., and S. H. Langer, *J. Electrochem. Soc.*, **111**, 438 (1964); **113**, 385 (1966).
183. Myüller, L., and L. N. Nekrasov, *Dokl. Akad. Nauk SSSR*, **157**, 416 (1964); *J. Electroanal. Chem.*, **9**, 282 (1965).
184. Nagel, K., and H. Dietz, *Electrochim. Acta*, **4**, 141 (1961).

185. Náray-Szabó, St. V., Z. Elektrochem., 33, 15 (1927).
186. Nernst, W., and H. von Wartenberg, Z. Physik. Chem., 56, 534 (1906).
187. Nesterova, V. I., and A. N. Frumkin, Zh. Fiz. Khim., 26, 1178 (1952).
188. Obrucheva, A. D., Zh. Fiz. Khim., 26, 1448 (1952).
189. Parks, L. R., and H. C. Beard, J. Am. Chem. Soc., 54, 856 (1932).
190. Pearson, J. D., and J. A. V. Butler, Trans. Faraday Soc., 34, 1163 (1938).
191. Pitman, A. L., M. Pourbaix, and N. De Zoubov, J. Electrochem. Soc., 104, 594 (1957); Proc. CITCE, 9, 32 (1959).
192. Pollard, W. B., J. Chem. Soc., 1926, 1347.
193. Rädlein, G., Z. Elektrochem., 61, 724, 727 (1957).
194. Randles, J. E. B., Trans. Faraday Soc., 44, 322, 327 (1948).
195. Rao, M. L. B., A. Damjanovic, and J. O'M. Bockris, J. Phys. Chem., 67, 2508 (1963).
196. Reddy, A. K. N., M. Genshaw, and J. O'M. Bockris, J. Electroanal. Chem., 8, 406 (1964).
197. Reddy, A. K. N., M. L. B. Rao, and J. O'M. Bockris, J. Chem. Phys., 42, 2246 (1965). References to earlier work are contained here.
198. Richards, W. T., J. Phys. Chem., 32, 990 (1928).
199. Roberts, E. J., and F. Fenwick, J. Am. Chem. Soc., 49, 2787 (1927); 50, 2125 (1928).
200. Roiter, V. A., et al., J. Phys. Chem. USSR, 4, 461, 465 (1933).
201. Roseveare, W. E., and T. F. Buehrer, J. Am. Chem. Soc., 49, 1221 (1927).
202. Ruff, O., and E. Vidic, Z. Anorg. Allgem. Chem., 136, 49 (1924).
203. Sand, H. J. S., Phil. Mag., 1, 45 (1901).
204. Sawyer, D. T., and R. J. Day, Electrochim. Acta, 8, 589 (1963).
205. Sawyer, D. T., and L. V. Interrante, J. Electroanal. Chem., 2, 310 (1961).
206. Schenck, R., and F. Finkener, Ber., 75, 1962 (1942).
207. Schmid, G. M., and N. Hackerman, J. Electrochem. Soc., 110, 440 (1963).
208. Schmid, G. M., and R. N. O'Brien, J. Electrochem. Soc., 111, 832 (1964).
209. Schoch, E. P., J. Phys. Chem., 14, 665 (1910).
210. Schuhmann. R., J. Am. Chem. Soc., 46, 52 (1924).
211. Schuhmann, R., J. Am. Chem. Soc., 46, 1444 (1924).
212. Schuldiner, S., and R. M. Roe, J. Electrochem. Soc., 110, 332 (1963).
213. Schuldiner, S., and R. M. Roe, J. Electrochem. Soc., 110, 1142 (1963).
214. Schuldiner, S., and R. M. Roe, J. Electrochem. Soc., 111, 369 (1964).
215. Schuldiner, S., and T. B. Warner, J. Electrochem. Soc., 112, 212 (1965).
215a. Schuldiner, S., B. J. Piersma, and T. B. Warner, J. Electrochem. Soc., 113, 573 (1966).
215b. Schuldiner, S., T. B. Warner, and B. J. Piersma, J. Electrochem. Soc., 114, 343 (1967).
216. Schulman, J. H., and W. C. Schumb, J. Am. Chem. Soc., 65, 878 (1943).
217. Schwabe, K., Z. Elektrochem., 53, 125 (1949).
218. Sevcik, A., Collection Czech. Chem. Commun., 13, 349 (1948).
219. Shriner, R. L., and R. Adams, J. Am. Chem. Soc., 46, 1683 (1924).
220. Shumilova, N. A., G. V. Zhutaeva, M. R. Tarasevich, and R. Kh. Burshtein, Zh. Fiz. Khim., 39, 1012 (1965).
221. Shutt, W. J., and A. Walton, Trans. Faraday Soc., 29, 1209 (1933).
222. Sidgwick, N. V., The Chemical Elements and Their Compounds, Oxford Univ. Press, London, 1950, p. 1581.

223. Slygin, A., and A. N. Frumkin, *Acta Physicochim. URSS*, **3**, 791 (1935).
224. Smale, F. J., *Z. Physik. Chem.*, *(Frankfurt)*, **14**, 577 (1894).
225. Smith, D. P., *Z. Physik*, **78**, 815 (1932).
226. Spielmann, P. E., *Trans. Faraday Soc.*, **5**, 88 (1909).
227. Stoch, J. T., W. C. Purdy, and L. M. Garcia, *Chem. Rev.*, **58**, 611 (1958).
228. Tamman, G., and F. Runge, *Z. Anorg. Allgem. Chem.*, **156**, 85 (1926).
229. Tarter, H. V., and M. Walker, *J. Am. Chem. Soc.*, **52**, 2256 (1930).
230. Tarter, H. V., and V. E. Wellman, *J. Phys. Chem.*, **32**, 1171 (1928).
231. Taube, M., *Ber.*, **15**, 2434 (1882).
232. Thacker, R., and J. P. Hoare, *Electrochem. Tech.*, **2**, 61 (1964).
233. Tikhomirova, V. I., V. I. Luk'yanycheva, and V. S.B agotskii, *Elektrokhimiya*, **1**, 645 (1965).
233a. Tikhomirova, V. I., A. I. Oshe, V. S. Bagotskii, and V. I. Luk'yanycheva, *Dokl. Akad. Nauk SSSR*, **159**, 644 (1964).
234. Tilley, G. S., and O. C. Ralston, *Trans. Am. Electrochem. Soc.*, **44**, 31 (1923).
235. Tomicek, O., and F. Poupe, *Collection Czech. Chem. Commun.*, **8**, 520 (1936).
236. Tourky, A. R., and A. A. Mousa, *J. Chem. Soc.*, **1948**, 752, 756, 759.
237. Tourky, A. R., and A. A. Mousa, *J. Chem. Soc.*, **1949**, 1297, 1302, 1305.
238. Trapnell, B. M. W., *Chemisorption*, Butterworths, London, 1955, p. 150.
238a. Tyurin, Yu. M., *Dokl. Akad. Nauk SSSR*, **126**, 827 (1959).
239. Uhl, A., and W. Kestranek, *Monatsh. Chem.*, **44**, 29 (1923).
239a. Van Muylder, J., and M. Pourbaix, *CEBELCOR Rapp. Tech.*, **59** (1958).
239b. Van Muylder, J., and M. Pourbaix, *CEBELCOR Rapp. Tech.*, **62** (1958).
240. Van Muylder, J., and M. Pourbaix, *Proc. CITCE*, **9**, 20 (1959).
240a. Van Muylder, J., N. De Zoubov, and M. Pourbaix, *CEBELCOR Rapp. Tech.*, **63** (1958).
241. Van Rysselberghe, P., *Electrochim. Acta*, **5**, 28 (1961); **8**, 543 (1963).
242. Vetter, K. J., and D. Berndt, *Z. Elektrochem.*, **62**, 378 (1958).
243. Vielstich, W., *Z. Instrumentenk.*, **71**, 29 (1963).
244. Visscher, W., and M. A. V. Devanathan, *J. Electroanal. Chem.*, **8**, 127 (1964).
245. Vogel, A. I., *Practical Organic Chemistry*, Longmans, London, 1948, p. 457.
246. Vogel, J., in *Progress in Polarography*, Vol. I, P. Zuman and I. M. Kolthoff, Eds., Interscience, New York, 1962, p. 429.
247. Von Wartenberg, H., *Ann.*, **440**, 102 (1924).
248. Voorhees, V., and R. Adams, *J. Am. Chem. Soc.*, **44**, 1397 (1922).
249. Wagner, C., and W. Traud, *Z. Elektrochem.*, **44**, 391 (1938).
250. Warner, T. B., and S. Shuldiner, *J. Electrochem. Soc.*, **112**, 853 (1965).
251. Waser, J., and E. D. McClanahan, *J. Chem. Phys.*, **19**, 413 (1951).
252. Watanabe, N., and M. A. V. Devanathan, *J. Electrochem. Soc.*, **111**, 615 (1964).
253. Weiss, J., *Trans. Faraday Soc.*, **31**, 1547 (1935).
254. Will, F. G., and C. A. Knorr, *Z. Elektrochem.*, **64**, 258 (1960).
255. Will, F. G., and C. A. Knorr, *Z. Elektrochem.*, **64**, 270 (1960).
256. Wilsmore, N. T. M., *Z. Physik. Chem.*, *(Leipzig)*, **35**, 291 (1900).
257. Winkelmann, D., *Z. Elektrochem.*, **60**, 731 (1956).
258. Witherspoon, R. R., H. Urbach, E. Yeager, and F. Hovorka, Tech. Rept. No. 4, ONR Contract Nonr. 581(00), Western Reserve Univ. (1954).
259. Wöhler, L., *Z. Anorg. Allgem. Chem.*, **40**, 423 (1904).
260. Wöhler, L., *Z. Elektrochem.*, **11**, 836 (1905).
261. Wöhler, L., P. Balz, and L. Metz, *Z. Anorg. Allgem. Chem.*, **139**, 213 (1924).

262. Wöhler, L., and L. Metz, *Z. Anorg. Allgem. Chem.*, **149**, 301 (1925).
263. Wöhler, L., and W. Frey, *Z. Elektrochem.*, **15**, 129 (1909).
264. Wöhler, L., and J. Konig, *Z. Anorg. Allgem. Chem.*, **46**, 323 (1905).
265. Wöhler, L., and F. Martin, *Z. Anorg. Allgem. Chem.*, **57**, 398 (1908).
266. Wöhler, L., and F. Martin, *Z. Elektrochem.*, **15**, 791 (1909).
267. Wöhler, L., and F. Martin, *Chem. Ber.*, **42**, 3326 (1909).
268. Wöhler, L., and F. Martin, *Chem. Ber.*, **42**, 3958 (1909).
269. Wöhler, L., and W. Müller, *Z. Anorg. Allgem. Chem.*, **149**, 125 (1925); L. Wöhler and N. Jochum, *Z. Physik. Chem.*, (*Leipzig*), **167**, 169 (1933).
270. Wöhler, L., and W. Witzmann, *Z. Anorg. Allgem. Chem.*, **57**, 323 (1907).
271. Wöhler, L., and W. Witzmann, *Z. Elektrochem.*, **14**, 97 (1908).
272. Wrotenbery, P. T., R. M. Hurd, and E. S. Snavely, ASTIA Rept. No. AD 415090, July (1963).
273. Yablakova, I. E., and V. S. Bagotskii, *Dokl. Akad. Nauk SSSR*, **85**, 599 (1952).
274. Yeager, E., P. Krause, and K. V. Rao, *Electrochim. Acta*, **9**, 1057 (1964).
275. Young, L., *Anodic Oxide Films*, Academic Press, New York, 1961, p. 1.
276. Zalkind, Ts. I., and B. V. Ershler, *Zh. Fiz. Khim.*, **25**, 565 (1951).
277. Zhukov, I. I., and G. P. Avseevich, *Z. Elektrochem.*, **35**, 349 (1929).

chapter **III**

The Anodic Evolution of Oxygen

A. PLATINUM

From rest potential studies, information may be obtained about the potential-determining reactions occurring at the electrode under open-circuit conditions and about the thermodynamics of the electrochemical system. Some understanding of the nature of the electrode surface, the type of adsorbed films, and the extent of surface coverage may be deduced from data obtained as constant current charging curves, potential sweep curves, and open-circuit potential-decay curves. To investigate the catalytic activity of the electrode surface for a given electrochemical reaction and to gain some knowledge of the kinetics of the reaction, steady-state polarization measurements (reaction rate studies) must be made. Such kinetic information is most useful only if one has some understanding of the state of the electrode surface before these polarization studies are undertaken. For this reason, so much stress has been placed in Chapter II on the work carried out under open-circuit conditions.

By choosing the proper experimental conditions (e.g., design of electrodes, vigorous stirring of solution,) the effects of mass transfer on the experimental parameters may be minimized and the system comes under kinetic control. (Mass transfer control processes will be studied in Chapter VI.) Under these conditions, the potential is a logarithmic function of the current density (see ref. 33, p. 96 et seq.), and regions in which such a relationship holds are known as Tafel regions.

A plot of the polarization data as E vs. log i is called an oxygen overvoltage curve only if the potential is recorded as the difference in potential between the measured and the reversible oxygen potential and if the current is expressed as current density. This is important because meaningful values of the exchange current density i_0, may be determined only from the overvoltage curve (see ref. 33, p. 158, for a discussion of i_0).

Since this chapter deals with anodic processes, the electrode reactions are written as oxidation reactions (electrons on the right-hand side of the equation).

1. Steady-State Polarization Curves

Much of the early polarization measurements obtained on Pt anodes for the evolution of oxygen (6,28,44,82,92,93,96,109,113,118,124) is difficult to interpret because only too often experimental details are missing and because experimental procedures were not carried out with sufficient care and preparation. Certainly, some of the lack of agreement among the various workers arises from the fact that the electrochemical properties of the Pt/O_2 system depend greatly on the previous history of preparation of the electrode surface, as seen from rest potential studies.

Picheta (96) noted that the overvoltage varied with time and depended on the electrode history. To get reproducible results he had to develop a definite and unvarying procedure. Yet the overall behavior of oxygen overvoltage on Pt was uncovered in these initial studies. Even in the very early work (6,124), Tafel regions were observed in the polarization curves. All investigators found that the oxygen overvoltage, η, is very high; and Roiter and Yampolskaya (104) showed that the η decreased with an increase in temperature.

The first to make a detailed investigation of the oxygen overvoltage on Pt anodes were Bowden (14) and Hoar (61). Since then, a number of workers (7,11,30,30a,35,37,38,51,57,67,84,85,102,111,112,117,125) have studied the evolution of oxygen on Pt anodes in acid and alkaline solutions.

Over a current density range from 10^{-7} to 10^{-2} A/cm², the shape of the galvanostatic, steady-state overvoltage curve on bright Pt anodes in O_2-saturated acid solution is similar to the one shown in Fig. 3.1, except that the point at which the curve deviates from the Tafel region may occur at higher overvoltages, depending on where the rest potential for the given electrode lies. The curve in Fig. 3.1 was obtained by Bockris and Huq (11) on an electrode at which the rest potential was very near the reversible potential. Under ordinary conditions, Pt electrodes in O_2-saturated solutions exhibit a rest potential of about a volt, and in this case, the deviation from Tafel behavior occurs at a potential of about 1.6 V ($\eta \sim 0.4$ V) or at a current density of about 10^{-6} A/cm² (see top curve of Fig. 5.1).

In any event, a Tafel region is observed in the anodic oxygen overvoltage curve over the current density range between 10^{-6} and 10^{-2} A/cm² with a slope having a value between 0.10 and 0.13. Extrapolation of this slope to $\eta = 0$ yields a value of i_0 between 10^{-10} and 10^{-9} A/cm² in systems in which rigorous control of impurities has been maintained. On the first run of a polarization curve taken at a Pt anode,

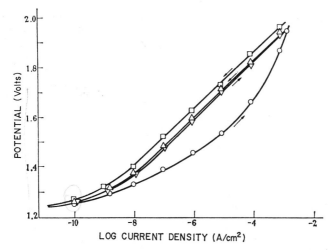

Fig. 3.1. A plot of the oxygen overvoltage on a Pt anode in a solution of $0.01N$ H_2SO_4 which had been cathodically preelectrolyzed for 24 hr at 10^{-2} A/cm² for 24 hr at 10^{-2} A/cm², followed by anodic preelectrolysis for 72 hr at 10^{-2} A/cm² (11). Two complete cycles of polarization are shown: increasing (○) and decreasing (△) current of first cycle; increasing (▽) and decreasing (□) current on second cycle. (By permission of The Royal Society.)

steady-state points are usually not obtained, and a curve without a Tafel region is found, as shown in Fig. 3.1. A probable reason for this behavior is the likelihood that the adsorbed oxygen layer is continually building up with anodic polarization. Once this adsorbed oxygen layer is formed, subsequent runs give reproducible results because now the catalytic surface on which the electrode reaction reaction takes place is fully formed and stable. The second run for decreasing current (Fig. 3.1) shows some hysteresis as compared to the corresponding data of the first run. Possibly, Bockris and Huq, who took their measurements very quickly, did not give the system enough time to form completely the adsorbed layer or the catalytic surface. In support of this viewpoint, it is seen that the slope did not change (i.e., the mechanism did not change) between the first and second decreasing Tafel lines, but the i_0 did (i.e., the catalytic activity of the surface did change). Provided a prepurified system is used, reproducible results (within ± 0.005 V) over at least three complete cycles of polarization may be obtained (67) if, before the η data are taken, the electrode is anodized initially at the highest current density investigated until a steady-state value is obtained. In this way, a stable catalytic surface is formed on the electrode, and the subsequent data will be reproducible.

Although rigorous control of impurities may not be important in some anodic processes, such as those reported (29) at nickel anodes in alkaline solutions, it is important in the investigations carried out at Pt anodes. This may be true because, even though some impurities are desorbed at such anodic potentials, those impurities which are not desorbed may be occluded in the formation of multilayers of nickel oxide. In the Pt case, multilayers are not formed, and the active sites on the electronically conducting unimolecular film may be impaired by the adsorption of impurities in the system.

Erdey-Gruz and Vajasdy (38) have shown that the oxygen evolution mechanism (Tafel slope) is not changed if Pt is hot rolled, cold rolled, etched, anodically polished, or mechanically polished. The polished surface gave higher overvoltages, which probably reflect the effect of a change in surface area. These results support the contention that the electrode process takes place on an adsorbed surface film.

It is reasonable to assume that the anodic evolution of oxygen on Pt is activation-controlled because a Tafel region is observed (e.g., 11,61, 67) and because the η decreases as the temperature increases while the Tafel slope remains independent of temperatures (57,103,104).

As one proceeds from alkaline to acid solutions, it is found (11,57,61, 102) that the anodic η for the evolution of O_2 on bright Pt electrodes decreases slightly as the pH decreases, but for very strong acids the η increases again.

Particularly for anodic processes, polarization studies carried out under controlled potential conditions give more information about changes in the nature of the surface while the polarization curve is taken than galvanostatic techniques. A typical steady-state, potentiostatic overvoltage curve for the evolution of O_2 on a bright Pt anode in O_2-saturated acid solution (71,72) is presented in Fig. 3.2. Various regions of the curve are marked by letters.

An interesting feature of Fig. 3.2 is the appearance of a limiting current region between B and C and between D and E. In the region BC it is concluded (71,72) that the film of Pt-O is being built up, and since this film is electronically conducting, the only change in the system as the potential becomes more anodic is an increase in the coverage of the electrode with the adsorbed film because the current-producing process reaches a limiting rate at B. Eventually, the potential reaches a point at C where the evolution of O_2 begins, and the current increases with the potential again. The conversion of some Pt-O sites to PtO_2 sites is considered to take place in the region DE. At E, a stable and completely formed catalytic surface (electronically conducting monolayer of adsorbed oxygen) is obtained, and between E and F, a reproducible (within ± 0.05 unit of log i) η curve is observed. A value of 0.14 is recorded (71) for the slope and 8.5×10^{-9} A/cm^2 for i_0, in good agreement with the galvanostatic results (11,61,67,102). The output of the potentiostat at F is virtually zero current density. As the potential is lowered, a measurable current is not observed until the point G has been reached.

Since an oxide reduction current was not observed between F and G, it was concluded (72) that PtO$_2$ is not present at potentials below 1.55 V. If PtO$_2$ is formed above 1.8 V, it must, therefore, be unstable in the presence of Pt at potentials below 1.6 V.

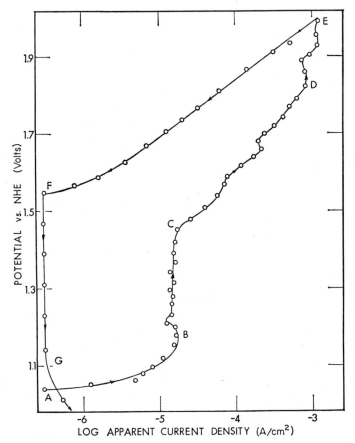

Fig. 3.2. Oxygen overvoltage curves obtained potentiostatically on a bright Pt anode in O$_2$-saturated $2N$ H$_2$SO$_4$ solution (72). Arrows indicate direction of polarization. (By permission of Elsevier.)

2. Mechanisms

Bockris (10) considered at least five different overall mechanisms for the evolution of oxygen from the discharge of OH$^-$ ions on Pt anodes in alkaline solutions and tabulated the kinetic parameters to be expected for various rate-determining steps. As pointed out recently by Milner (89) in an analysis of the oxygen reaction, an unbelievable number of reaction paths are possible if all intermediates are considered. By applying the most drastic of assumptions, still this number may only be reduced to about eleven mechanisms. Since many of these reaction paths are not unique in the value of the kinetic parameters, evidence other than that obtained from Tafel slopes is required to identify unambiguously the complete mechanism. To date this has not been accomplished.

However, some information about the mechanism of O$_2$ evolution at a Pt anode is available. From steady-state measurements it is possible to learn about the rate-determining step, but as cautioned by Bockris and Huq (11), information about the rapid steps preceding or following the slow step cannot be obtained from such studies. Certain fast pulse techniques (e.g., 107) may eventually provide a means for sorting out a well-founded overall mechanism.

Most workers agree that the rate-limiting step is an electron transfer reaction, and the most likely step is the discharge of H$_2$O molecules from acid or OH$^-$ ions from alkaline solutions to form adsorbed OH radicals on the electrode surface. The observed Tafel slopes between 0.1 and 0.13 agree with Bockris's analysis (10), since mechanisms in which the electron transfer step is rate-limiting produce polarization curves with a calculated Tafel slope of 0.12. Using Hickling and Hill's η data (57), Rüetschi and Delahay (106) were able to demonstrate that variations in η from one surface to another could be accounted for by variations in the energy of the M—OH bond determined theoretically.

It may be speculated as to what happens next, i.e., whether OH radicals combine to form adsorbed oxygen atoms and H$_2$O molecules, or whether another electron transfer step takes place to form an adsorbed O atom and a solvated H$^+$ ion in solution from the discharge of the adsorbed OH radical. In either case, two adsorbed O atoms eventually combine to form an adsorbed O$_2$ molecule which may then be desorbed. The mechanism in which two adsorbed OH radicals combine to form adsorbed H$_2$O$_2$ is rejected. In fact, Glasstone and Hickling (50) proposed a mechanism of electrolytic oxidation in which adsorbed H$_2$O$_2$ molecules which were formed by the combination of adsorbed OH radicals react with the electrode reactant in the oxidation reaction. This peroxide theory of oxidation has been severely criticized by Butler and Leslie (27) and by Walker and Weiss (123), who point out not only that peroxide is not observed at the anode but also that it is unlikely that two OH radicals would combine to form H_2O_2.

Theoretical calculations made by Hickling and Hill (58) show that a mechanism in which the combination of oxygen atoms to produce O_2 is the rate-limiting step demands an oxygen partial pressure of 10^{14} atm, a situation which is most unlikely. Using the concept that the adsorbed oxygen on the surface affects only the Tafel slope, Nassonov (91) has also concluded from theoretical considerations that the $O + O \rightarrow O_2$ reaction is not likely to be rate-determining.

Taking into account the results of these various investigations, the following mechanism for the evolution of oxygen on Pt is favored. First of all, the reaction takes place on a Pt surface covered with an electronically conducting monolayer of adsorbed oxygen. This adsorbed layer acts only as the catalytic surface on which the electrode reaction takes place. Absorbed OH radicals are produced by the discharge of OH^- ions or H_2O molecules as

$$(H_2O)_{ads} \xrightarrow{\text{slow}} (OH)_{ads} + H^+ + e \qquad (3.1)$$

followed by a fast reaction either

$$(OH)_{ads} + (OH)_{ads} \rightleftharpoons (O)_{ads} + H_2O \qquad (3.2)$$

or

$$(OH)_{ads} \rightleftharpoons (O)_{ads} + H^+ + e \qquad (3.3)$$

Finally, O_2 is evolved by the reactions

$$(O)_{ads} + (O)_{ads} \rightleftharpoons (O_2)_{ads} \qquad (3.4)$$

and

$$(O_2)_{ads} \rightleftharpoons O_2 \qquad (3.5)$$

This mechanism can account for the behavior of the overvoltage curves in the Tafel region (e.g., the region EF in Fig. 3.2), but cannot in the low current density region (e.g., the region AB in Fig. 3.2).

Even in rigorously purified solutions, Bockris and Huq (11) found that a routine of certain preparation steps was required to get reproducible results at low current densities. Hickling and Hill (57) and Efimov and Izgaryshev (35), among others (67), have also observed this phenomenon. An explanation for such behavior may be understood (67) from a polyelectrode point of view.

Since the rest potential is a mixed potential, the value of which depends on the ratio of cathodic to anodic sites which in turn depends on the history of the electrode preparation, the polarization curve at low current densities will reflect variations in the preparation of the electrode surface.   At low current densities, as seen in Fig. 3.3, the polarization curve describes the deviation of the mixed potential from the rest potential value as a function of the current density.   Only under open-circuit conditions at the mixed potential value, E_m, are the local cell anodic and the cathodic currents equal, $I_m = I' = I''$ (cathodic current I' is negative; anodic current I'', positive).   When the system is polarized in the anodic direction to the arbitrary value of E_2, the total observable current is the algebraic sum of the local cell partial currents, $I_2 = I_2' + I_2''$, which is an anodic current.   In a similar manner the cathodic current, I_1, resulting from the cathodic polarization to E_1, is $I_1' + I_2''$. Eventually a point is reached where one of the two partial currents becomes negligible to the other, depending on the direction of polarization, and the system is no longer a polyelectrode.

    In this low current density region (mixed potential region), any electrode preparation procedure which produces a reproducible surface will

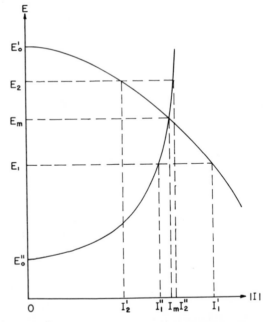

Fig. 3.3. Schematic diagram of the local cell polarization curves of the two half-reactions of the polyelectrode system, Pt/Pt-O/O$_2$ showing effect of external application of anodic or cathodic polarization (67).   Potential is plotted along the ordinate, the absolute value of the local current, along the abscissa.   At $E = E_m$, $I_m = I' = I''$; at all other potentials $I = I' + I''$ where I' is a negative (cathodic) current. (By permission of The Electrochemical Society.)

give reproducible polarization data. However, it must be remembered that this set of data may not agree with another set of reproducible data which has been obtained on electrodes prepared by a different procedure. The catalytic surfaces and the ratio of the anodic to cathodic sites are not the same in the two cases, and as a result the polarization behavior of the polyelectrode system is different in each instance. This viewpoint may explain why the results of two careful investigations of a polyelectrode system may not agree. A kinetic equation which is compatible with the experimental data has been derived (67).

In Fig. 3.3, as the potential becomes more noble, a point is reached at which the mixed potential can no longer maintain control of the kinetics. Here, the kinetics are controlled by the Pt/Pt-O reaction which has reached a limiting current situation. Under galvanostatic conditions, the potential rises, as in Fig. 3.1, until a new process, the evolution of O_2 on a layer of Pt-O, can take place. Under potentiostatic conditions a limiting current region is observed (Fig. 3.2) until O_2 is evolved which is interpreted (67) to mean that the layer of Pt-O is electronically conducting. The current goes into the formation of Pt-O, but the coverage of the surface depends on the potential (12,39,48,95,108). The Pt-O formed in excess of the steady-state value governed by the potential decomposes chemically by dissolving either in the solution or in the surface layers of the metal.

Finally, in the region EF of Fig. 3.2, a stable surface is obtained, and the evolution of O_2 takes place on the electronically conducting layer of Pt-O, possibly containing some PtO_2. Using ^{18}O tracer techniques, Rozental' and Veselovskii (105) have shown that the adsorbed layer of Pt-O does not participate in the electrode reaction but only serves as the surface on which the reaction proceeds. Breiter (18) has reached similar conclusions. This region in which a Tafel relationship holds is referred to as the oxygen evolution region. As mentioned before, the anodization procedure causes oxygen to be dissolved in the surface layers of the Pt, but this does not affect the evolution reaction which occurs on the adsorbed film of Pt-O. This effect is demonstrated by the observation that the η curve obtained on a bright Pt anode (67) is virtually the same as that obtained on bright Pt-O-alloy anode (67a).

3. Effects of pH

There is some question about whether the mechanism is the same in acid and in alkaline solutions. Hoar (62) studied the evolution of oxygen at Pt anodes in $0.1N$ NaOH solutions and interpreted the results in terms of Tafel slopes. Because the observed slopes for the anodic evolution and the cathodic reduction of oxygen on Pt in alkaline solutions did not correspond to any one of the Bockris mechanisms (10) and because a value of the stoichiometric number, ν, equal to 2 was determined, he suggested the step

$$(OH)_{ads} + OH^- \rightarrow (H_2O_2^-)_{ads} \qquad (3.6)$$

which does not involve electron transfer, as the rate-determining step. This reaction, Eq. 3.6, is different from the one, Eq. 3.1, proposed as rate-limiting in acid solution.　Besides, Bockris and Huq (11) has estimated a value of 4 for ν from the values of the anodic and cathodic Tafel slopes in H_2SO_4 solutions.　Consequently, the mechanism of O_2 evolution at Pt anodes is different in acid and alkaline solutions according to the interpretation of these data.

Activation energies for the evolution of O_2 at a Pt anode determined by Stout (114) from the dependence of η on temperature have a value of 25.3 kcal/mole at a potential of 1.23 V in 0.1N NaOH.　When these results were compared (114) with similar determinations of the activation energy carried out by Bowden (15) at a potential of 1.23 V in 0.2N H_2SO_4 (18.7 kcal/mole), Stout concluded that the mechanism for oxygen evolution is different in acid and alkaline solutions because the two energy values did not agree.　However, later values of 14 to 16 kcal/mole obtained by Yoneda (125) from η measurements in 0.1N NaOH solutions agree well with Bowden's value, and he concluded that the mechanism is the same in acid and alkaline solutions.

Since the oxygen overvoltage value varies with time initially before a steady state was reached, Bockris and Huq (11) analyzed the η–time plots and suggested that the so-called electrochemical oxide path (10),

$$(OH)_{ads} + OH^- \rightarrow (O)_{ads} + H_2O + e \qquad (3.7)$$

is the rate-determining step.　As mentioned before, the steady-state measurements and the value of $\nu = 4$ (10) indicate that the discharge of OH^- ion,

$$OH^- \rightarrow (OH)_{ads} + e \qquad (3.8)$$

is rate determining.

Recently, Riddiford (100) objected to this idea that the mechanism under steady-state conditions is different from that under transient conditions.　Although he used an analysis involving the anodic and cathodic Tafel slopes to get a value of 2 for ν as Hoar (62) did, he concluded that the mechanism of O_2 evolution at Pt anodes is the same in all solutions.　Riddiford was careful to point out that too much faith in Tafel slopes may be a dangerous procedure for complicated systems, such as the evolution of oxygen, but from the analysis of his data of Hoar and of Bockris and Huq, he favored Eq. 3.7 as the rate-determining step.

It is important to realize, as Parsons (94) has pointed out, that reliable values of ν may be obtained from the equation

$$\nu = 1 \bigg/ \left[\frac{RT}{nF} \left(\frac{1}{b_a} - \frac{1}{b_c} \right) \right] \tag{3.9}$$

where b_a and b_c are the Tafel slopes for the anodic and cathodic polarization data only if the activated complex in the rate-determining reactions is the same for both the oxidation and reduction processes. This appears to be unlikely (67) in the case of the oxygen electrode, since the electron transfer occurs in the O_2 evolution process between the electrode and a singly bonded oxygen, $(OH)_{ads}$, and in the O_2 reduction between the electrode and a doubly bonded oxygen, $(O_2)_{ads}$. This problem will be discussed more fully in Chapter V. The concept of the stoichiometric number seems to be of little help in the studies of the evolution of oxygen on noble metals. It seems improbable that a reaction such as Eq. 3.6 is rate determining since H_2O_2 is not observed (27) at the anode during the evolution of oxygen.

In the recent literature, Bockris and co-workers (30a) find that the O_2 overvoltage curve for the evolution of O_2 on Pt anodes changes slope from 0.055 below 10^{-4} A/cm² to 0.110 above. This change of slope is interpreted as a change in the rate-limiting step, rather than a change in the overall mechanism. Hydroxyl ion adsorption may play an important role in the electrode kinetics at Pt anodes in alkaline solution.

Most likely the overall mechanism of oxygen evolution, Eqs. 3.1–3.5, is the same in acid and alkaline solutions as suggested by Krasil'shchikov (84) with the rate-determining step as Eq. 3.1 in acid and as Eq. 3.8 in alkaline solutions.

4. Effects of Adsorbed Ions

When the overvoltage curves shown in Figs. 3.1 and 3.2 are extended beyond the indicated current density range to higher anodic current densities, it is found (42) that the η increases abruptly to a higher curve as seen in Fig. 3.4 for $HClO_4$ solutions. As the $HClO_4$ concentration is increased, the jump is shifted to lower current density values and becomes less abrupt. Similar results were observed (78) in H_2SO_4 solutions, but the break in the curve seemed to occur at higher current densities than in $HClO_4$ solutions.

It is interesting that Gerovich and co-workers (47) found that tagged $H_2S^{18}O_4$ exchanged oxygen with the water, whereas tagged $HCl^{18}O_4$ did not. In $H_2^{35}SO_4$, Balashova (2) observed rapid adsorption of SO_4^{2-} ions on Pt black and a slow adsorption which was considered to be an interaction with the surface. For these reasons and the fact that anodization

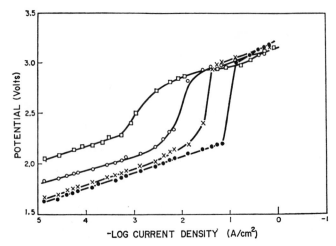

Fig. 3.4. Anodic polarization curves on Pt in solutions of $HClO_4$ of various concentrations (42); ($\bullet$) $1.34N$; ($\times$) $2.98N$; ($\bigcirc$) $5.0N$; ($\square$) $9.8N$. (By permission of Pergamon.)

of Pt in high concentrations of H_2SO_4 produces $H_2S_2O_8$ and H_2SO_5 (75, 80) which becloud the effects of adsorbed anions on the evolution of oxygen, $HClO_4$ makes a more desirable acid to use in these studies at high current densities and high acid concentrations.

This break in the anodic polarization curve obtained on Pt electrodes at high current densities was first reported by Kaganovich and co-workers (77,78) and later confirmed by isotope studies (47,105) using tagged $HCl^{18}O_4$. Frumkin (41) has summarized the Russian work using the ^{18}O isotope. It was found that ^{18}O enrichment was not detected in the oxygen evolved in the lower Tafel region but was detected in the upper Tafel region. This behavior has been interpreted in terms of the specific adsorption of anions. Both Frumkin (42) and Breiter (17) have assembled reviews of the role of ion adsorption on the kinetics of anodic processes.

Grube and Mayer (53) anodized Pt electrodes in $11N$ $HClO_4$ solution and observed that more oxygen was evolved than could be accounted for from the discharge of water. They concluded that ClO_4^- ions must be discharged to produce adsorbed ClO_4 radicals which decompose according to the reaction

$$ClO_4 \rightarrow ClO_2 + O_2 \tag{3.10}$$

In concentrated $HClO_4$ solutions, Beck and Moulton (5) also found two Tafel regions separated by a limiting current region. They observed that this limiting current was the same on rotating and stationary electrodes from which they deduced that the limiting current was not the result of diffusion control but was produced by the adsorption of ClO_4^- ions. Since the η had to increase to a point where ClO_4^- ions could be discharged, a break in the polarization curve was observed. ClO_2 was detected only in the region above the limiting current and was considered to be produced by Eq. 3.10.

The presence of ClO_3^- ions was detected at high anodic potentials by all workers (5,47,53). Both Grube and Moulton suggest that it arises from the reaction of ClO_2 with OH^- ions,

$$ClO_2 + OH^- \rightarrow HClO_3 + e \qquad (3.11)$$

Since the process occurs in strong acids, Eq. 3.11 does not seem as likely to occur as the reaction proposed by Gerovich et al. (47),

$$3ClO_4 + H_3O^+ + e \rightarrow 3HClO_3 + 2O_2 \qquad (3.12)$$

Incidentally, Eq. 3.12 is incorrectly written in the report (47).

Gerovich and co-workers (47) disagree with Beck and Moulton (5) since they distinguish between two mechanisms in the upper Tafel region. At the beginning of the upper η curve, oxygen was first thought to be produced by the interaction of ClO_4 radicals with water, according to Eq. 3.12, where ClO_2 was not detected. Above current densities at least 10 times greater than those at the beginning of the upper curve, ClO_2 was detected, and O_2 was evolved by the breaking up of ClO_4 radicals according to Eq. 3.10.

However, from the isotope studies at current densities before the point where ClO_2 was detected, ClO_3^- ions were not found, even though the mass spectroscopic data correspond to the decomposition of 67% of the $Cl^{18}O_4$ radicals. This is interpreted (41,42,47) to mean that the adsorbed ClO_4^- ions interact with the oxides formed on Pt at these high potentials. It is thought that the $Cl^{18}O_4^-$ ion is distorted at the high overvoltage values, and this situation permits an exchange of oxygen with the adsorbed platinum oxides. In this way, ^{18}O may appear in the evolved O_2 without the discharge of a ClO_4^- ion because oxygen could be evolved by the chemical decomposition of a higher oxide of Pt, such as PtO_2 or PtO_3, which may be formed at the high potentials (about 3 V) where such observations are made. Krasil'shchikov (84) has suggested such an oxide mechanism for the anodic evolution of oxygen on a number of metal anodes.

Consequently, it is now believed (42,47) that oxygen is evolved through a surface oxide complex rather than according to Eq. 3.12 at current densities corresponding to the first part of the upper η curve of Fig. 3.4. In the second part of the upper Tafel region (above 3×10^{-1} A/cm^2), the potential is high enough to discharge ClO$_4^-$ ions, and both ClO$_2$ and ClO$_3^-$ ions are detected.

Kheifets and Rivlin (80) found two Tafel regions separated by a limiting current region on Pt anodes. From a study of the effect of organic additives on the oxygen η (81), they favor the suggestion that the evolution of oxygen proceeds through the reaction

$$(O)_{ads} + H_2O \rightarrow O_2 + 2H^+ + 2e \qquad (3.13)$$

As the temperature is lowered, the oxygen overvoltage on a Pt anode in strong HClO$_4$ solutions at high current densities increases, and ozone is detected (5) in the gases evolved at the anode. For some time it has been known that O$_3$ appears at the anode when strong solutions of H$_2$SO$_4$ are electrolyzed at high current densities. Fischer and Massenez (40) used ice–salt cooling mixtures to lower the temperature of a Pt-tube electrode to $-14°$C. Using high current densities (60–80 A/cm^2), they found that the yield of O$_3$ was better for bright than for black Pt anodes and increased with decreasing temperature. Ozone was detected by Briner and co-workers (21) in the electrolysis of concentrated solutions of H$_3$PO$_4$ and KOH besides H$_2$SO$_4$ and HClO$_4$. Briner and Yalda (21a) studied the evolution of O$_3$ in concentrated solutions of H$_2$SO$_4$ at $-40°$ to $-50°$C at Pt anodes and observed that the rest potential of a Pt electrode in O$_3$ saturated acid solution was always more noble than one in O$_2$-saturated acid solution. Both a reduction in pressure and a reduction in temperature increased the current efficiency for the evolution of O$_3$ as determined by Putnam and co-workers (98).

Beck and Moulton (5) used Pt and Pt-Ir-alloy anodes in $5M$ HClO$_4$ solution and showed that O$_3$ was not formed at current densities corresponding to the lower Tafel region. At current densities corresponding to the limiting current region at temperatures of $-40°$ to $-50°$C, a maximum was observed in the evolution of O$_3$, but at much higher current densities, the yield of O$_3$ increased again. The evolution of O$_3$ in the first region was considered (5) to arise from the discharge of H$_2$O. As soon as ClO$_4^-$ ions begin to discharge, the evolution of O$_3$ falls off, producing a maximum in the yield curve with increasing potential. At higher current densities (>1 A/cm^2) where the current efficiency for O$_3$ formation reaches 30% (40,98), it was suggested (5) that ClO$_4$ free radicals are involved.

From ^{18}O tracer studies of the evolution of O$_3$ in concentrated solutions of HClO$_4$ at temperatures between $-40°$ and $-50°$C, Gerovich and co-workers (46) concluded that ClO_4^- ions take part in the reaction. The actual evolution of O_3 takes place by the decomposition (40a,46) of adsorbed oxygen layers on PtO,

$$PtO(nO)_{ads} + H_2O \rightarrow PtO[(n - 2)O]_{ads} + O_{\frac{3}{2}} + 2H^+ + 2e \qquad (3.14)$$

The ^{18}O enters the adsorbed oxygen layers by exchange with the adsorbed anions.

Recently, Kasathin and Rakov (79) electrolyzed $10N$ H_2SO_4 solutions cooled to $-60°C$ in the potential range from 3 to 12 V potentiostatically. The resulting polarization curve shows several negative resistance regions (zones of inhibition) which are associated with the inhibition of $H_2S_2O_8$ formation ($\sim$4 V), O_2 formation ($\sim$5 V), and O_3 formation (8 V). They reason that the formation of a given product depends on the nature and the accumulation of a given intermedite absorbed on the PtO-covered surface. By lowering the temperature, the reactivity of certain intermediates is lowered, and a given end product is favored. At potentials above $\sim$3 V, where the formation of $H_2S_2O_8$ occurs through the discharge of adsorbed HSO_4^- ions, lowering of the temperature favors O_3 formation. Studies of the electrolysis of H_2SO_4 in $H_2^{18}O$-enriched solutions (86) indicate that O_3 and $H_2S_2O_8$ are produced simultaneously.

Reports of other instances of the interference of ion adsorption on anodic processes may be found in the literature. It has been reported that peroxides were detected at Pt anodes in KOH solutions if fluoride ions were present (101) and that the presence of F^- ions increased the overvoltage (57,76,90). Additions of reducing agents containing sulfur or nitrogen, such as thiourea, Na_2S, and $NaNO_2$, raise the η (59); but the addition of SO_4^{2-} ions as Na_2SO_4 lowers the η (11). Frumkin (43) has reported that the adsorption of alkali cations at high anodic potentials influences the electrode processes at Pt anodes in strong $HClO_4$ and H_2SO_4 solutions, but he finds (42) that definite conclusions about the mechanisms of these phenomena are difficult to make because of the complex nature of these effects.

Kozawa (83) has discussed the effect of halide ions in acid solutions and alkaline-earth ions in alkaline solutions on the evolution and reduction of O_2 at Pt electrodes. The anodic overvoltage increased with the addition of Sr^{2+} ions. He explains the observed results in terms of an ion-exchange adsorption of cations and anions on a PtO_2 covered surface in the potential range from 0.7 to 0.9 V. It is believed (72) that adsorbed PtO_2 does not exist in the presence of metallic Pt on Pt anodes at potentials below 1.5 V. Effects of the adsorption of alkali metals ions on oxygen adsorption have been investigated by Kazarinov et al. (79a) using radioactive tracers.

An account of ion adsorption effects is also given by Delahay (ref. 33, p. 269 et seq.)   In summary, Vetter (121) points out that a theoretical understanding of the effects of ion adsorption on anodic processes is still unclear.

5. Effect of Nonaqueous Media

Such studies are rare in the reviewed literature.   Bockris (9) has studied the effect of nonaqueous media on the oxygen evolution reaction at Pt electrodes in acetic acid– and dioxane–water mixtures.   The solutions were always $1N$ in H_2SO_4.   He found that the η was greater in the mixtures than in the pure water solutions, and the Tafel slope increased as the nonaqueous component increased.   Because it was assumed that the combination of oxygen atoms was rate determining, he reasoned that the η increased because the nonaqueous component inhibited this process.   This work represents a step further in the complexity of an already complex system, and since the purity of these systems may be questioned, it is difficult to comment on these studies.

6. Photo-Effects

The effects on the polarization of electrodes when irradiated with light have been investigated by several workers (1,16,34,49,55).   Audubert (1) attributed the depolarization of Pt and Au anodes illuminated with ultraviolet light to the depolarizing effect of the presence of H and O atoms produced when H_2O molecules which had adsorbed photons were decomposed.   The rapid depolarization of a W anode by ultraviolet radiation was reported by Duclaux (34).

These investigations (16,49,55) were carried out on Pt foil anodes in H_2SO_4 solutions which were irradiated with light of wavelengths between 2000 and 4000 Å.   The threshold limit of activity was found to be at 4000 Å; any light of lower frequency produced no effect.   The photo-current or the anodic depolarization increased with the frequency of the illuminating light.   In general, the procedure used by these workers involved anodizing the electrode to a given potential in the dark, and after a steady state had been reached, illuminating the electrode with the desired monochromatic light.   In all cases, a lowering of the overvoltage was observed.   Ginzburg and Veselovskii (49) noted that for a given light intensity and frequency the depolarization effect increased as the potential value to which the electrode was preanodized in the dark was increased.   They also recorded that the double layer capacity was the same, $300 \ \mu F/cm^2$, whether the electrode was in the dark or illuminated.

Hillson and Rideal (55) and Bowden (16) explain their results in terms of the adsorption of the light by the oxide layer produced on the electrode surface by anodization. It is assumed that the combination of O atoms is rate controlling in the anodic evolution of oxygen. Adsorption of light by the oxide layer causes a decomposition of the oxide which produces O atoms. These extra O atoms produced by the light adsorption are responsible for the photocurrent or depolarization effects. Ginzburg and Veselovskii (49) view the situation in a slightly different manner. They also believe that the anodic process produces an oxide layer which adsorbs the light, but they conclude that the adsorption of light produces a defect structure. An electron is liberated in the oxide which migrates to the metal, and the resulting "hole" migrates to the solution side of the oxide layer where an electron transfer reaction from H_2O or OH^- ions in the double layer is facilitated. This explanation (49) seems to fit the experimental facts better than the atom combination idea (55).

B. GOLD

It is impossible to study galvanostatically the evolution of O_2 on an Au-O surface (70) because the potential drifts until a point is reached where the surface is covered with a layer of Au_2O_3. As pointed out by Tsinman (119), O_2 is not evolved until the surface is covered with Au_2O_3, and reproducible results are obtained only after the electrode has been preanodized at constant current until a steady current is reached. Preanodization was required in other investigations (4,70,87). In the current density range 10^{-4} to 1 A/cm^2, most investigators (4,57,70,87,112, 119) observed a Tafel region of low slope (between 0.04 and 0.05) in the overvoltage curve for the anodic evolution of oxygen on a Au anode in acid solutions. This slope yields the extremely low value of 10^{-22} A/cm^2 for i_0 (4,87).

The overvoltage is independent of pH (4,87,119) in acid solutions, a fact which led MacDonald and Conway (87) to the conclusion that the rate-determining step is the discharge of water, Eq. 3.1. Barnartt observed that the Tafel slope did not fit any of the schemes listed by Bockris (10) and concluded that more reliable data were required before a definite mechanism could be suggested.

Rather complicated mechanisms involving the higher oxides of gold have been suggested by Tsinman (119), and a similar viewpoint is held by Krasil'shchikov (84), where oxygen is evolved from the chemical decomposition of Au_2O_4 produced at high anodic potentials. In a later report, however, Tsinman (120) favors Eq. 3.1 as the rate-determining step for the evolution of O_2 at Au anodes. According to Bockris's schemes, the 0.045 slope is too low to correspond to a mechanism in which Eq. 3.1 is the rate-controlling step.

To account for the low Tafel slope of 0.045 and the resulting low value of the transfer coefficient ($\alpha = 0.25$ instead of 0.5), MacDonald and Conway (87) proposed a twin barrier model in which a potential drop occurs not only across the Helmholtz double layer but also across the oxide layer adsorbed on the Au surface. It may be shown (88) that in such cases an overall α may be determined equal to $\alpha_1\alpha_2/(\alpha_1 + \alpha_2)$. If it may be assumed that α_1 for the process across the oxide film and α_2 for the one across the Helmholtz double layer are each 0.5, a value of 0.25 is obtained for α; and the experimental value of $2.3(4RT/5F)$ for the Tafel slope may be explained.

In the current density range from 10^{-7} to 10^{-4} A/cm^2, a Tafel region with a slope equal to 0.118 was reported (70) for the steady-state galvanostatic η studies, and at about 10^{-4} A/cm^2, the slope changed to a value of 0.047. The open-circuit potential of this anodized electrode was near 1.36 V, the E_0 value for the Au/Au_2O_3 reaction,

$$Au_2O_3 + 6H^+ + 6e \rightleftharpoons Au + 3H_2O \qquad E_0 = 1.36 \text{ V} \qquad (3.15)$$

and when the current was reversed, an oxide-reduction current was observed. The extrapolated anodic and cathodic Tafel slopes crossed at about 1.36 V as seen in Fig. 3.5. In this low current density range, the η data were independent of P_{O_2}. It was concluded (70) that in this current density range the electrode system is not a polyelectrode but a true metal–metal oxide electrode with the overall electrode reaction being Eq. 3.15. Consequently, in this η range, the current goes into the formation or reduction of Au_2O_3. Similar findings were obtained with steady-state potentiostatic overvoltage measurements (69).

Brumer (23,24) has studied the reduction of Au_2O_3 in $HClO_4$ solution as a function of the forming potential and the time of formation. He has presented a mechanism in which the reduction of Au^{II} is rate determining. If this were so, one would have expected a buildup of Au^{II} in the system, but Hickling (56) could not detect its presence. As a matter of fact, Sidgwick (110) casts doubts on its existence. These studies are complicated by the fact that at constant potential, Au_2O_3 not only continues to build up with time but also continuously decomposes in the acid solutions. Grüneberg (54) favors a mechanism involving Au^I, but this has not been detected (25,45,97) either.

From an analysis of potential sweep curves taken on bright Au electrodes, Grüneberg (54) observed that Au_2O_3 is formed above 1.37 V and

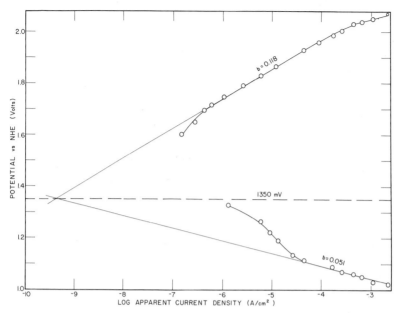

Fig. 3.5. Cathodic and anodic polarization curves taken galvanostatically on the same Au/Au_2O_3 electrode (70). The extrapolated Tafel lines cross at 1.355 V, very near the reversible potential of the Au/Au_2O_3 reaction. (By permission of Pergamon.)

O_2 is evolved above 1.8 V. With such observations, one would expect to observe a break in the anodic η curve similar to the one observed experimentally (69,70). Because the Au_2O_3 film is a poor electronic conductor (64,66), it is likely that the oxygen reactions proceed with difficulty on oxide-covered sites. It was reasoned (70) that since the Au_2O_3 film is poorly adherent and easily flakes off (3,64) there is always present a number of electronically conducting "bare" gold sites on which O_2 may be evolved with Eq. 3.1 rate determining as suggested in the literature (87,120). In parallel to the O_2 evolution reaction on uncovered sites, the Au_2O_3 builds up on clusters of ionically conducting oxide-covered sites. If such parallel reaction mechanisms are accepted, one would expect unusual values for the Tafel slope as well as for i_0, since the reference state for the determination of the η cannot be established.

Steady-state polarization data obtained potentiostatically on Au anodes (69) are shown in Fig. 3.6. From an open-circuit potential of about 0.85 V, the current increases with increasing potential until a value of about 0.9 V is reached. This region is the mixed potential

region, and the current goes into the reduction of oxygen and the formation of a partial layer of adsorbed oxygen atoms, Au-O.   Between 0.9
and 1.35 V, a limiting current region is observed, during which the partial layer of Au-O grows with increasing potential.   In this potential
range, the electrode surface is electronically conducting. Up to this point,
the system is similar to the Pt-O_2 system, except that the Au-O layer oc-
cupies at the most a small fraction of the total electrode surface.   Above
1.35 V, Au_2O_3 begins to form, and the current increases slightly.   Because Au_2O_3 sites are poorly conducting sites, a negative resistance or
passivation region appears between 1.36 and 1.6 V.   Finally, above 1.6
V, O_2 evolution begins, and the current increases with the potential.   A
point is eventually reached where reproducible results may be obtained
and where a polarization curve with two Tafel regions is found similar
to the galvanostatic measurements (Fig. 3.5).   It is interpreted (69,70)
that the upper region of low slope (0.05) corresponds to the parallel mechanisms of O_2 evolution and Au_2O_3 formation, whereas the lower region
of high slope (0.15) corresponds to Au_2O_3 formation alone.   When the
potential of this system is lowered below 1.36 V, a Au_2O_3-reduction curve
is obtained with a slope of 0.05.   As in the galvanostatic case, the extrapolated Tafel slopes cross at about 1.36 V.   Brummer and Makrides (24)
also observed a low slope (0.04) for the reduction of Au_2O_3.

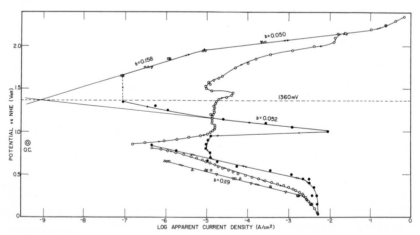

Fig. 3.6. Cathodic and anodic polarization curves taken potentiostatically on
bright Au in O_2 saturated $2N$ H_2SO_4 solution (69).   (O) Increasing current, first
cycle; (△) decreasing current, first cycle; (▽) increasing current, second cycle; (□)
decreasing current, second cycle; (●) cathodic polarization of Au/Au_2O_3 electrode.
The rest potential is given at the extreme left by symbol O O.C..   (By permission
of Pergamon.)

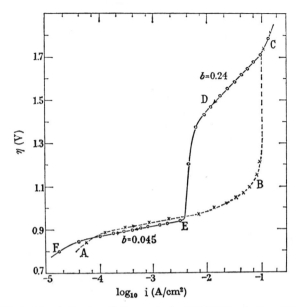

Fig. 3.7. Polarization behavior of Au anode in $1M$ KOH solution (87).   (By permission of The Royal Society.)

Apparently the electrode processes occurring at Au anodes in acid solutions are very complex, and one must await the further accumulation of fundamental observations before firm conclusions about the kinetic mechanisms may be made.

It appears that the anodic processes in alkaline solutions are even more complicated.   In general, the η for O_2 evolution is higher in alkaline than in acid solutions.   Several investigators (57,87,120) observed two Tafel regions in the anodic η curves separated by a limiting current region.   A large hysteresis is observed in the beginning of the limiting current region for increasing and decreasing currents as presented in Fig. 3.7.   For increasing currents it was found (87) that the Au surface darkened at the beginning of the limiting current region and became deep brown in the upper Tafel region.   With decreasing currents, the color did not disappear until the beginning of the limiting current region was reached.   The Tafel slope is about five times larger in the upper curve than in the lower. MacDonald and Conway (87) explain these observations with the twin barrier model in terms of the degree of coverage of the metal-oxide surface with adsorbed OH radicals.   Because the η is independent of pH, they favor the formation of adsorbed OH radicals from the discharge of

H_2O molecules, Eq. 3.1, as the rate-limiting step. However, Tsinman (120) reports a dependency of the η on the activity of OH^- ion and prefers the discharge of OH^- ions, Eq. 3.8, as the rate-determining step.

C. PALLADIUM

Steady-state galvanostatic oxygen overvoltage curves have been obtained for the evolution of oxygen at Pd anodes in acid (68,87) and in alkaline solutions (57,87). MacDonald and Conway (87) record a slope of 0.1 and an i_0 of about 10^{-10} A/cm^2 in acid solutions and similar values in alkaline solutions. It was concluded that the mechanism for the anodic process is the same in both acid and alkaline solutions with Eq. 3.1 rate-controlling in either case. Although Hickling and Hill (57) report an upper and lower Tafel region in the anodic η curve on Pd, MacDonald and Conway (87) did not observe such a phenomenon.

In another investigation of the evolution of O_2 at a Pd anode in O_2 saturated acid solutions (68), the overvoltage at low current densities was described in terms of a polyelectrode model. The current goes into the local cell processes responsible for the mixed potential (65) observed under open-circuit conditions. Finally, a limiting current density is reached, as seen in Fig. 3.8, where the local cell process can no longer control the kinetics. At this point the potential drifts to a value where another process, the evolution of O_2, takes place. In the higher potential region above 1.45 V ($\eta = -0.22$ V), a reproducible η curve is obtained with the high Tafel slope of 0.198. When the current is reversed, an oxide reduction current is observed until the active material (oxide) has been consumed, after which the system goes to the reduction of O_2, as shown by the broken vertical line in Fig. 3.8.

More information about these processes is obtained from the steady-state potentiostatic polarization data given in Fig. 3.9. In agreement with the galvanostatic data of Fig. 3.8, a mixed potential region is observed followed by a negative resistance region in Fig. 3.9, corresponding to the limiting current in Fig. 3.8. This observation was interpreted (68) that in the potential range between 0.98 V ($\eta = 0.25$ V) and 1.43 V ($\eta = 0.2$ V) a layer of poorly conducting PdO is formed on the electrode surface. At potentials above 1.45 V ($\eta = 0.22$ V), the evolution of O_2 takes place, and a reproducible Tafel region ($b = 0.2$) is found. It is believed (68) that above 1.45 V the PdO sites are converted to PdO_2 sites.

It has been observed (8,60,122) that oxygen is first evolved in a potential region in the neighborhood of 1.47 V, which is considered (65) to be the standard potential of the Pd/PdO_2 electrode. Since it is known that PdO_2 is unstable (8,26,36,60), O_2 may be liberated by the decomposition of PdO_2 to PdO (36,60,68).   It was proposed (68) that O_2 is evolved at Pt anodes by the chemical decomposition of PdO_2,

$$PdO_2 \rightarrow PdO + \tfrac{1}{2}O_2 \qquad (3.16)$$

and the external current is used to convert more PdO to PdO_2,

$$PdO + H_2O \rightarrow PdO_2 + 2H^+ + 2e \qquad (3.17)$$

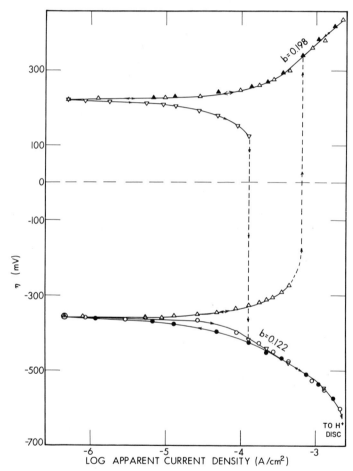

Fig. 3.8. Oxygen overvoltage curves for the reduction of oxygen ($\bigcirc$) and the evolution of oxygen ($\triangle$) on bright Pd electrodes in O_2-saturated $2N$ H_2SO_4 solution, obtained galvanostatically (68).   The arrows give direction of polarization, and in general, open symbols represent data obtained with increasing current and filled symbols those with decreasing current.   The reduction curve for a Pd/PdO_2 electrode ($\triangledown$) is also shown. η is determined from 1.23 V.   (By permission of The Electrochemical Society.)

In the potential region in the vicinity of 1.45 V, the system behaves
as a metal–metal oxide electrode with the current being used in either
the formation or reduction of PdO_2 similar to the case of the Au/Au_2O_3
electrode, except that at high anodic potentials, O_2 is evolved by the
decompostion of the oxide at a Pd/PdO_2 electrode instead of by a side
reaction to oxide formation at an Au/Au_2O_3 electrode. In both cases,
at potentials cathodic to the metal–metal oxide potential, an oxide-reduction current is obtained until the active material is used up. After

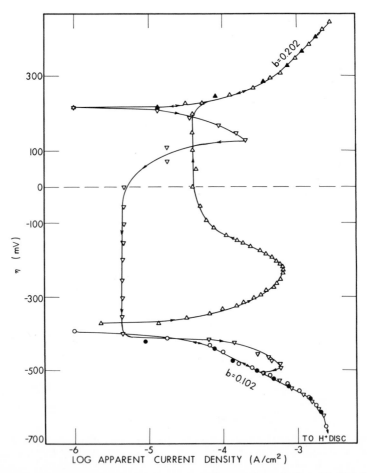

Fig. 3.9. Oxygen overvoltage curves obtained potentiostatically on bright Pd in
O_2-saturated $2N$ H_2SO_4 solution (68). Symbols have the same meaning as in Fig.
3.8. (By permission of The Electrochemical Society.)

reaching this point, the system goes to the reduction of oxygen. Mac-Donald and Conway (87) pointed out that the polarization curves of many anodic reactions, in which surface oxides participate, exhibit high values of the Tafel slope, as observed in Figs. 3.8 and 3.9 for Pd anodes.

To identify the rate-determining step uniquely and to sort out the preceding and following rapid steps in these anodic processes, more information than is available in the literature is required.

D. RHODIUM

Early polarization studies of the evolution of O_2 at Rh anodes (6,124) gave results in which the E–log i curves possessed Tafel regions. Steady-state galvanostatic overvoltage measurements obtained on Rh anodes in acid solution (73) indicate that a reproducible polarization curve cannot be attained until the system has been very strongly ano-dized, as shown by the broken and solid anodic curves in Fig. 3.10. Once a stable catalytic surface was prepared by the preanodization pro-cedure, a reproducible η curve was obtained with a Tafel slope of 0.106. It was also observed (Figs. 3.10 and 3.11) that the extrapolated anodic and cathodic Tafel slopes crossed near the reversible potential of 1.23 V ($\eta = 0$). Damjanovic and co-workers (30b) have obtained similar results.

The steady-state potentiostatic data of Fig. 3.11 show a short mixed potential region before the curve enters a limiting current situation. The kinetic processes may be somewhat similar to those operating in the Pt–O_2 system. In the limiting current region from 0.96 V ($\eta = -0.27$) to 1.38 V ($\eta = 0.15$), the surface is electronically conducting, and oxygen is not only adsorbed on the surface but is dissolved in the metal to form a surface layer of Rh-O alloy as noted in Chapter II. The concentration of dissolved oxygen increases with potential until a point is reached when a stable steady-state concentration is obtained. Under these conditions a reproducible Tafel curve with a slope of 0.085 is observed. Krushcheva and co-workers (81a) observed that below 1.2 V the adsorbed oxygen was in the form Rh-O; above 1.2 V, in the form of Rh_2O_3. This oxide may correspond to the Pt-O alloy.

When the potential was lowered below 1.03 V ($\eta = 0.2$), a reduction current was observed until a cathodic limiting current density was reached equal to the anodic limiting current density. This lack of hysteresis in the positive overvoltage region is unique with Rh (Pt and Ir exhibit no reduction current, and Au and Pd show an oxide reduction current that disappears when the oxide is used up). Apparently (73) the Rh-O alloy can support a small current (about 10^{-5} A/cm^2) at these potentials, involving a reaction similar to Eq. 3.18,

$$Rh(O)_x + 2xH^+ + 2xe \rightarrow Rh + xH_2O \qquad (3.18)$$

where the symbol $Rh(O)_x$ represents the concentration of oxygen dissolved in the Rh-O alloy. Over this potential region it is interpreted that the catalytic activity of the electrode surface is the same for the anodic processes responsible for the limiting current whether the electrode is in the reduced or preanodized state. This statement does not apply to any one of the other noble metals.

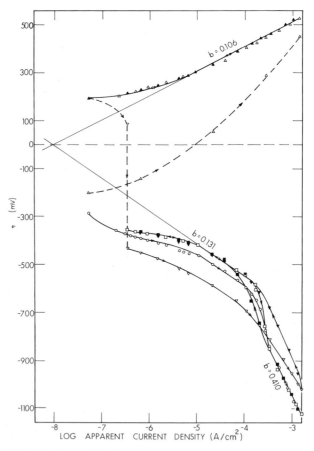

Fig. 3.10. Oxygen overvoltage curves obtained galvanostatically on bright Rh in O$_2$-saturated $2N$ H$_2$SO$_4$ solution (73). Notation same as in Fig. 3.8. (By permission of Pergamon.)

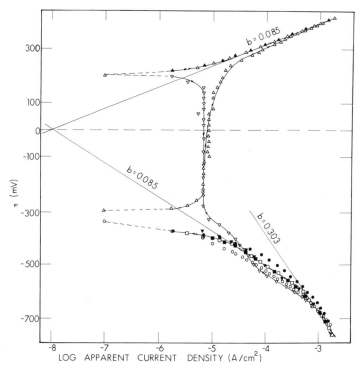

Fig. 3.11. Oxygen overvoltage curves obtained potentiostatically on bright Rh in O_2-saturated $2N$ H_2SO_4 solution (73)   Notation same as in Fig. 3.8. (By permission of Pergamon.)

The mechanism of O_2 evolution on Rh anodes in acid solutions is considered (73) to be the same as the one on Pt anodes, Eqs. 3.1-3.5 with Eq. 3.1 as rate determining.   The anodic process contributing to the current in the limiting current region is the formation of the adsorbed layer of oxygen atoms, Rh-O.   A steady state is reached when the anodic rate of formation of Rh-O,

$$Rh + H_2O \rightarrow Rh\text{-}O + 2H^+ + 2e \qquad (3.19)$$

equals the chemical decomposition of Rh-O which either dissolves in solution or in the metal.   The steady-state coverage of electronically conducting Rh-O is determined by the potential.

E. IRIDIUM

Tafel regions were also observed in the O_2-evolution polarization curves obtained in the early work (6,124) on Ir anodes.   Galvanostatic, steady-state η measurements obtained (74) on Ir anodes show (Fig. 3.12) that reproducible data cannot be obtained until the electrode has been preanodized to produce a stable catalytic surface on which the evolution reaction may take place.   After such preparation, a reproducible Tafel curve with a slope of 0.077 is found.   Steady-state potentiostatic data in Fig. 3.13 yield an anodic η curve similar to that found for Pt with the mixed potential region followed by a limiting current region.   The curve in the oxygen evolution region is nearly identical to the one obtained galvanostatically.

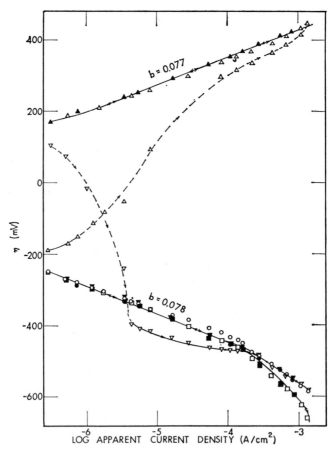

Fig. 3.12. Oxygen overvoltage curves obtained galvanostatically on bright Ir in O_2-saturated 2N H_2SO_4 solution (74).   Notation same as in Fig. 3.8.   (By permission of Elsevier.)

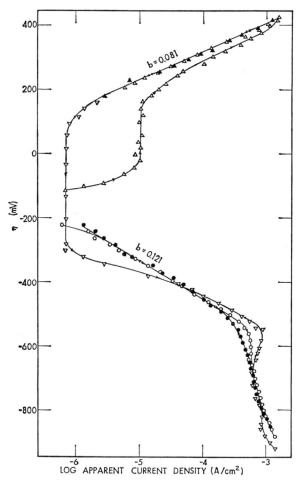

Fig. 3.13. Oxygen overvoltage curves obtained potentiostatically on bright Ir in O_2-saturated $2N$ H_2SO_4 solution (74). Notation same as in Fig. 3.8. (By permission of Elsevier.)

As in the Pt and Rh cases, the current in the limiting current region goes into the formation of the electronically conducting layer of Ir-O, the electrode coverage of which is determined by the potential. At potentials above about 1.4 V($\eta = 0.21$ V), some Ir-O sites may be converted to IrO_2 sites. With prolonged anodization the oxide layer on Ir anodes does not build up much beyond a monolayer thick (13,26), indicating that the layer of IrO_2 is a good electronic conductor. Oxygen evolution takes place on the surface which probably contains a mixture of Ir-O and IrO_2 sites.

When the potential in Fig. 3.13 is lowered below 1.4 V ($\eta = 0.21$ V), a reduction current is not observed as in the Pt case. Böld and Breiter (13) observed that the formation and reduction of the adsorbed oxygen layers on Ir occurred reversibly. The data of Fig. 3.13 are interpreted (74) that the number of Ir-O sites converted to IrO_2 sites depends on the potential (number increases with increasing potential) because an oxide reduction current is not observed when the potential is lowered below 1.4 V ($\eta = 0.21$ V) in Fig. 3.13. As the potential becomes less noble, the IrO_2 sites formed at higher potentials are reconverted to Ir-O sites with the result that IrO_2 is no longer present on the electrode surface at potentials below 1.4 V. Consequently, a reduction current is not found at the Ir electrode.

In the potential range where a Tafel region with a slope of 0.08 is found, the anodic current goes into both the evolution of O_2 and the conversion of Ir-O to IrO_2 sites by parallel reactions as proposed for Au anodes. Such a situation may possibly account for the low value of the Tafel slope and for the fact that the extrapolated anodic and cathodic Tafel lines do not cross at 1.23 V. Since a Tafel relationship is observed, kinetic control is assumed; and the evolution of O_2 by the mechanism of Eqs. 3.1-3.5 is favored to occur with Eq. 3.1 as the rate-controlling step. Evolution of O_2 by the decomposition of IrO_2 (as in the Pd case) is not likely since IrO_2 is a stable compound and the Ir/IrO_2 potential (0.93 V) is nowhere near 1.4 V, the point above which O_2 is liberated on an Ir anode. Damjanovic and co-workers (30b) have interpreted galvanostatic polarization data obtained on Ir anodes in terms of Tafel parameters.

F. NOBLE METAL ALLOYS

Some studies of the dependence of the oxygen overvoltage on the composition of various alloys of Ni (52,115,116) were made, but either there was no correlation or the η was independent of the alloy composition. MacDonald and Conway (87) studied the evolution of oxygen on a series of Au-Pd alloys in acid and alkaline solutions. Although the Tafel slopes and i_0 values varied with composition, any d-band effects of the electronic structure of the alloys were not demonstrated, most likely because such effects were obscured by the adsorbed oxide films. Recently, Brooman and Hoar (22) reported on galvanostatic polarization studies carried out on anodes made of Ru-Pt, Ir-Pt, Rh-Pt, and Pd-Pt alloys. The onset of oxide formation and the η for the evolution of O_2 were different for each set of alloys. For the anodic evolution of O_2, the η on the alloys was lower than on pure Pt, and the Rh-Pt alloy exhibited the lowest η. An explanation for this synergistic effect was not offered. It was also observed by Damjanovic et al. (30b) that the η was lower on a Pt-Rh anode than on a Pt anode.

However, if the interaction of oxygen with the surfaces of a series of alloys could be studied before an interfering adsorbed film could be formed, the dependence on the electronic structure might be detected. Rao, Damjanovic, and Bockris (99) investigated the adsorption of oxygen on the oxide-free surfaces of Pt, Pd, Rh, Ir, and Au wires in O_2 saturated, H_2SO_4 solutions. The amount of O_2 adsorbed as a function of time was determined from constant current cathodic stripping pulses. These experiments were carried out in periods of time less than one hour. In the first 10 or 15 min the potential changed rapidly with time and then tapered off quickly to a very slow variation with time, at which point the surface had reached maximum coverage with adsorbed oxygen dipoles (31). It was found (99) that the maximum coverage of the metal surface with oxygen dipoles was about 25% for metals such as Pt and Pd to which a value of 0.6 may be assigned as the number of holes in the d band. For those metals, Rh, Ir, and Ru, to which a value of almost 2 may be estimated for the number of holes in the d band, a coverage of nearly 100% was found. Gold which does not have any holes in the d band had a coverage of less than 3%.

The Bockris group reasoned that if two valences could be assigned to the oxygen atom, it would require four Pt atoms with 0.5 valences to satisfy each adsorbed O atom and a coverage of 25% would be expected. Those metals with two valences should accept an O atom for each surface Pt atom. As a check on this viewpoint, the Bockris group (32) studied the adsorption of oxygen on a series of Rh-Pt alloys. They observed that the maximum coverage increased with the Rh content, and hence, with the number of holes in the d band. Consequently, in this dipole region, there is a correlation between the electronic structure of the electrode metal and the adsorption of O dipoles.

Over periods of time greater than an hour, a local cell may be set up by electrode reactions which may occur on the covered and uncovered Pt sites. Under the driving force of the local cell, the Pt surface will be nearly covered with a layer of Pt-O (63) after about 24 hr. The rest potential is no longer determined by a dipole- or χ-potential mechanism, which was assumed (31,32) during the first 30 min of adsorption, but by a mixed potential mechanism (63). Under these conditions the correlation between electronic structure of the Rh-Pt alloy and the surface coverage with Pt-O should not be observed. This experiment has not been performed.

Breiter (20) studied the formation and reduction of the adsorbed oxygen layers on a series of Au-Pt alloys in acid solutions with the potential sweep method. The oxygen reduction trace contained two peaks, the first of which increased and the second decreased with increasing gold content. He assigned the reduction of the oxygen layer on Au sites to the first peak and that on Pt sites to the second. These data seem to indicate that the ratio of Pt to Au sites on the electrode surface is a function of the alloy composition and the interaction of these sites with oxygen is independent of the presence of one or the other. In fact, it was found that the superposition of the sweep curve obtained for pure Pt upon that for pure Au produced a combined curve which approximated the shape of the curve obtained on the 40 at % Au alloy. Similar measurements were carried out on a series of Pt-Cr alloys (19), and the influence of the passive Cr sites on the oxygen region was pointed out.

If effects of the electronic structure of the electrode material on the catalytic activity of the electrode surface for the oxygen electrode reactions are to be observed, one must choose experimental conditions where the surface is not completely covered with adsorbed oxygen films.

References

1. Audubert, R., *Compt. Rend.*, **189**, 1265 (1929).
2. Balashova, N. A., *Dokl. Akad. Nauk SSSR*, **103**, 639 (1955).
3. Barnartt, S., *J. Electrochem., Soc.* **106**, 722 (1959).
4. Barnartt, S., *J. Electrochem. Soc.*, **106**, 991 (1959).
5. Beck, T. R., and R. M. Moulton, *J. Electrochem. Soc.*, **103**, 247 (1956).
6. Bennewitz, K., *Z. Physik. Chem. (Leipzig)*, **72**, 202 (1910).
7. Bianchi, G., and F. Mazza, *Electrochim. Acta*, **1**, 198 (1959).
8. Blackburn, T. R., and J. J. Lingane, *J. Electoranal. Chem.*, **5**, 216 (1963).
9. Bockris, J. O'M., *Nature*, **159**, 401 (1947); *Discussions Faraday Soc.*, **1**, 229 (1947).
10. Bockris, J. O'M., *J. Chem. Chem. Phys.*, **24**, 817 (1956).
11. Bockris, J. O'M., and A. K. M. S. Huq, *Proc. Roy. Soc. (London)*, **A237**, 277 (1956).
12. Böld, W., and M. W. Breiter, *Electrochim. Acta*, **5**, 145 (1961).
13. Böld, W., and M. W. Breiter, *Electrochim. Acta*, **5**, 169 (1961).
14. Bowden, F. P., *Proc. Roy. Soc., (London)*, **A125**, 446 (1929).
15. Bowden, F. P., *Proc. Roy. Soc. (London)*, **A126**, 107 (1929).
16. Bowden, F. P., *Trans. Faraday Soc.*, **27**, 505 (1931).
17. Breiter, M. W., in *Advances in Electrochemistry and Electrochemical Engineering*, Vol. 1, P. Delahay, Ed., Interscience, New York, 1961, p. 123.
18. Breiter, M. W., *Electrochim. Acta*, **9**, 441 (1964).
19. Breiter, M. W., *J. Electroanal. Chem.*, **10**, 191 (1965).
20. Breiter, M. W., *Electrochim. Acta*, **10**, 543 (1965); *J. Phys. Chem.*, **69**, 901 (1965).
21. Briner, E., H. Haefeli, and H. Pillard, *Helv. Chim. Acta*, **20**, 1510 (1937).
21a. Briner, E., and A. Yalda, *Helv. Chim. Acta*, **24**, 109, 1328 (1941).

22. Brooman, E. W., and T. P. Hoar, *Platinum Metals Rev.*, **9**, 122 (1965).
23. Brummer, S. B., *J. Electrochem, Soc.*, **112**, 633 (1965).
24. Brummer, S. B., and A. C. Makrides, *J. Electrochem. Soc.*, **111**, 1122 (1964).
25. Buehrer, T. F., F. S. Wartman, and R. L. Nugent, *J. Am. Chem. Soc.*, **49**, 1271 (1927).
26. Butler, J. A. V., and G. Drever, *Trans. Faraday Soc.*, **32**, 427 (1936).
27. Butler, J. A. V., and W. M. Leslie, *Trans. Faraday Soc.*, **32**, 435 (1936).
28. Cassel, H. M., and E. Krumbein, *Z. Physik. Chem. (Leipzig)*, **171**, 70 (1935).
29. Conway, B. E., and P. L. Bourgault, *Can. J. Chem.*, **37**, 292 (1959).
30. Criddle, E. E., *Electrochim. Acta*, **9**, 853 (1964).
30a. Damjanovic, A., A. Dey, and J. O'M. Bockris, *Electrochim. Acta*, **11**, 791 (1966).
30b. Damjanovic, A., A. Dey, and J. O'M. Bockris, *J. Electrochem. Soc*, **113**, 739 (1966).
31. Damjanovic, A., M. L. B. Rao, and M. Genshaw, ASTIA Rept. No. AD 414675, Mar. (1963).
32. Damjanovic, A., M. L. B. Rao, and M. Genshaw, ASTIA Rept. No. AD 431148, Sept. (1963).
33. Delahay, P., *Double Layer and Electrode Kinetics*, Interscience, New York, 1965.
34. Declaux, P. E., *Compt. Rend.*, **200**, 1938 (1935).
35. Efimov, E. A., and N. A. Izgaryshev, *Zh. Fiz. Khim*, **39**, 1606 (1956).
36. El Wakkad, S. E. S., and A. M. S. El Din, *J. Chem. Soc.*, **1954**, 3094.
37. Erdey-Gruz, T., and O. Golopenczane-Bajgr, *Magy. Kim. Folyoirat*, **67**, 435 (1961); *Acta Chim. Acad. Sci. Hung.*, **34**, 281 (1962).
38. Erdey-Gruz, T., and I. Vajasdy, *Magy. Kem. Folyoirat*, **67**, 90 (1961).
39. Ershler, B. V., and A. N. Frumkin, *Trans. Faraday Soc.*, **34**, 464 (1939).
40. Fischer, F., and K. Massenez, *Z. Anorg. Allgem. Chem.*, **52**, 208, 229 (1907).
40a. Flisskii, M. M., and L. M. Surova, *Elektrokhim.*, **1**, 1995 (1965).
41. Frumkin, A. N., *Proc. CITCE*, **9**, 396 (1959).
42. Frumkin, A. N., *Electrochim. Acta*, **5**, 265 (1961).
43. Frumkin, A. N., A. I. Kaganovich, E. V. Yakoleva, and V. V. Sobol, *Dokl. Akad. Nauk SSSR*, **141**, 1416 (1961).
44. Garrison, A. D., and J. F. Lilly, *Trans. Faraday Soc.*, **65**, 275 (1934).
45. Gerke, R. H., and M. D. Rourke, *J. Am. Chem. Soc.*, **49**, 1855 (1927).
46. Gerovich, M. A., R. I. Kaganovich, Yu. A. Mazitov, and M. A. Ghorokhov, *Dokl. Akad. Nauk SSSR*, **137**, 634 (1961).
47. Gerovich, M. A., R. I. Kaganovich, V. A. Vergelesov, and L. N. Gorokhov, *Dokl. Akad. Nauk SSSR*, **114**, 1049 (1957).
48. Giner, J., *Z. Elektrochem.*, **63**, 386 (1959).
49. Ginzburg, V. I., and V. I. Veselovskii, *Zh. Fiz. Khim.*, **24**, 366 (1950).
50. Glasstone, S., and A. Hickling, *Chem. Rev.*, **25**, 407 (1939).
51. Gorodetski, Yu. S., *Zh. Fiz. Khim.*, **38**, 2717 (1964).
52. Grube, G., and W. Gaupp, *Z. Elektrochem.*, **45**, 290 (1930).
53. Grube, G., and K. H. Mayer, *Z. Elektrochem.*, **43**, 859 (1937).
54. Grüneberg, G., *Electrochim. Acta*, **10**, 339 (1965).
55. Hillson, P. J., and E. K. Rideal, *Proc. Roy. Soc. (London)*, **199A**, 295 (1949).
56. Hickling, A., *Trans. Faraday Soc.*, **42**, 518 (1946).
57. Hickling, A., and S. Hill, *Discussions Faraday Soc.*, **1**, 236 (1945); *Trans. Faraday Soc.*, **46**, 550 (1950).
58. Hickling, A., and S. Hill, *Trans. Faraday Soc.*, **46**, 557 (1950).

59. Hickling, A., and W. H. Wilson, *Nature*, **164**, 673 (1949).
60. Hickling, A., and G. G. Vrjosek, *Trans. Faraday Soc.*, **57**, 123 (1961).
61. Hoar, T. P., *Proc. Roy. Soc. (London)*, **A142**, 628 (1933).
62. Hoar, T. P., *Proc. CITCE*, **8**, 439 (1958).
63. Hoare, J. P., *J. Electrochem. Soc.*, **109**, 858 (1962).
64. Hoare, J. P., *J. Electrochem. Soc.*, **110**, 245 (1963).
65. Hoare, J. P., *J. Electrochem. Soc.*, **111**, 610 (1964).
66. Hoare, J. P., *Electrochim. Acta*, **9**, 1289 (1964).
67. Hoare, J. P., *J. Electrochem. Soc.*, **112**, 602 (1965).
67a. Hoare, J. P., *J. Electrochem. Soc.*, **112**, 849 (1965).
68. Hoare, J. P., *J. Electrochem. Soc.*, **112**, 1129 (1965).
69. Hoare, J. P., *Electrochim. Acta*, **11**, 203 (1966).
70. Hoare, J. P., *Electrochim. Acta*, **11**, 311 (1966).
71. Hoare, J. P., *J. Electrochem. Soc.*, **113**, 846 (1966).
72. Hoare, J. P., *J. Electroanal. Chem.*, **12**, 260 (1966).
73. Hoare, J. P., *Electrochim. Acta*, in press.
74. Hoare, J. P., *J. Electroanal. Chem.*, in press.
75. Izgaryshev, N. A., and E. A. Efimov, *Zh. Fiz. Khim.*, **27**, 130 (1953); **31**, 1141 (1957).
76. Izgaryshev, N. A., and D. Stepanov, *Z. Elektrochem.*, **30**, 138 (1924).
77. Kaganovich, R. I., and M. A. Gerovich, *Zh. Fiz. Khim.*, **32**, 957 (1958).
78. Kaganovich, R. I., M. A. Gerovich, and E. Kh. Enikeev, *Dokl. Akad. Nauk SSSR*, **108**, 107 (1956).
79. Kasathin, E. V., and A. A. Rakov, *Electrochim. Acta*, **10**, 131 (1965).
79a. Kazarinov, V. E., N. A. Balashova, and M. I. Kulesneva, *Elektrokhimiya*, **1**, 867 (1965).
80. Kheifets, V. L., and I. Ya. Rivlin, *Zh. Prikl. Khim.*, **28**, 1291 (1955).
81. Kheifets, V. L., and I. Ya. Rivlin, *Zh. Prikl. Khim.*, **29**, 69 (1956).
81a. Khrushcheva, E. I., N. A. Shumilova, and M. R. Tarasevich, *Elektrokhimiya*, **2**, 277 (1966).
82. Knobel, M., P. Caplan, and M. Eiseman, *Trans. Faraday Soc.*, **43**, 55 (1923).
83. Kozawa, A., *J. Electroanal. Chem.*, **8**, 20 (1964).
84. Krasil'shchikov, A. I., *Zh. Fiz. Khim.*, **37**, 531 (1963).
85. Llopis, J., and F. Colom, *Proc. CITCE*, **8**, 414 (1958).
86. Lunenok-Burmakins, V. A., A. P. Potemskaya, and A. I. Brodsky, *Dokl. Akad. Nauk SSSR*, **137**, 634 (1961).
87. MacDonald, J. J., and B. E. Conway, *Proc. Roy. Soc. (London)*, **A269**, 419 (1962).
88. Meyer, R. E., *J. Electrochem. Soc.*, **107**, 847 (1960).
89. Milner, P. C., *J. Electrochem. Soc.*, **111**, 438 (1964).
90. Müller, E., *Z. Elektrochem.*, **10**, 776 (1904).
91. Nassonov, P. M., *Zh. Fiz. Khim.*, **36**, 2628 (1962).
92. Newberry, E., *J. Chem. Soc.*, **109**, 1066 (1916); *Proc. Roy. Soc. (London)*, **A114**, 103 (1927).
93. Onada, T., *J. Chem. Soc. Japan*, **42**, 782 (1922); *Z. Anorg. Allgem. Chem.*, **165**, 79 (1927).
94. Parsons, R., *Trans. Faraday Soc.*, **47**, 1332 (1951).
95. Pearson, J. D., and J. A. V. Butler, *Trans. Faraday Soc.*, **34**, 1163 (1938).
96. Picheta, V. V., *J. Gen. Chem. USSR*, **1**, 377 (1931).

97. Pollard, W. B., *J. Chem. Soc.*, **1926**, 1347.
98. Putnam, G. L., R. W. Moulton, W. N. Fillmore, and L. H. Clark, *Trans. Am. Electrochem. Soc.*, **93**, 211 (1948).
99. Rao, M. L. B., A. Damjanovic, and J. O'M. Bockris, *J. Phys. Chem.*, **67**, 2508 (1963).
100. Riddiford, A. C., *Electrochim. Acta*, **4**, 170 (1961).
101. Rius, A., *Helv. Chim. Acta*, **3**, 355 (1920).
102. Rius, A., J. Llopis., and P. Gandia, *Anales Real Soc. Espan. Fis. Quim.*, **46B**, 225, 279 (1950); **49B**, 329 (1953).
103. Rius, A., J. Llopis, and J. Giner, *Anales Real Soc. Espan. Fis. Quim.*, **49B**, 329 (1953).
104. Roiter, V. A., and R. B. Yampolskaya, *J. Phys. Chem. USSR*, **9**, 763 (1937); *Acta Physicochim. URSS*, **7**, 247 (1937).
105. Rozental', K. I., and V. I. Veselovskii, *Dokl. Akad. Nauk SSSR*, **111**, 637 (1956).
106. Rüetschi, P., and P. Delahay, *J. Chem. Phys.*, **23**, 556 (1955).
107. Schuldiner, S., and C. H. Presbrey, *J. Electrochem. Soc.*, **111**, 457 (1964); **108**, 986 (1961).
108. Schuldiner, S., and R. M. Roe., *J. Electrochem. Soc.*, **110**, 332 (1963).
109. Sederholm, P., and C. Benedecks, *Trans. Am. Electrochem. Soc.*, **56**, 169 (1929).
110. Sidgwick, N. V., *The Chemical Elements and Their Compounds*, Oxford Univ. Press, London, 1950, p. 178.
111. Solc, M., *Collection Czech. Chem. Commun.*, **26**, 1749 (1961).
112. Solc, M., and V. Srb, *Chem. Prumsyl*, **11**, 290 (1961).
113. Spitalski, E., and V. V. Picheta, *J. Phys. Chem. USSR*, **60**, 1351 (1928).
114. Stout, H. P., *Discussions Faraday Soc.*, **1**, 246 (1947).
115. Thompson, M. de K., and A. Kaye, *Trans Am.. Electrochem. Soc.*, **60**, 229, (1931).
116. Thompson, M. de K., and G. H. Sistare, *Trans. Am. Electrochem. Soc.*, **78**, 259 (1940).
117. Tödd, F., S. Kahan, and W. Schwartz, *Z. Elektrochem.*, **56**, 19 (1952).
118. Tomoshov, N. D., *Compt. Rend. Acad. Sci. URSS*, **52**, 601 (1946).
119. Tsinman, A. I., *Ukrain. Khim. Zh.*, **26**, 454 (1960).
120. Tsinman, A. I., *Elektrokhimiya*, **1**, 409 (1965).
121. Vetter, K. J., *Elektrochemische Kinetik*, Springer-Verlag, Berlin, 1961, p. 502.
122. Vetter, K. J., and D. Berndt, *Z. Elektrochem.*, **62**, 378 (1958).
123. Walker, O. J., and J. Weiss, *Trans. Faraday Soc.*, **31**, 1011 (1935).
124. Westhaver, J. B., *Z. Physik. Chem. (Leipzig)*, **51**, 65 (1905).
125. Yoneda, Y., *Bull. Chem. Soc. Japan*, **22**, 266 (1949); *J. Electrochem. Soc. Japan*, **17**, 247 (1949); *J. Chem. Soc. Japan*, **71**, 216 (1950).

chapter IV

The Cathodic Reduction of Oxygen

A. PLATINUM

In recent years, a worldwide interest in fuel cells has promoted a vigorous investigation of the mechanisms of the electrochemical reduction of oxygen. The electroanalytical chemist has always been interested in these processes, if for no other reason than that the electrochemical reduction of dissolved oxygen represents complications to the interpretation of polarographic analyses or potentiometric titrations. One of the most difficult problems to be faced in studying the reduction of oxygen is the presence of a stable intermediate, H_2O_2.

Since the presence of H_2O_2 is observed whenever O_2 is reduced, it is reasonable to assume that the reduction process occurs by two two-electron steps, the reduction of O_2 to H_2O_2 and the reduction of H_2O_2 to H_2O. Consequently, the cathode surface must be at the same time not only a good electron transfer catalyst, but also a good peroxide-decomposing catalyst. These conclusions are borne out by the experimental evidence described in this chapter.

Most electrochemical reactions are written in this chapter as reduction reactions (electrons on the left-hand side of the equation) because attention is paid chiefly to the cathodic processes occurring on oxygen electrodes.

1. Steady-State Polarization Curves

A large portion of the work concerned with the reduction of oxygen on Pt cathodes in both acid and alkaline electrodes has appeared in the last twenty years (1,2,8,10,13,14b,16,20,24,25,27,38,39,44–46,48–50,52,53,56, 58,61,62,64–66,69,71,72). From steady-state polarization studies on Pt cathodes (1,8,10,14b,24,25,27,61,71) which provide information about the rate-determining step, a cathodic overvoltage curve similar to the one plotted in Fig. 4.1 is obtained. In general, these curves contain a Tafel region in the current density range from 10^{-6} to 10^{-3} A/cm^2 with a slope in the neighborhood of 0.1. Extrapolation of the Tafel slope to zero overvoltage gives a value between 10^{-9} and 10^{-10} A/cm^2 for i_0. Roiter and Yampolskaya (61) found that η decreased with an increase in temperature but that the value of the Tafel slope remained constant. Since a plot of the results of the dependence of the η on P_{O_2} gave an η–

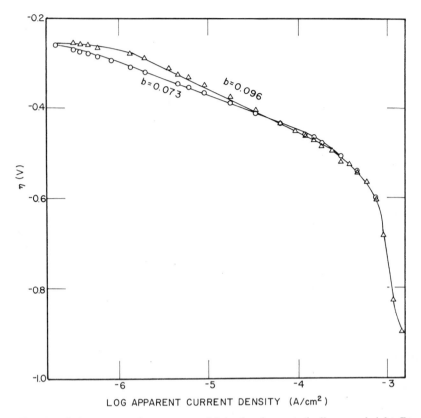

Fig. 4.1. Oxygen overvoltage curves obtained galvanostatically on a bright Pt cathode in O_2-saturated $2N$ H_2SO_4 solution (27). Circles represent data of first cycle of polarization and triangles, the average over four additional cycles. η is determined from 1.23 V. (By permission of The Electrochemical Society.)

$\log P_{O_2}$ curve with a slope of about 60 mV, it was concluded (27) that the rate-limiting step is an electron transfer step involving molecular O_2. Above 10^{-3} A/cm², the curve of Fig. 4.1 enters a limiting current region which leads to a potential at which the reduction of H^+ ions takes place.

Polarographic studies of the reduction of oxygen at Hg electrodes (e.g., 41) give curves containing two waves. It is generally agreed (40) that the reduction of O_2 to H_2O_2 corresponds to the first wave and the reduction of H_2O_2 to OH^- ions or water molecules takes place in the second. Yeager and co-workers (15) observed from ^{18}O, tagged [18]O_2 and H_2[18]O experiments on active carbon electrodes in $1M$ KOH solutions that the reduction of oxygen leads to H_2O_2 and the O—O bond is not broken.   At Pt cathodes, H_2O_2 is not detected at low current densities, but with higher current densities H_2O_2 is always detected.   With prolonged cathodic polarization at Pt cathodes (8,10,24,27,38), the concentration of H_2O_2 in solution builds up to a constant value between 10^{-6} and $10^{-5}M$.

At very low current densities (below 10^{-6} A/cm²), it has been found (27,71) that the polarizarion curve was shifted to larger η values with successive cycles of polarization (see Fig. 4.2) until a point was reached where a reproducible η curve was obtained.   It was concluded (27) that a reproducible η curve (within $\pm$ 6 mV) could be obtained on a Pt cathode only after the electrode had been polarized at a high cathodic current density ($>10^{-4}$ A/cm²) until the steady-state concentration of H_2O_2 in solution had been reached.   The lowering of the polarization curve in this very low current density range with repeated cycles of polarization is caused primarily by the lowering of the rest potential due to the presence of increasing concentrations of H_2O_2 produced in the electrolyte by

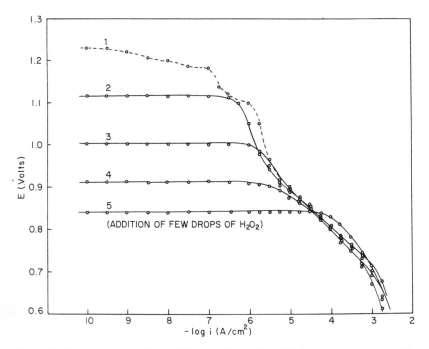

Fig. 4.2. Cathodic polarization of Pt in acid solution (71) for successive runs of increasing current as indicated by the numbers showing influence of H_2O_2 in solution on polarization behavior near open circuit.   (By permission of The Electrochemical Society.)

cathodization at high current densities (11a). Supporting evidence for this conclusion rests on the observation (71) that the curves are parallel in the region below 10^{-6} A/cm² but coalesce to a more or less common plot above 10^{-6} A/cm².

As seen in Fig. 4.1, the first polarization run at intermediate current densities (between 10^{-6} and 10^{-4} A/cm²) lies at higher η values than the following runs. This phenomenon is interpreted that the surface of a Pt electrode which had not previously been cathodized has an adsorbed layer of oxygen on it. In agreement with the findings of Breiter (13) and Vielstich (70) from potential sweep studies and of Myuller and Nekrasov (52) from rotating ring studies, the presence of the adsorbed oxygen film inhibits the reduction of O_2. However, after the electrode is cathodized at high current densities, the adsorbed oxygen layer is removed, the overvoltage is reduced, and the succeeding curve lies above the preceding one, as in Fig. 4.1.

During the first polarization run at current densities above 10^{-4} A/cm² on a Pt cathode, the η values are less than the η values of succeeding runs. This phenomenon is more pronounced in the lower half of Fig. 4.3 than in Fig. 4.1. Apparently, at the high current densities where H_2O_2 formation is important, the oxygen η is lowered. This observation is in agreement with the findings of Bianchi and co-workers (6,7) plotted in Fig. 4.3 that the presence of adsorbed oxygen films accelerates the reduction of H_2O_2 to H_2O. Once the adsorbed oxygen film is consumed, the reduction occurs on a "bare" surface with increased η. In Fig. 4.4,

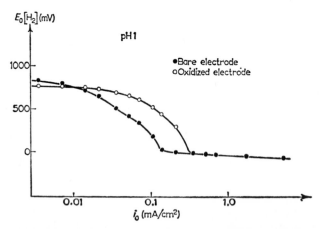

Fig. 4.3. Cathodic polarization curves (6) on bare (●) and oxidized (○) Pt in solution 0.5M in H_2SO_4 and 5 × $10^{-4}M$ in H_2O_2. (By permission of Pergamon.)

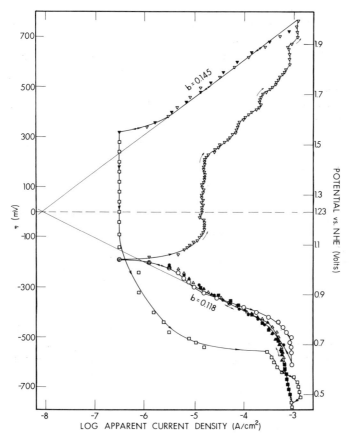

Fig. 4.4. Oxygen overvoltage curves obtained potentiostatically on a bright Pt cathode in O_2-saturated $2N$ H_2SO_4 solution (35).   Arrows indicate direction of polarization; double-headed arrow indicates reproducible regions. (By permission of The Electrochemical Society.)

the cathodic η curve for the Pt electrode which had been strongly anodized (open squares) lies at a much higher η values between 10^{-6} and 10^{-3} A/cm² than the curve (triangles) which had not been previously anodized.   Also, above 10^{-3} A/cm² the reverse situation occurs.   At high enough cathodic overvoltages (> -0.7 V), the oxide is removed, and the following curves (filled squares) are virtually identical to the film-free curve (triangles).   The η data obtained on a Pt cathode in O_2 saturated acid solutions give the same results whether obtained galvanostatically, Fig. 4.1. or potentiostatically, Fig. 4.4, except that the value of the Tafel slope in the potentiostatic case is closer to 0.12, the value which is associated with an electron-transfer-controlled process (9).

Consequently, irreproducibility of steady-state polarization measurements may be traced to an unstable catalytic surface on which the electrode process is taking place. To obtain reproducible results at a Pt cathode, the electrode should be cathodized at current densities approaching 10^{-3} A/cm^2 until a steady potential is reached. Under these conditions, the H_2O_2 in solution has reached a steady-state concentration, and the adsorbed oxygen film on the electrode surface has been removed.

2. Mechanisms

The polarization behavior at low current densities (below 10^{-6} A/cm^2) represents (27) the variation of the mixed potential, E_m, with changes in the externally applied current (galvanostatic) or potential (potentiostatic), as seen in the diagram of Fig. 3.3. As the potential is shifted to more cathodic values, a point is reached ($E_m = E_0''$ in Fig. 3.3), where the local cell mechanism no longer controls the kinetics. For cathodic potentials beyond this point, the electrode kinetics are controlled by a single process, the reduction of O_2, and the polarization curve enters a Tafel region.

Since the cathodic η curves on Pt in acid solutions exhibit Tafel regions with a slope of about 0.1 and since the η depends on temperature and on the partial pressure of O_2, it is reasonable to assume that a one-electron transfer reaction involving molecular oxygen is the rate-determining step for the reduction of O_2 on Pt in acid solutions at current densities $> 10^{-6}$ A/cm^2. The most likely step is

$$(O_2)_{ads} + e \xrightarrow{\text{slow}} (O_2^-)_{ads} \tag{4.1}$$

Without doubt, a reaction with a solvated hydrogen ion follows, producing an adsorbed HO_2 radical,

$$(O_2^-)_{ads} + H^+ \rightleftarrows (HO_2)_{ads} \tag{4.2}$$

With the transfer of another electron,

$$(HO_2)_{ads} + e \rightleftarrows (HO_2^-)_{ads} \tag{4.3}$$

followed again by the reaction with a H^+ ion,

$$(HO_2^-)_{ads} + H^+ \rightleftarrows (H_2O_2)_{ads} \tag{4.4}$$

an adsorbed molecule of peroxide is formed which may either desorb or react further by another electrode process.

From Yeager's observation (15) that the O—O bond is not broken on carbon cathodes in KOH, the H_2O_2 molecule must be a very stable intermediate and is the end product under these conditions. Since a steady-state concentration of H_2O_2 in the electrolyte is observed at Pt cathodes, H_2O_2 is probably not the end product of the reduction of O_2 in this system. Platinum is known to be a good peroxide-decomposing catalyst, and in all likelihood the H_2O_2 is decomposed by an electrochemical local cell mechanism (Eqs. 2.31 and 2.36) such as that described by Gerischer and Gerischer (21) on a Pt surface free of an adsorbed oxygen film. If the adsorbed oxygen film is present, the H_2O_2 may be decomposed chemically, as suggested by Bianchi and co-workers (7) by the reaction of H_2O_2 with the film,

$$H_2O_2 + Pt\text{-}O \longrightarrow Pt + O_2 + H_2O \qquad (4.5)$$

until the film has been destroyed, after which the Gerischer mechanism (21) sets in.

In any event, the adsorbed H_2O_2 is decomposed catalytically by the overall reaction,

$$2(H_2O_2)_{ads} \xrightarrow{\text{cat}} 2H_2O + O_2 \qquad (4.6)$$

The liberated oxygen may then be reduced to H_2O_2 by repeating the reaction scheme from Eqs. 4.1 to 4.4. Effectively, four electrons are transferred, and H_2O is the end product. Since the current involved in the decomposition of H_2O_2 is a local cell current, it is lost to the external circuit, and as a result, the polarization curve is not affected by the H_2O_2 decomposition current.

Roiter and Yampolskaya (61) suggested that the rate-controlling step is $O_2 + e \rightarrow O^- + O$, which is unlikely, since it would be difficult with such a mechanism to account for the presence of H_2O_2 detected experimentally in the reduction of O_2. Two Tafel regions were found by Akopyan (1) in the steady-state polarization curves on Pt in $1N$ H_2SO_4 solution. He associated the first region of low slope with the reduction of O_2 to H_2O_2 and the second region of high slope with the reduction of H_2O_2 to H_2O. More likely, the region of high slope corresponds to the point where the system goes into H^+ reduction. It is difficult to see how the Tafel regions can be treated as polargraphic half-wave regions.

From constant current charging curves taken in rather impure systems, Sawyer and co-workers (65,66) concluded that the reduction of O_2 proceeds by the electrochemical decomposition of a film of hydrated PtO produced by the interaction of dissolved O_2 with surface sites on the Pt electrode. Such a mechanism is not considered favorably for the following reasons. Breiter (13) made potential sweep measurements in both O_2 stirred and He stirred solutions from which he determined the ohmic and capacitive components of the electrode impedance as a function of potential. It turned out that the impedance curves were the same in the presence or absence of O_2 reduction. He concluded from the results that any mechanism in which adsorbed oxygen atoms participate may be eliminated from consideration. Also according to Sawyer's mechanism, it would be predicted that the presence of adsorbed oxygen should increase the rate of reduction on O_2 to H_2O_2. This is contrary to the experimental results recorded in the literature (13,35,36, 52,70).

From a study of the transition times of constant current charging curves obtained on Pt cathodes in O_2 saturated acid and alkaline electrolytes, Lingane (46) proposed that the reduction of oxygen goes directly to H_2O. These curves showed only one plateau if the Pt electrode was preanodized. On a reduced Pt surface, the potential fell quickly to the value where the reduction of H^+ ions takes place. This observation, of course, is not surprising since the charging current used by Lingane was about 10^{-3} A/cm², which is in the region of the overvoltage curve where the system goes to the reduction of H^+ ions (27), as seen in Fig. 4.1. He observed a plateau in the charging curve only when "oxide" was present on the electrode surface; however, the plateau decreased in length with repeated runs. It appears, then, that Lingane's measurements had little to do with the reduction of oxygen, since he worked in the limiting current region of the oxygen η curve, and what was measured by these cathodic stripping studies was the amount of adsorbed oxygen present on the electrode produced by preanodization or adsorption from solution. It is unlikely that O_2 is reduced electrochemically to water by the direct transfer of four electrons.

Although Myuller and Nekrasov (51) present good evidence that H_2O_2 is an intermediate in the reduction of O_2 on Pt, Liang and Juliard (45) have concluded that H_2O_2 is formed by a side reaction. Using a Pt rotating ring electrode (1a,20,60), Myuller and Nekrasov (51) investigated the intermediates formed in the reduction of oxygen. They found that the reduction occurred in two stages:

$$O_2 + 2H^+ + 2e \xrightarrow{k_1} H_2O_2 \tag{4.7}$$

and

$$H_2O_2 + 2H^+ + 2e \xrightarrow{k_2} 2H_2O \tag{4.8}$$

As an important point in this work, the specific rate for the reduction of H_2O_2 to H_2O, k_2, is an order of magnitude larger than the specific rate for the reduction of O_2 to H_2O_2, k_1, on oxidized Pt; but $k_1 > k_2$ on reduced Pt.

These observations are in good agreement with the steady-state over-voltage data obtained on Pt cathodes in Figs. 4.1 and 4.4, which indicate that the adsorbed oxygen film on Pt impedes the rate of reduction of O_2 to H_2O_2 (increases η) in the region (10^{-6} to 10^{-4} A/cm^2) where H_2O_2 does not accumulate in the electrolyte but accelerates the reduction of H_2O_2 to H_2O (decreases η) in the region (10^{-4} to 10^{-2} A/cm^2) where H_2O_2 accumulates in the electrolyte. Sandler and Pantier (64) also observed this effect.

It is believed (35) that the reduction of O_2 occurs by the mechanism Eqs. 4.1–4.4 and Eq. 4.6 over the complete range of current densities studied, so that H_2O_2 is always formed at all current densities. Peroxides do not accumulate in the low current density region (below 10^{-4} A/cm^2) because the peroxide-decomposing capabilities of the Pt electrode surface are adequate for the demanded current. At high current densities (above 10^{-4} A/cm^2), the H_2O_2 is produced at a higher rate than the rate at which it can be decomposed, and the result is an accumulation of the highly soluble H_2O_2 in solution. Proposed kinetic schemes which involve a change in mechanism from one (in which H_2O_2 is not an intermediate) at low current densities to another (in which H_2O_2 is an intermediate) at high current densities are not considered favorably.

From a recent analysis of the data obtained on rotating Pt ring electrodes for the reduction of O_2 in acid solution, Bockris and co-workers (14c) have concluded that H_2O_2 is not an intermediate in the reduction of O_2 but is produced by a side reaction. The side reaction arises from the adsorption of impurities from solution on the Pt surface. As supporting evidence, they note that H_2O_2 is not detected when the impurities are removed from solution. However, another equally likely explanation exists. In the absence of impurities, the activity of the Pt surface is adequate to decompose the H_2O_2 as fast as it is formed at the current densities demanded; and so, H_2O_2 is not detected. From this viewpoint, the impurities inhibit the catalytic activity of Pt for the decomposition of H_2O_2.

It is interesting to note that the presence of adsorbed oxygen on a Pt surface inhibits the reduction of O_2 to H_2O_2 (13,39,51,70) but promotes the catalytic decomposition of H_2O_2 to H_2O (6–8,39,52). An explanation may lie in the possibility that the reduction of O_2 to H_2O_2 occurs only through an electron transfer-controlled process, which is retarded by the negative dipoles of the adsorbed O atoms on the surface, but that the reduction of H_2O_2 to H_2O can occur by a chemical reaction with the adsorbed oxygen layer similar to Eq. 4.5, as suggested by Bianchi (7).

The reduction of O_2 was carried out on Pt-O-alloy cathodes (29) which were made by a HNO_3 treatment (26), and the results are presented in Fig. 4.5. Beginning with an electrode exhibiting the reversible potential of 1.23 V, the first polarization run (circles) had a very low η, but as successive runs were taken, the η increased until a reproducible curve (inverted triangles) was obtained. The η of this final curve is much lower than the η of a pure Pt cathode, Fig. 4.1. Evidently, the Pt-O-alloy surface is a better catalyst for the reduction of O_2, and the presence of oxygen dissolved in the surface layers (subsurface oxygen) accelerates

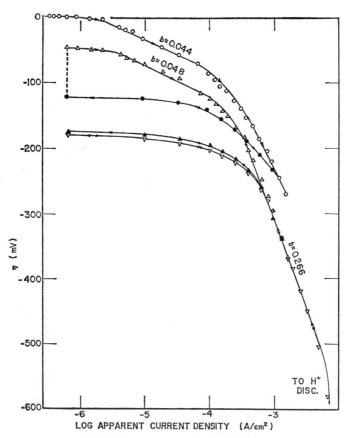

Fig. 4.5. Cathodic overvoltage data obtained galvanostatically on a Pt-O-alloy cathode in O_2-saturated $2N$ H_2SO_4 solution (29). Arrows indicate direction of polarization: ($\bigcirc$), ($\bullet$), first cycle of polarization; ($\triangle$), ($\blacktriangle$), second; ($\triangledown$) final reproducible polarization curve. η is determined from 1.23 V. (By permission of The Electrochemical Society.)

the reduction of O_2 probably by facilitating the transfer of electrons. This observation is to be contrasted with the presence of oxygen adsorbed on the surface which inhibits the reduction of O_2.

In addition, it was found (29) that the Pt-O-alloy surface is a better catalyst for the decomposition of H_2O_2. After prolonged cathodic polarization, the presence of H_2O_2 could be detected with $TiSO_4$ reagant at a Pt cathode but not at a Pt-O-alloy cathode under equivalent experimental conditions.

The steady-state polarization curves for the evolution of oxygen are nearly the same on bright Pt anodes (27) and on bright Pt-O-alloy anodes (29) which stands in contrast to the large differences in the curves for the reduction of O_2 on bright Pt (Fig. 4.1) and Pt-O-alloy (Fig. 4.5) cathodes. In the case of the anodic evolution of O_2, the electrode process takes place on a surface covered with an electronically conducting layer of adsorbed oxygen which masks any effects of the dissolved or subsurface oxygen on the electrode kinetics. On the other hand, if the reduction of O_2 were to take place on a surface free of adsorbed oxygen, the effects of dissolved oxygen on the electrode kinetics would be detected. Since such experimental observations were found on Pt and Pt-O-alloy cathodes (27,29), it is believed that the steady-state reduction of O_2 takes place on a surface free of adsorbed oxygen.

Along with Myuller and Nekrasov (52), Breiter (13) also considers that H_2O_2 is reduced directly to H_2O by the transfer of two electrons, according to Eq. 4.8. But these mechanisms do not account for the observation made by Yeager and co-workers (15) from tracer experiments that the O—O bond is not broken on carbon cathodes. It is generally known that carbon is a very poor catalyst for the decomposition of H_2O_2 which accounts for Yeager's findings. In the case of Pt, which is a good peroxide-decomposing catalyst, the O—O bond is broken by a local cell process (21) in the absence of adsorbed oxygen films or by a chemical process (7) in its presence. With extended cathodic polarization, the magnitude of the steady-state concentration of H_2O_2 detected in the electrolyte is a function of the catalytic activity of the electrode surface for the decomposition of H_2O_2 (28). This concentration of H_2O_2 is high for carbon (15,73), less for Au (32,7a), small for Pt (27), and below detection for Pt-O-alloy (29) cathodes.

Both Bagotskii and co-workers (56) and Sandler and Pantier (64) find that the electrode pretreatment greatly affects the kinetics of the reduction of O_2 on Pt cathodes.

For studies of the interaction of adsorbed hydrogen with adsorbed oxygen on Pt electrodes, one may consult Breiter and co-workers (12,

14,19), Belina and Krasil'shchikov (4) and Rosental' and Veselovskii
(62,63). It appears that adsorbed oxygen on Pt (12) is reduced initially
by the dissolved H_2 in solution by a chemical process, but the process
afterwards continues electrochemically by local cell action.

3. Effects of pH

The experimental results (e.g., 24,14b) show that the overvoltage for
the reduction of O_2 is lower in alkaline than in acid solutions and the
value of the Tafel slope in alkaline solutions is about one-half as large
(0.05) as it is in acid solutions. Winkelmann (72) found that the η
depended on the P_{O_2}. Probably the decrease in the η in going from
acid to alkaline electrolytes rests in the fact that H_2O_2 is more stable in
acid than in alkaline solutions (67). As a result it is easier to exchange
the equivalent of four electrons per O_2 molecule in alkaline solutions
because of the more labile nature of H_2O_2 in these solutions. Balashova
(3) has shown that SO_4^{2-} ions are adsorbed on Pt at potentials between
0.4 and 0.8 V, and the Russian investigators (39,56) conclude that the η
is higher in acid solutions because adsorbed anions interfere with the
oxygen reduction processes occurring at Pt cathodes.

A reasonable mechanism for the reduction of O_2 on Pt cathodes in
alkaline solutions is the following. As in acid solutions, the discharge
of an electron according to Eq. 4.1 produces an adsorbed $(O_2^-)_{ads}$ ion
which may react with a water molecule,

$$(O_2^-)_{ads} + H_2O \rightarrow (H_2O)_{ads} + OH^- \tag{4.9}$$

This is followed by the discharge of another electron,

$$(HO_2)_{ads} + e \longrightarrow (HO_2^-)_{ads} \tag{4.10}$$

so that the overall reaction is

$$O_2 + H_2O + 2e \longrightarrow OH^- + HO_2^- \tag{4.11}$$

The peroxide is catalytically decomposed as in the acid case,

$$2H_2O^- \longrightarrow O_2 + 2OH^- \tag{4.12}$$

to give an O_2 molecule which may undergo the electron transfer steps,
Eqs. 4.1 and 4.10, so that the equivalent of four electrons per O_2 molecule
has been transferred. Without further basic information, it is difficult
to identify the rate-determining step. Because the η depends on P_{O_2},
Eq. 4.1 may be rate controlling even though the observed Tafel slope
value deviates strongly from the usual value of 0.12, which is associated
(9) with an electron transfer step as the rate-limiting step.

4. Effect of Adsorbed Ions

Several authors (2,42,53,56) have studied the effects of the presence of foreign ions added to the electrolyte on the kinetics of O_2 reduction. Kozawa (42) found that the η was increased by additions of alkaline earth ions to alkaline solutions and by additions of halide ions to acid solutions. The effects of alkaline earth ions in acid solutions are not expected to be found since Obrucheva (55) observed that the alkaline earth ions are not adsorbed by Pt in acid solutions. He (42) explained the lowering of the current efficiency for H_2O_2 production at Pt cathodes in the presence of Ca^{2+}, Ba^{2+}, and Sr^{2+} ions by an ion-exchange mechanism.

From studies of the rate of H_2O_2 formation at a rotating ring electrode in KOH solutions, Myuller and Sobel (53) found that Ba^{2+} reduced the rate of reduction of O_2 to H_2O_2 but accelerated the rate of H_2O_2 to OH^- ion. They explain these effects by suggesting that the presence of the adsorbed ions strengthens the bond between the Pt and O atoms.

Using radioactive ^{45}Ca and ^{140}Ba, Balashova and Kulezneva (2) studied the reduction of O_2 at stationary Pt cathodes. On Pt electrodes free of adsorbed oxygen, very small amounts of Ca^{2+} or Ba^{2+} ions were adsorbed, but the quantity of adsorbed ions increased in the presence of adsorbed oxygen. Oxygen reduction on Pt cathodes was retarded twice as much by Ba^{2+} ions as by Ca^{2+} ions, and it was concluded that Ca^{2+} and Ba^{2+} ions inhibit the reduction of O_2 by stabilizing the Pt—O bond.

B. GOLD

In all cases, investigations of the reduction of oxygen at gold electrodes in acid and alkaline solutions (1,7a,18,31,32,43,57,62,71) show that the η is higher on Au than on Pt cathodes and that peroxide was detected as a stable intermediate. The Tafel slope of the steady-state η curves on Au cathodes in acid solutions (Fig. 4.6) is about 0.11 (1,31,32,43), and the η is a function of the oxygen partial pressure, P_{O_2}. Between 10^{-3} and 10^{-2} A/cm^2, the curve reaches a limiting current where the system goes to the reduction of H^+ ions.

As in the case with Pt, steady-state η measurements could not be obtained on Au cathodes (32) until a steady-state concentration of H_2O_2 was built up. With Au cathodes, the steady-state concentration of H_2O_2 is about $10^{-4}M$, which is to be expected, since Au seems to be a poorer catalyst for the decomposition of H_2O_2 (7,7a). It is interesting that the reproducible steady-state η curve for Au cathodes lies below

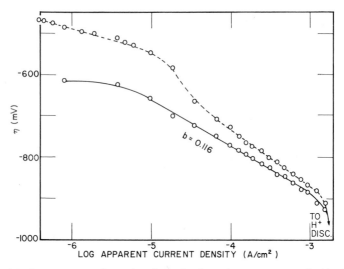

Fig. 4.6. Oxygen overvoltage for the reduction of oxygen on an Au/Au-O electrode (32).   The dashed curve is a plot of the first cycle of polarization from open-circuit conditions.   The average of four additional cycles of polarization is shown by the solid curve which was reproducible within ± 10 mV. η is determined from 1.23 V.   (By permission of Pergamon.)

(at larger η values) the initial curve (see Fig. 4.6), a situation which is opposite to the one found for Pt cathodes (see Fig. 4.1).   The amount of adsorbed oxygen on Au is so low that the reduction of oxygen is not hindered as for Pt, but the higher steady-state concentration of H_2O_2 in solution depresses the potential to less noble values.   Hence, the η increases with each polarization run until the H_2O_2 in solution builds up to about $10^{-4}M$.

Polarization curves obtained potentiostatically on Au cathodes (31) agree very well with those obtained galvanostatically (compare Figs. 3.6 and 4.6).   When a Au/Au_2O_3 electrode (filled circles of Fig. 3.6) is cathodized, an oxide reduction curve is obtained until the Au_2O_3 is removed.   Then a cathodic η curve for the reduction of O_2 is found which has the same Tafel slope as the η curve taken on a reduced Au surface but which lies at lower η values.   The decrease in η is accounted for by the increase in surface area caused by the reduction of Au_2O_3.   Again, inhibition of the reduction of O_2 on Au by the presence of an anodized surface was not observed in Fig. 3.6 as for Pt in Fig. 4.4.   In agreement with the results of investigations by Rao and co-workers (59), the coverage of the Au surface with adsorbed oxygen is probably no greater than about 3% which is not enough to produce a measurable influence on the polarization curves.

In the light of the experimental observations, the reduction of O_2 probably takes place on Au cathodes in acid solutions by the mechanism Eqs. 4.1–4.4 followed by Eq. 4.6. According to this scheme, Eq. 4.1 is rate determining. A likely cause of the higher η on Au cathodes than on Pt cathodes is the poorer peroxide-decomposing properties of the Au surface. In the reduction of O_2, Krasil'shchikov (43) feels that the effect of the adsorption of oxygen on anion adsorption is a much more important consideration than the presence of oxide films on electrode kinetics. Recently, Bianchi and co-workers (7a) have concluded from current efficiency studies that the presence of adsorbed HO_2^- ions blocks the Au surface for the reduction of O_2 to H_2O_2 in solutions of high H_2O_2 concentrations, and therefore, oxygen is reduced directly to H_2O by a four-electron process.

In alkaline solutions the reduction of oxygen takes place at lower η values, and the Tafel slope is about a half as large ($\sim$0.06) as that found in acid solutions (1,58). No doubt the mechanism is different in acid and alkaline solutions, but one must await the accumulation of fundamental investigative results before firm conclusions may be made.

C. PALLADIUM

Steady-state overvoltage measurements have been obtained galvanostatically on Pd cathodes (30), and the results are presented in the bottom curves of Fig. 3.8. The experimental observations of the dependence of η on P_{O_2} and the finding of a Tafel slope of 0.12 support the contention that the mechanism of oxygen reduction on Pd cathodes in acid solutions is the same as that on Pt cathodes with Eq. 4.1 rate controlling. Steady-state reduction of O_2 occurs on a "bare" Pd surface.

As seen in Fig. 3.9, the η is increased by the presence of adsorbed oxygen on the electrode surface at low current densities but decreased at high current densities. An explanation which is the same as the one accounting for similar results on preanodized Pt cathodes, Fig. 4.4, may be used for Pd cathodes. Sobol and co-workers (68) studied the reduction of O_2 on a rotating Pd disc with a Pt-Pd-alloy ring and came to the conclusion that the reduction of O_2 occurred in two steps with H_2O_2 as a stable intermediate. The first step, the reduction of O_2 to H_2O_2, is hindered by the presence of adsorbed oxygen on the Pd surface, but the second step, reduction of H_2O_2 to water, is promoted by the adsorbed O_2 layer. Reasons for this behavior are the same as those offered by the similar behavior observed on Pt electrodes.

In alkaline solutions (68) the η is lower than in acid solutions. Sobol and co-workers suggest that the interference of adsorbed anions with the oxygen reduction processes at Pd cathodes in acid solutions causes the η to be higher than in alkaline solutions.

D. RHODIUM

Steady-state η curves obtained galvanostatically (bottom curves of Fig. 3.10) or potentiostatically (bottom curves of Fig. 3.11) for the reduction of O_2 on Rh cathodes in acid solutions show two Tafel regions (34) with a b value of about 0.1 below 10^{-4} A/cm² and about 0.35 above. The Tafel slope of the low current density region (0.1) corresponds to similar regions in the O_2-reduction curves of other noble metals, and the mechanism is the same as that described for Pt cathodes. The high value of the slope of the high current density region, which extends into the range of potentials where H^+ ion reduction is expected, may reflect the interaction of the H^+ ion discharge reactions with oxygen which is strongly held to the Rh surface. Of all the noble metals, Rh makes the strongest bond with the adsorbed oxygen on the surface as noted from the overlap of the hydrogen and oxygen reduction regions of the potential sweep curves taken (11,69a) on Rh electrodes. The available data do not allow a full explanation of these phenomena. Damjanovic and co-workers (14d) have interpreted their η data entirely in terms of Tafel parameters, but this procedure has been critized (34).

Preanodization of the Rh electrode produces a surface on which the reduction of O_2 to H_2O_2 is inhibited but on which the reduction of H_2O_2 to water is increased (inverted triangles in Fig. 3.10). The explanation is the same as that given for preanodized Pt cathodes (Fig. 4.4).

Nekrasov and co-workers (54) studied the reduction of O_2 in KOH solutions on a rotating Rh disc with a Rh-Pt-alloy ring. They concluded that the reduction occurred in two steps with H_2O_2 as an intermediate and the presence of an adsorbed oxygen film (preanodized) lowered the rate of the first step $(O_2{\rightarrow}H_2O_2)$ but increased the rate of the second step $(H_2O_2{\rightarrow}H_2O)$.

The high η values obtained on Rh cathodes agree with the viewpoint that the Rh surface is a poor peroxide-decomposing catalyst.

E. IRIDIUM

In Figs. 3.12 and 3.13, the steady-state η curves obtained (34a) on Ir cathodes in acid solutions both galvanostatically and potentiostatically may be found. The mechanism for the reduction of O_2 on Ir cathodes in acid solution is considered to be the same as on Pt cathodes since the polarization behavior is the same in the two cases. Preanodization increases the η at low current densities but lowers it at high current densities, just as for preanodized Pt cathodes. Peroxide is detected, and the higher values of the η on Ir than on Pt cathode probably result from Ir being a poor peroxide-decomposing catalyst as reported by Bianchi et al. (7). Tafel parameter analysis of polarization data of Ir cathodes has been made (14d).

Polarization curves for the reduction of O_2 on bright Ru or Os cathodes do not exist, to our knowledge, in the published literature.

F. ALLOYS

Very little work is reported in the reviewed literature on the reduction of O_2 on bright cathodes composed of the alloys of noble metals. Recently Hoar and Brooman (25) studied the reduction of O_2 galvanostatically on a series of Pt-Ru and Pt-Rh alloys in O_2 saturated H_2SO_4 solutions. They found a Tafel slope of about 0.1 between 10^{-6} and 10^{-5} A/cm^2 and that the system went to the reduction of H^+ ion between 10^{-3} and 10^{-2} A/cm^2. In all cases, the η was lower for the alloy than for pure Pt cathodes, but the overall mechanism is probably unchanged. Damjanovic et al. (14d) obtained similar results on Pt-Rh cathodes.

These data support the contention that in the steady-state reduction of oxygen the process takes place on a "bare" (absence of an adsorbed oxygen or oxide film) metal surface because in this case the electrode reaction can "see" the changes in electronic structure. In anodic processes where the reaction occurs on an adsorbed oxygen or oxide film (47), no effect of the electronic structure can be observed because the film masks the "bare" surface. In contrast to the results of Hoar and Brooman (25), Damjanovic et al. (14a) found that the reduction of O_2 on a Pt-Rh alloy cathode was slower than on a Pt or Rh cathode in acid solution, but the same on a Pt cathode in alkaline solutions. They consider the reduction process to take place mainly on Pt oxides.

Several possibilities may be called forth to account for the better polarization characteristics of the alloy cathodes (25). The alloy may facilitate the transfer of electrons; it may improve the peroxide-decomposing ability of the electrode surface; it may decrease the amount of inhibiting anion adsorption. However, not enough data are available to enable one to make an educated guess for the cause of the polarization behavior of these systems.

G. PEROXIDE-STABILIZED SYSTEMS

It has been pointed out by Berl (5) and Yeager (73) that the addition of H_2O_2 to the electrolyte produced a reversible oxygen electrode system. Yeager referred to such systems as "peroxide-stabilized" systems. Both the anodic and cathodic polarization of these systems with noble metal electrodes is discussed here because peroxides are detected only with the reduction of O_2.

Polarization studies on Pt electrodes in H_2O_2 stabilized solutions over wide ranges of current density, pH, P_{O_2}, and H_2O_2 concentration were carried out by Giner (22) and Gerischer and Gerischer (21), who came to the conclusion that an explanation for all the observed phenomena may not be possible at this time. If one accepts the viewpoint that at low H_2O_2 concentrations (0 to $10^{-5}M$) the Pt/O_2 system may be described in terms of a local cell mechanism composed of the reactions

$$O_2 + 4H^+ + 4e \longrightarrow 2H_2O \tag{4.13}$$

and

$$Pt + H_2O \longrightarrow Pt\text{-}O + 2H^+ + 2e \tag{4.14}$$

and at high concentrations ($>10^{-3}M$) by a local cell mechanism composed of the reactions

$$H_2O_2 + 2H^+ + 2e \longrightarrow 2H_2O \tag{4.15}$$

ánd

$$H_2O_2 \longrightarrow O_2 + 2H^+ + 2e \tag{4.16}$$

it is not difficult to see that in the transition region the competition between the two mechanisms for kinetic control would produce a frightfully complex system. Obviously, a simple explanation should not be expected. Yet, in the two extreme cases where one mechanism may be assumed to be dominant, there is hope for a successful interpretation.

1. Low Current Density Studies

This section is devoted to a consideration of the polarization studies carried out in strong solutions of H_2O_2. It is to be noted that H_2O_2 is continually consumed in the local cell mechanism of Eqs. 4.15 and 4.16, accounting for the great instability of H_2O_2 solutions in the presence of noble metals. Steady-state anodic and cathodic polarization curves have been obtained on Pt electrodes (21,22,28) in strong solutions of H_2O_2 in H_2SO_4. At very low current densities (28), an E–i curve (Fig. 4.7) is obtained, which passes from cathodic to anodic current densities

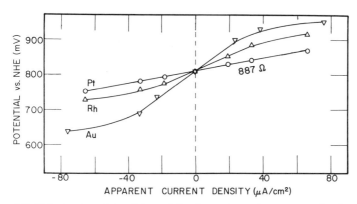

Fig. 4.7. Polarization of Au, Rh, and Pt electrodes at low current density in O_2-saturated, $0.1M$ H_2O_2-stabilized $2N$ H_2SO_4 solution (28). All electrodes have the same rest potential (810 mV) but different rates of reaction $(-di/d\eta)$. (By permission of The Electrochemical Society.)

through the rest potential without a change of slope. As noted in Chapter II, the rest potential of the H_2O_2 stabilized system is a mixed potential, which is diagrammed in Fig. 2.19. Since the H_2O_2/H_2O reaction, Eq. 4.15, is in a limiting current region at the rest potential, the kinetics are controlled by the O_2/H_2O_2 reaction, Eq. 4.16, and the system behaves like an O_2/H_2O_2 electrode with an E_0 of about 0.81 V, the value of the rest potential. As a result, the polarization curve is linear on both sides of 0.81 V at low current densities because of contributions to the back reactions (17).

Under similar experimental conditions, Au and Rh electrodes were studied (28) in H_2O_2 stabilized acid solutions at low current densities. The linear region was shorter for these metals than for Pt, as seen in Fig. 4.7, a situation which may reflect the poorer catalytic activity of these metals for the H_2O_2 reactions.

2. Anodic Polarization Studies

When the system was anodized to higher potentials, the polarization curve left the linear region and entered a logarithmic or Tafel region with a slope of about 0.12. However, a reproducible curve could not be obtained since the curve was shifted to more noble potentials with each cycle of polarization, but the value of the Tafel slope did not change, as seen in Fig. 4.8. Both Giner (22) and Gerischer (21) demonstrated that the η curves depend on the H_2O_2 concentration. With continued anodization, H_2O_2 was consumed, and the rest potential rose to more

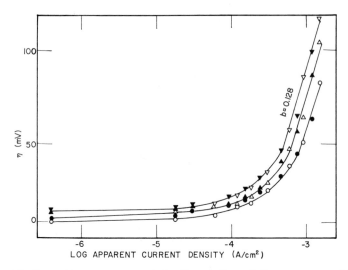

Fig. 4.8. Overvoltage measurements for the anodic evolution on oxygen obtained galvanostatically on bright Pt in O_2-saturated, $0.1M$ H_2O_2-stabilized, $2N$ H_2SO_4 solution (28). η is determined from the rest potential. Open symbols represent increasing current, and filled symbols, decreasing current. Three cycles of polarization are presented, the polarization increasing with each cycle. (By permission of The Electrochemical Society.)

noble potential values (28), shifting in turn the polarization curve to more noble potentials with each cycle of polarization. Similar observations were made on Au anodes (33).

In this Tafel or logarithmic region, it is believed that O_2 is evolved from the discharge of an adsorbed H_2O_2 molecule. The first electron transfer step,

$$(H_2O_2)_{ads} \longrightarrow (HO_2)_{ads} + H^+ + e \qquad (4.17)$$

is assumed (28) to be rate controlling followed by a second electron-transfer step,

$$(HO_2)_{ads} \longrightarrow O_2 + H^+ + e \qquad (4.18)$$

Huq and Makrides (37) also found Tafel regions on Au anodes and cathodes in H_2O_2 stabilized $HClO_4$ solutions at low current densities (below 10^{-2} A/cm²).

Gerischer (21) polarized a Pt anode galvanostatically in strong H_2O_2 stabilized H_2SO_4 solutions to high current densities ($>10^{-2}$ A/cm²). As the current density was increased (see Fig. 4.9), a potential was reached at which a limiting current was established. The magnitude of the limiting current increased as the concentration of H_2O_2 was increased, but the potential at which the limiting current occurred was independent of the H_2O_2 concentration.   Similar observations were made by Hickling and Wilson (23).

The potentiostatic polarization curves were obtained by Giner (22) on Pt anodes in H_2O_2-stabilized acid solutions; and these curves, shown in Fig. 4.10, exhibit a typical passivation region in the same potential range where the limiting current appeared in the galvanostatic curves. These curves also depend on the H_2O_2 concentration.   The limiting current in Fig. 4.9 and the current maximum in Fig. 4.10 occur above 1 V where an adsorbed oxygen film can form on Pt.   Apparently, the presence of the film inhibits the evolution of O_2 from adsorbed H_2O_2 molecules.   This may be true because H_2O_2 molecules are not easily adsorbed on a surface covered with adsorbed oxygen, or more likely, because the H_2O_2 molecules form a stable complex with the surface adsorbed oxygen which cannot be decomposed until a very high potential is reached where the current rises once more.

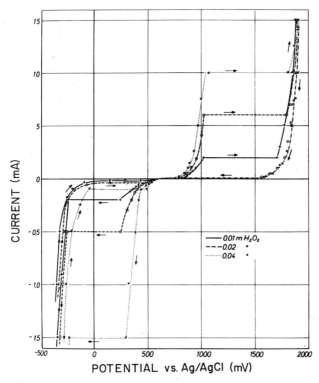

Fig. 4.9. Complete polarization curves for bright Pt in $1N$ H_2SO_4 solutions which contain various concentrations of H_2O_2 (21).   (By permission of Akademische Verlagsgesellschaft.)

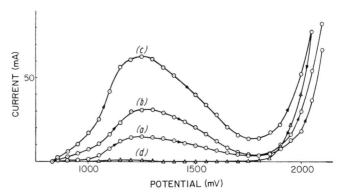

Fig. 4.10. The anodic current–potential curves obtained potentiostatically on bright Pt in $1N$ H_2SO_4 (22) containing various concentrations of H_2O_2; (a) $10^{-2}M$, (b) $2 \times 10^{-2}M$; (c) $4 \times 10^{-2}M$ for increasing potential, and (d) $4 \times 10^{-2}M$ for decreasing potential. (By permission of Bunsengesellschaft.)

Evidence to support the latter conclusion is the observation from both Figs. 4.9 and 4.10 that a large hysteresis is found (21–23) after decreasing the current density from the highest value studied. The formation of a passivating complex of the surface oxygen and H_2O_2 molecules renders the electrode inactive for the evolution of O_2 and the current falls to a nearly zero value. To activate the electrode, the complex must be removed by reduction at cathodic potentials.

A similar set of results were obtained by Hickling and Wilson (23) on Au and Pt anodes in both H_2O_2-stabilized acid and alkaline solutions except that the limiting current does not appear on Au in acid solutions until a potential greater than 1.3 V is reached. The observation that a potential at which an oxide is formed must be reached before the electrode passivates is offered as further support for the passivating complex point of view.

3. Cathodic Reduction Studies

With increasing cathodic current on a Pt electrode in H_2O_2-stabilized H_2SO_4 solutions (28), the system leaves the linear region and enters a logarithmic or Tafel region with a slope of about 0.12. In contrast to the results obtained on Pt anodes, the cathodic η curves were reproducible, as shown in Fig. 4.11. The preferred mechanism (28) for the reduction of O_2 in H_2O_2-stabilized systems of O_2 saturated acid solutions involves the discharge of the first electron,

$$(O_2)_{ads} + e \longrightarrow (O_2^-)_{ads} \tag{4.1}$$

as the rate-determining step followed by Eqs. 4.2–4.4 so that the over-all cathodic reaction is

$$O_2 + 2H^+ + 2e \longrightarrow H_2O_2 \qquad (4.19)$$

The end product is H_2O_2.

In parallel with the cathodic reduction of O_2 to H_2O_2, which contributes current to the external circuit, H_2O_2 may be consumed by a local cell process, Eqs. 4.15 and 4.16, which does not contribute current to the external circuit. Consequently, over the period of time required to record the polarization curve in Fig. 4.11, the H_2O_2 concentration remains relatively steady because of the counteracting effects of cathodic and local cell processes. With anodic polarization both processes consume H_2O_2, the H_2O_2 concentration consequently changes with time, and hence, a reproducible polarization curve cannot be obtained at a Pt anode (see Fig. 4.8).

The oxygen overvoltage for both Pt anodes and cathodes in O_2 saturated, H_2O_2-stabilized acid solutions is much lower than for the corresponding η in H_2O_2-free solutions. A possible explanation is that O_2 is evolved from (Eq. 4.16) or reduced to H_2O_2 (Eq. 4.19) and the O—O bond does not have to be formed or broken, thus producing a more reversible system. This is shown by the high i_0 value, 4×10^{-5} A/cm², obtained (28) from Fig. 4.8.

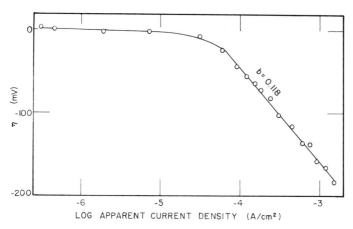

Fig. 4.11. Overvoltage measurements for the cathodic reduction of oxygen on bright Pt in O_2-saturated, $0.1M$ H_2O_2-stabilized, $2N$ H_2SO_4 solution (28). The η is determined fron the rest potential. (By permission of The Electrochemical Society.)

The results obtained (33) on Au cathodes are virtually the same as those obtained on Pt cathodes, except that the η is larger on Au. In this case i_0 is 10^{-7} A/cm^2, showing that Au is a poorer catalyst for the H_2O_2 reactions.

At high enough current densities, depending on the H_2O_2 concentration (see Fig. 4.9), a limiting current density is also obtained (21,72) on Pt cathodes in acid solutions. Since the limiting current occurs at potentials ($\sim$0.3 V) where H^+ ion may be discharged, the curves probably represent the reduction of H^+ ions at potentials below 0.3 V. The hysteresis in the cathodic curves is interesting, but a simple explanation is not available.

As shown by the curves in Fig. 4.7, the rest potential is the same for Pt, Au, or Rh electrodes in O_2-saturated, H_2O_2-free H_2SO_4 solutions, indicating (28) that surface oxides are not involved in the potential-determining mechanism. It is likely that surface films are not involved in the electrode processes responsible for the curves in Figs. 4.8 and 4.11 since a break in the anodic curve did not occur until a potential was reached where surface oxides are known to be produced. At cathodic potentials, any surface oxide layer would be removed by the reaction with H_2O_2 similar to Eq. 4.5.

Bianchi and co-workers (7) have shown that polarization curves obtained on Au, Pt, Pd, and Ir cathodes in N_2 stirred, H_2O_2-stabilized acid solutions do not have Tafel regions. In this case, only the reduction of H_2O_2 to H_2O occurs which is a more difficult process since the O—O bond must be broken. These results were explained (7) in terms of oxide formation produced on the metal surface by the attack of surface atoms by H_2O_2 molecules followed by an oxide reduction process similar to Eq. 4.5.

References

1. Akopyan, A. U., *Zh. Fiz. Khim.*, **33**, 82 (1959).
1a. Albery, W. J., S. Bruckenstein et al., *Trans. Faraday Soc.*, **62**, 1915, 1920, 1932, 1938, 1946, 2584, 2596 (1966).
2. Balashova, N. A., and M. I. Kulezneva, *Elektrokhimiya*, **1**, 155 (1965).
3. Balashova, N. A., and N. S. Merkulova, *Proceedings of the 4th Conference on Soviet Electrochemistry* (*English translation*), Consultants Bureau, New York, 1961, Vol. I., p. 23.
4. Belina, T. N., and A. I. Krasil'shchikov, *Zh. Fiz. Khim.*, **28**, 1286 (1954).
5. Berl, W. G., *Trans. Am. Electrochem. Soc.*, **83**, 253 (1943).
6. Bianchi, G., and G. Caprioglio, *Electrochim. Acta*, **1**, 18 (1959).
7. Bianchi, G., F. Mazza, and T. Mussini, *Electrochim. Acta*, **7**, 457 (1962).
7a. Bianchi, G., F. Mazza, and T. Mussini, *Electrochim. Acta*, **11**, 1509 (1966).
8. Bianchi, G., and T. Mussini, *Electrochim. Acta*, **10**, 445 (1965).

9. Bockris, J. O'M., *J. Chem. Phys.*, **24**, 817 (1956).
10. Bockris, J. O'M., and A. K. M. S. Huq, *Proc. Roy. Soc. (London)*, **A237**, 277 (1956).
11. Böld, W., and M. W. Breiter, *Electrochim. Acta*, **5**, 169 (1961).
11a. Bowen, R. J., H. B. Urbach, and J. H. Harrison, *Nature*, **213**, 592 (1967).
12. Breiter, M. W., *J. Electrochem. Soc.*, **109**, 425 (1962).
13. Breiter, M. W., *Electrochim. Acta*, **9**, 441 (1964).
14. Breiter, M. W., and M. Becker, *Z. Elektrochem.* **60**, 1080 (1965).
14a. Damjanovic, A., A. Dey, and J. O'M. Bockris, *Catalysis*, **4**, 721 (1965).
14b. Damjanovic, A., A. Dey, and J. O'M. Bockris, *Electrochim. Acta*, **11**, 791 (1966).
14c. Damjanovic, A., M. A. Genshwa, and J. O'M. Bockris, *J. Phys. Chem.*, **70**, 3761 (1966).
14d. Damjanovic, A., A. Dey, and J. O'M. Bockris, *J. Electrochem. Soc.*, **113**, 739 (1966).
15. Davies, M. O., M. Clark, E. Yeager, and F. Hovorka, *J. Electrochem. Soc.*, **106**, 56 (1959).
16. Delahay, P., *J. Electrochem. Soc.*, **97**, 198 (1950).
17. Delahay, P., *Double Layer and Electrode Kinetics*, Interscience, New York, 1965, p. 160.
18. Evans, D. H., and J. J. Lingane, *J. Electroanal. Chem.*, **6**, 283 (1963).
19. Franke, K. F., C. A. Knorr, and M. W. Breiter, *Z. Elektrochem.*, **63**, 226 (1959).
20. Frumkin, A. N., L. Nekrasov, V. B. Levich, and Yu. Ivanov, *J. Electroanal. Chem.*, **1**, 84 (1959).
21. Gerischer, R., and H. Gerischer, *Z. Physik. Chem. (Frankfurt)*, **6**, 178 (1956).
22. Giner, J., *Z. Elektrochem.*, **64**, 491 (1960).
23. Hickling, A., and W. H. Wilson, *J. Electrochem. Soc.*, **98**, 425 (1951).
24. Hoar, T. P., *Proc. Roy. Soc. (London)*, **A142**, 628 (1933).
25. Hoar, T. P., and E. W. Brooman, *Electrochim. Acta*, **11**, 545 (1966).
26. Hoare, J. P., *J. Electrochem. Soc.*, **110**, 1019 (1963).
27. Hoare, J. P., *J. Electrochem. Soc.*, **112**, 602 (1965).
28. Hoare, J. P., *J. Electrochem. Soc.*, **112**, 608 (1965).
29. Hoare, J. P., *J. Electrochem. Soc.*, **112**, 849 (1965).
30. Hoare, J. P., *J. Electrochem. Soc.*, **112**, 1129 (1965).
31. Hoare, J. P., *Electrochim. Acta*, **11**, 203 (1966).
32. Hoare, J. P., *Electrochim, Acta*, **11**, 311 (1966).
33. Hoare, J. P., *Electrochim. Acta*, **11**, 549 (1966).
34. Hoare, J. P., *Electrochim. Acta*, in press.
34a. Hoare, J. P., *J. Electroanal. Chem.*, in press.
35. Hoare, J. P., *J. Electrochem. Soc.*, **113**, 846 (1966).
36. Hoare, J. P., *J. Electroanal. Chem.*, **12**, 260 (1966).
37. Huq, A. K. M. S., and A. C. Makrides, *J. Electrochem. Soc.*, **112**, 756 (1965).
38. Jacq, J., and O. Bloch, *Electrochim. Acta*, **9**, 551 (1964).
39. Khrushcheva, E. I., N. A. Shumilova, and M. R. Tarasevich, *Elektrokhimiya*, **1**, 730 (1965).
40. Kolthoff, I. M., and J. J. Lindane, *Polarography*, Interscience, New York, 1952, p. 552.
41. Kolthoff, I. M., and C. S. Miller, *J. Am. Chem. Soc.*, **63**, 1013 (1914).
42. Kozawa, A., *J. Electroanal. Chem.*, **8**, 20 (1964).
43. Krasil'shchikov, A. I., *Zh. Fiz. Khim.*, **23**, 322 (1949); **26**, 216 (1952).

44. Laitinen, H. A., and I. M. Kolthoff, *J. Phys. Chem.*, **45**, 1061 (1941).
45. Liang, C. C., and A. L. Juliard, *J. Electroanal. Chem.*, **9**, 390 (1965); *Nature*, **207**, 629 (1965).
46. Lingane, J. J., *J. Electroanal. Chem.*, **2**, 296 (1961).
47. MacDonald, J. J., and B. E. Conway, *Proc. Roy. Soc. (London)*, **A269**, 419 (1962).
48. Maget, H. J., and R. Roethlein, *J. Electrochem. Soc.*, **112**, 1034 (1965).
49. Mazitov, Iu. A., N. A. Fedotov, and N. A. Aladzhalova, *Zh. Fiz. Khim.*, **39**, 218 (1965).
50. Mochizuki, M., *J. Electrochem. Soc. Japan*, **28**, E232 (1961).
51. Myuller, L., and L. N. Nekrasov, *Dokl. Akad. Nauk SSSR*, **157**, 416 (1964); *J. Electroanal. Chem.*, **9**, 282 (1965).
52. Myuller, L., and L. N. Nekrasov, *Dokl. Akad. Nauk SSSR*, **154**, 437, 2815 (1964); **149**, 1107 (1963); *Electrochim. Acta*, **9**, 1015 (1964).
53. Myuller, L., and V. V. Sobol, *Elektrokhimiya*, **1**, 111 (1965).
54. Nekrasov, L. N., E. I. Khrushcheva, N. A. Shumilova, and M. R. Tarasevich, *Elektrokhimiya*, **2**, 363 (1966).
55. Obrucheva, A. D., *Dokl. Akad. Nauk SSSR*, **120**, 1072 (1958).
56. Oshe, A. I., V. I. Tikhomirova, and V. S. Bagotskii, *Elektrokhimiya*, **1**, 688 (1965).
57. Palous, S., and R. Buvet, *Bull. Soc. Chim. France*, **1962**, 1602; **1963**, 2490.
58. Peters, D. G., and R. A. Mitchell, *J. Electroanal. Chem.*, **10**, 306 (1965).
59. Rao, M. L. B., A. Damjanovic, and J. O'M. Bockris, *J. Phys. Chem.*, **67**, 2508 (1963).
60. Riddiford, A. C., in *Advances in Electrochemistry and Electrochemical Engineering*, Vol. 4, C. Tobias, Ed., Interscience, New York, 1966, p. 47.
61. Roiter, V. A., and R. B. Yampolskaya, *J. Phys. Chem. USSR*, **9**, 763 (1937); *Acta Physicochim. URSS*, **7**, 247 (1937).
62. Rozental', K. I., and V. I. Veselovskii, *Zh. Fiz. Khim.*, **31**, 1555 (1956).
63. Rozental', K. I., and V. I. Veselovskii, *Zh. Fiz. Khim.*, **35**, 2256 (1961).
64. Sandler, Y. L., and E. A. Pantier, *J. Electrochem. Soc.*, **112**, 928 (1965).
65. Sawyer, D. T., and R. J. Day, *Electrochim. Acta*, **8**, 589 (1963).
66. Sawyer, D. T., and L. V. Interrante, *J. Electroanal. Chem.*, **2**, 310 (1961).
67. Sidgwick, N. V., *The Chemical Elements and Their Compounds*, Oxford Univ. Press, London, 1950, p. 869.
68. Sobol, V. V., E. I. Khrushcheva, and V. A. Dagaeva, *Elektrokhimiya*, **1**, 1332 (1965).
69. Songina, O. A., and B. K. Toibaev, *Izv. Akad. Nauk Kaz. SSR*, **1963**, 8.
69a. Tyurin, Yu. M., *Dokl. Akad. Nauk SSSR*, **126**, 827 (1959).
70. Vielstich, W., *Z. Instrumentenk.*, **71**, 29 (1963).
71. Watanabe, N., and M. A. V. Devanathan, *J. Electrochem. Soc.*, **111**, 615 (1964).
72. Winklemann, D., *Z. Elektrochem.*, **60**, 731 (1956).
73. Witherspoon, R. R., H. Urbach, E. Yeager, and F. Hovorka, Tech. Rept. No. 4, ONR Contract Nonr. 581(00), Western Reserve Univ., 1954.

chapter **V**

The Reversible Oxygen Electrode

Although the reversible oxygen potential, 1.23 V, has been observed (1,3,11,18), the question may be raised whether a reversible oxygen electrode has been obtained. For an electrode process consisting of a series of consecutive steps to be truly reversible, not only must the overall electrode reaction be the same in both the anodic and cathodic directions, but also each individual step of the forward process must be reversed in the back reaction. Since this question is considered a central point in the discussion of the electrochemistry of oxygen, this chapter is devoted to a consideration of the prerequisites for attaining a reversible oxygen electrode and to an account of how near to this ideal have the investigators in the field actually approached.

For the sake of review, the overall electrochemical reaction for this oxygen electrode is

$$O_2 + 4H^+ + 4e \rightleftharpoons 2H_2O \tag{5.1}$$

in acid solutions, or

$$O_2 + 2H_2O + 4e \rightleftharpoons 4OH^- \tag{5.2}$$

in alkaline solutions. The accepted value for the E_0 of Eq. 5.1 is 1.229 V obtained thermodynamically from thermochemical data.

The first requisite for a reversible oxygen electrode is the ability to maintain a potential of 1.23 V under open-circuit conditions. In ordinary circumstances, this cannot be done at noble metal electrodes since the rest potential is a mixed potential. When a complete layer of electronically conducting adsorbed oxygen is formed on a Pt surface in pure systems (3,11,18), the reversible O_2 potential is obtained since the local cell mechanism cannot occur (11,13). The coefficient $dE/d \log P_{O_2}$ is 0.015 V, and dE/d pH is 0.06 V, which satisfy the Nernst relationship as expected if Eq. 5.1 is potential determining.

In addition to this requirement, the mechanisms of the anodic and cathodic reactions must be the reverse of one another for the same overall electrode process. This prerequisite does not seem to be achieved in most studies of the oxygen electrode on noble metals.

First of all, when O_2 is reduced at a Pt cathode in O_2-saturated acid solutions, H_2O_2 is detected (3,12) in the electrolyte, but when O_2 is evolved at a Pt anode, H_2O_2 is not detected (6). Such an observation indicates that the mechanism for the reduction is different from the one

for the evolution of O_2 at a Pt electrode since H_2O_2 is a stable intermediate in one case but not in the other.   Yet, from steady-state polarization curves, Hoar (9) was the first to show that the extrapolated
Tafel lines for the cathodic reduction and anodic evolution of O_2 cross
at the theoretically calculated potential of 1.23 V, an observation which
has since been confirmed (3,7,7a,12), as seen in Fig. 5.1.   The data of
Fig. 5.1 support the conclusion that the overall electrode process (Eq.
5.1) is the same for Pt/O_2 anodes and cathodes even if the mechanisms
are different in the two cases.

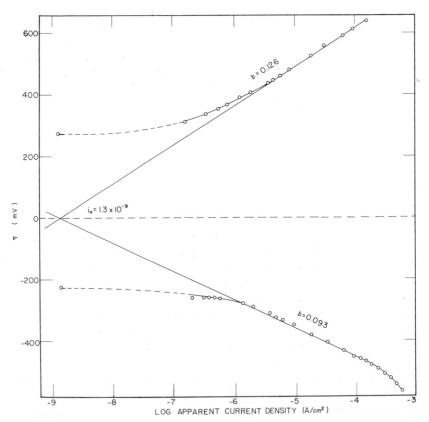

Fig. 5.1. A plot of the anodic and cathodic overvoltage obtained galvanostatically
on the same bright Pt electrode in O_2-saturated, $2N$ H_2SO_4 solution (12).   The cathodic curve was obtained first, after which the anodic curve was determined, and
the extrapolated Tafel slopes cross at $\eta = 0$. η is determined from 1.23 V.   (By
permission of The Electrochemical Society.)

One must distinguish between two types of Pt/O$_2$ electrodes: the one normally obtained (e.g., 4,9,10) when a clean Pt electrode is placed in O$_2$-saturated acid solution and which has a rest potential of about 1.05 V; and the other in which the Pt surface is covered by a complete layer of adsorbed oxygen produced by anodization and heating in pure O$_2$ (1,3,18) or by treating in HNO$_3$ (11) and which has a rest potential of 1.23 V. In this discussion, the former is designated by the symbol Pt/O$_2$, and the latter by Pt(O)/O$_2$.

Although Bockris and Huq (3) do not show the overvoltage curves for the reduction of O$_2$ on a Pt(O)/O$_2$ cathode, they do record values for the cathodic Tafel slopes and state that the extrapolated Tafel lines cross at 1.23 V. Recently (7a) such data have been reported.

Using the relationships for determining the stoichiometric number, ν (8),

$$\nu = \frac{nFi_0}{RT}\left(\frac{d\eta}{di}\right)_{\eta\to 0} \tag{5.3}$$

and (16)

$$\nu = \frac{nF}{2.3RT}\left(\frac{b_a b_c}{b_a + b_c}\right) \tag{5.4}$$

where n is the number of electrons exchanged, i_0 the exchange current, $(d\eta/di)_{\eta\to 0}$, the slope of the $E\text{–}i$ curve in the linear, low current density range, b_a and b_c the anodic and cathodic Tafel slope values, and F, R, and T have their usual significance; Bockris and co-workers (3,7a) calculated a value of about 4 for ν. Since b_a and b_c were nearly equal on the average and the value of ν is 4, they concluded that the electrode system Pt(O)/O$_2$ was a reversible electrode, and any H$_2$O$_2$ formed at the cathode was produced by a side reaction.

It was found (13) that the polarization curve on a Pt(O)/O$_2$ electrode in the low current linear region passed from the cathodic branch through the reversible potential of 1.23 V to the anodic branch without a change of slopes, as demonstrated in Fig. 5.2. This steady-state curve was reproducible (within $\pm$ 2 mV) provided a current density greater than about 40 μA/cm was not exceeded. Above these current density limits, the complete passive layer of adsorbed oxygen was impaired, and the system was no longer a Pt(O)/O$_2$ electrode. In the low current density range, the Pt(O)/O$_2$ electrode behaves as a reversible oxygen electrode which means that the cathodic process is just the reverse of the anodic process.

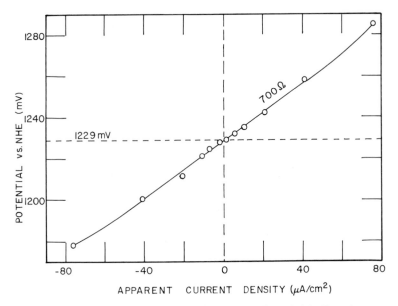

Fig. 5.2. Polarization data for the reversible O_2 electrode on bright Pt at low current densities (13). (By permission of The Electrochemical Society.)

If it is assumed that the evolution of oxygen proceeds on a bright Pt anode in acid solutions as follows,

$$H_2O \longrightarrow OH + H^+ + e \qquad (5.5)$$

$$OH + H \longrightarrow H_2O + O \qquad (5.6)$$

$$O + O \longrightarrow O_2 \qquad (5.7)$$

then the cathodic reduction should proceed by the reverse of Eqs. 5.7, 5.6, and 5.5, in that order. Under these conditions the O—O bond is broken before the electron transfer step takes place, and H_2O_2 is not an intermediate in this mechanism.

Support for this mechanism would be obtained if steady-state polarization data could be obtained in a current density range where a Tafel region exists. In Fig. 5.3 is a plot for the first cathodic run on a Pt(O)/ O_2 electrode and the first anodic run on a second Pt(O)/O_2 electrode. The arrowheads indicate the beginning of drifting points. Unfortunnately, a Tafel region cannot be determined in the anodic curve, and only a short region appears in the cathodic curve. A value of i_0 may be determined from the data in the linear range according to the relationship (ref. 8, p. 160)

$$i_0 = \frac{RT}{nF} (di/d\eta) \tag{5.8}$$

Using the value of 5.1×10^{-3} mho for $di/d\eta$ obtained from the data in Fig. 5.2 over the linear range, i_0 is calculated as 3.2×10^{-5} A/cm^2, which is much larger than the value 1.3×10^{-9} A/cm^2 (12) obtained on Pt/O$_2$ electrodes (Fig. 5.1). From the extrapolation of the Tafel line in Fig. 5.3 to $\eta = 0$, an i_0 value of 0.30×10^{-5} A/cm^2 is determined which is in fair agreement with the value obtained from Eq. 5.8.

Ideally, the Tafel slope should be 0.12 if Eq. 5.5 is rate determining. However, as soon as significant amounts of current (>40 μA/cm^2) are drawn, reduction of the passivating adsorbed oxygen layer contributes to the kinetics. This may offer some explanation for the low slope observed in Fig. 5.3. If the breaking of the O—O bond (reverse of Eq. 5.7) is the rate-determining step in the reduction of O$_2$ at a Pt(O)/O$_2$ cathode, a low value for the Tafel slope may be expected (2). Because

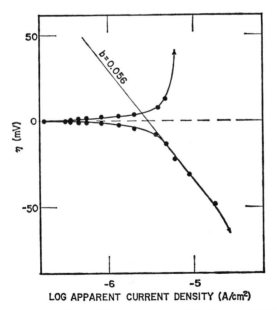

Fig. 5.3. Oxygen overvoltage obtained galvanostatically on a Pt(O)/O$_2$ electrode on O$_2$-saturated $2N$ H$_2$SO$_4$ solution. Arrowheads indicate beginning of drifting points. The $i_0 = 0.30 \times 10^{-5}$ A/cm^2. η was determined from 1.23 V.

the reduction process on the cathode removes the passivating film and the oxidizing process on the anode produces higher oxides, the $Pt(O)/O_2$ electrode cannot sustain a significant anodic or cathodic current without being converted to a Pt/O_2 electrode.

It is interesting that the i_0 (between 10^{-9} and 10^{-10} A/cm^2) determined from the data of Bockris and co-workers (3,7a) corresponds to the i_0 (9, 12) of a Pt/O_2 electrode (Fig. 5.1). Apparently the thermal treatment (3,7a,18) does not produce as stable a passivating film on the Pt as the HNO_3 treatment (11), and since this complete film is unstable in acid solutions (10,11), the complete layer of adsorbed oxygen on the surface of the electrodes used by Bockris and Huq were damaged very quickly with polarization. Consequently, these non-steady-state measurements (3) give i_0 values corresponding to a Pt/O_2 electrode because the i_0 value was determined from measurements made in the Tafel region at current densities where the passive film is damaged. In support of this conclusion, the thermally treated electrodes (3,18) maintained a potential of 1.23 V for periods of time no greater than an hour, whereas HNO_3-treated electrodes (11) maintained the reversible potential for periods of time greater than 24 hr.

Probably the reversible O_2 electrode does exist on a $Pt(O)/O_2$ electrode at very low current densities where the complete film of adsorbed oxygen has the capability of breaking the O—O bond before the electron transfer step occurs. In this case, H_2O_2 is not an intermediate, the anodic and cathodic processes are the same, and the possible rate-limiting steps are Eq. 5.5 for the anodic evolution of O_2 and the reverse of Eq. 5.7 for the cathodic reduction of O_2. At higher cathodic current densities, the reverse of Eq. 5.7 cannot support the demanded current, and a new mechanism sets in, as described by Eqs. 4.1–4.6, where H_2O_2 is an intermediate and the electron transfer precedes the catalytic breaking of the O—O bond. Although Bockris and Huq (3) do not show curves for the cathodic η, Watanabe and Devanathan (18) do. They (18) found that the curve lies at higher η values for each cycle of polarization (see Fig. 4.2) and H_2O_2 was detected in solution.

It is doubtful if the oxygen electrode on Pt behaves reversibly once a current density greater than about 40 μA/cm^2 is reached. A probable reason why the Tafel slopes of the steady-state anodic and cathodic η curves of a Pt/O_2 electrode cross at 1.23 V (2,5), as seen in Fig. 5.1, is the possibility that the anodic reaction takes place on a surface completely covered with Pt-O and PtO_2 and the cathodic reaction on a "bare" Pt surface. In this situation the electrode surface is chemically inert in each instance, and the overall electrochemical process is the

O_2/H_2O reaction, Eq. 5.1, even though the mechanism is different on the anode, Eqs. 5.5–5.7 than on the cathode, Eqs. 4.1–4.6.

Recently, Damjanovic and Bockris (7) determined polarization curves in acid solutions on Pt anodes and on Pt cathodes, the surfaces of which were either covered with an adsorbed oxygen film or were free of adsorbed oxygen.   In agreement with the previous findings, they (7) found (see Fig. 5.4) that the film-covered cathode had a higher η for the reduction of O_2 than the film-free cathode (5,14,15,17), that the anodic and cathodic Tafel lines cross at 1.23 V (3,9,12), that the i_0 is about 10^{-9} A/cm²

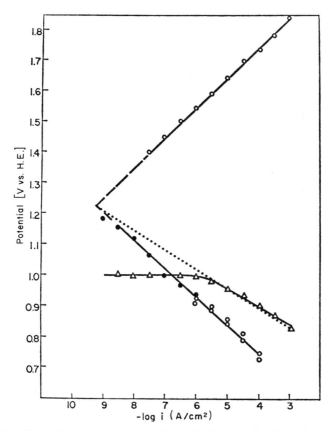

Fig. 5.4. Cathodic and anodic Tafel lines obtained on Pt/O_2 electrodes in acid solutions (7).   (●) is cathodic curve from reversible potential; (○) cathodic curve by reversing anodic current followed by fast measurements; (△) cathodic curve on film-free Pt surface.   The dotted line was added to the original drawing. (By permission of Pergamon.)

(3,9,12), that the kinetics of the film-free cathode at low current densities (below 10^{-6} A/cm^2) is controlled by a local cell mechanism (12), and that the reversible potential is observed only on the film-covered Pt surface (11). Since the Tafel lines for the film-covered anodes and cathodes cross at 1.23 V, Damjanovic and Bockris conclude that the measurements obtained by Bockris and Huq (3) and Hoar (9) were obtained on film-covered cathodes. This may not be true for the data obtained by Bockris and Huq, since these were not steady-state measurements, but it is doubtful for the steady-state data obtained by Hoar.

Damjanovic and Bockris (7) neglected to point out that the extrapolated Tafle slope for the film-free cathode (triangles in Fig. 5.4) also crosses the anodic Tafel slope at 1.23 V, just as in Fig. 5.1. As a result, the steady-state data for reduction of O_2 obtained by Hoar doubtlessly were taken on a film-free Pt surface. It is important to note that the overall mechanism on both the film-covered and film-free cathodes (Eq. 5.1) is the same (in both cases Tafel lines cross at 1.23 V). The electrode system corresponding to the data presented in Fig. 5.4 is not a Pt(O)/O_2 but a Pt/O_2 electrode since the i_0, $\sim 10^{-9}$ A/cm^2, corresponds to the Pt/O_2 electrode of Fig. 5.1 and not to the Pt(O)/O_2 electrode of Fig. 5.3.

Determinations of ν (3,7a) are of little use in elucidating the mechanism of oxygen electrodes. For a Pt(O)/O_2 electrode, the values on i_0, b_a, and b_c to be used in Eqs. 5.3 and 5.4 must be determined from kinetic data in a current density range (Tafel region) where the system is no longer a Pt(O)/O_2 but a Pt/O_2 electrode. Parsons (16) pointed out that Eq. 5.4 gives reliable values for ν only if the activated complexes are the same in the anodic and cathodic processes. This is not likely for a Pt/O_2 electrode, since the rate-determining step for the reduction process (Eq. 4.1) involves a doubly bonded oxygen, O_2, whereas the corresponding step for the oxidation process (Eq. 5.5) involves a singly bonded oxygen, H_2O. Also, the mechanism at low and at high current densities is different on a Pt/O_2 electrode; in the low range, the kinetics are governed by a polyelectrode or mixed potential mechanism, but in the high range by a single electrode mechanism.

To obtain a truly reversible oxygen electrode, the Pt surface must be made inert to the surroundings at all current densities, for example, by a stable, electronically conducting, passive film. In this case, the electrode surface acts only as a source or sink for electrons and as the catalytic surface on which the electrochemical process takes place. It is possible that this situation is realized (11,13) at the very low current densities (below 40 μA/cm^2) on a HNO$_3$-passivated Pt electrode (Fig. 5.2).

It seems that a reversible oxygen electrode may be made on Rh (11) but not on Au (11,18). Palladium and iridium make oxides that are electrochemically active, and the 1.23-V potential is not observed (11) on these electrodes.

In conclusion, two types of oxygen electrodes are distinguished. The irreversible Pt/O_2 electrode, which is normally prepared (Fig. 5.1) is a polyelectrode; the rest potential is a mixed potential; the anodic mechanism, Eqs. 5.5–5.7, is different from the cathodic mechanism, Eqs. 4.1–4.6; and in the reduction of O_2, electrons are transferred before the O—O bond is broken, producing the stable intermediate, H_2O_2. The reversible $Pt(O)/O_2$ electrode prepared only by special techniques (1,3,11,18) is a single electrode (Fig. 5.3); the rest potential is determined by the O_2/H_2O reaction, Eq. 5.1; the anodic Eqs. 5.5–5.7, and the cathodic Eqs. 5.7–5.5 are the reverse of one another; in the reduction of O_2, electrons are transferred after the O—O bond is broken, and the only intermediate involved is the unstable adsorbed OH radical.

References

1. Bianchi, G., and T. Mussini, *Electrochim. Acta*, **10**, 445 (1965).
2. Bockris, J. O'M., *J. Chem. Phys.*, **24**, 817 (1956).
3. Bockris, J. O'M., and A. K. M. S. Huq, *Proc. Roy. Soc. (London)*, **A237**, 277 (1956).
4. Bowden, F. P., *Proc. Roy. Soc. (London)*, **A125**, 446 (1929); **A126,** 107 (1929).
5. Breiter, M. W., *Electrochim, Acta*, **9**, 441 (1964).
6. Butler, J. A. V., and W. M. Leslie, *Trans. Faraday Soc.*, **32**, 435 (1936).
7. Damjanovic, A., and J. O'M. Bockris, *Electrochim. Acta*, **11**, 376 (1966).
7a. Damjanovic, A., A. Dey, and J. O'M. Bockris, *Electrochim. Acta*, **11**, 791 (1966).
8. Delahay, P., *Double Layer and Electrode Kinetics*, Interscience, New York, 1965, p. 183.
9. Hoar, T. P., *Proc. Roy. Soc. (London)*, **A142**, 628 (1933).
10. Hoare, J. P., *J. Electrochem. Soc.*, **109**, 858 (1962).
11. Hoare, J. P., *J. Electrochem. Soc.*, **110**, 1019 (1963).
12. Hoare, J. P., *J. Electrochem. Soc.*, **112**, 602 (1965).
13. Hoare, J. P., *J. Electrochem. Soc.*, **112**, 849 (1965).
14. Hoare, J. P., *J. Electrochem. Soc.*, **113**, 846 (1966).
15. Myuller, L., and L. N. Nekrasov, *Dokl. Akad. Nauk SSSR*, **154**, 437, 2815 (1964); **149**, 1107 (1963); *Electrochim. Acta*, **9**, 1015 (1964).
16. Parsons, R., *Trans. Faraday Soc.*, **47**, 1332 (1951).
17. Vielstich, W., *Z. Instrumentenk.*, **71**, 29 (1963).
18. Watanabe, N., and M. A. V. Devanathan, *J. Electrochem. Soc.*, **111**, 615 (1964).

chapter VI

Electroanalytical Chemistry

Electroanalytical Chemistry may be defined as that field of electro-chemistry in which the nature of a chemical species (qualitative analysis) present in a given system and its concnetration in solution (quantitative analysis) are determined from the measurement of the electrochemical parameters of the system. As pointed out by Lingane (166), most elec-troanalytical techniques are based on electrochemical principles which have been known for some time such as Faraday's laws and the Nernst relationship. Some techniques had to await the advent of modern developments in the electronic circuitry, such as the potentiostat (21, 56,111,230,242), before full exploitation of the methods could be realized.

Any electrochemical parameter which relates to the nature of the given chemical species under consideration (e.g., the electrode potential) or to its concentration in solution (e.g., the diffusion current) may be used as the basis of an electrochemical method of analysis. Arbitrarily, one may divide the various electroanalytical techniques into three cate-gories: those carried out under open-circuit conditions, those under constant-current polarization (galvanostatic) and those under con-trolled potential polarization (potentiostatic).

Although detailed discussions of the various electroanalytical methods are not given here, since excellent comprehensive treatments of these subjects are available in books by Lingane (166), Delahay (56), and Kolthoff and Lingane (149), a brief account will be presented in the nature of review.

According to the Nernst relationship, the electrode potential, E, is related to the activity, a, of a given ion in solution by

$$E = E_0 - (RT/nF) \ln a \qquad (6.1)$$

where E_0 is the standard potential, n is the number of electrons trans-ferred in the overall electrode reaction, and the other symbols have their usual significance. Since Eq. 6.1 is derived from thermodynamic prin-ciples, it applies only to equilibrium conditions (at open circuit). If an electrode system is constructed which responds specifically to the pres-ence of a given ion, the concentration of that ion in solution may be de-termined from the potential measured against a suitable reference elec-trode. A good example is the glass electrode, the potential of which varies linearly with H^+ ion concentration over a certain pH range, depending on structure of the glass membrane (131).

Another technique based on Eq. 6.1 is known as potentiometric titration in which the potential of an indicating electrode is measured against a suitable reference electrode as the titrant is added to the sample solution. The potential of the indicating electrode (usually a Pt or Au wire) responds to the nature of the ions present through the value of E_0, which has a unique value for each electrode couple (55), and to their concentration through the ln a term. Initially, the potential corresponds to the electrochemical system of the sample. As titrant is added, the number and kinds of ions present changes until finally a new electrochemical system with a different potential is obtained which corresponds to the electrochemical system of the titrant, provided that the product of the titration process is an electrochemically indifferent salt. In any case, the familiar S-shaped curve is obtained in a plot of potential against volume of titrant added. The end point or equivalence point occurs where the change in potential per unit of volume of titrant added has a maximum value.

Because the resistance or its reciprocal, the conductance, is a function of the concentration and mobility of the ions present in solution, conductometric titration procedures are possible. In this case, the resistance of the sample solution is measured with a bridge circuit such as a Wheatstone bridge while the titrant is added. During the titration process, ions may be added or removed from solution, or ions of different mobilities may replace one another in solution. A plot of the resistance or conductance as a function of the volume of titrant added consists of two straight lines. The point at which the lines meet is the end point. For example, the titration of a strong acid with a strong base produces a V-shaped conductometric curve because the H^+ ions are removed from solution, causing a decrease in the conductance until the end point is reached at the bottom of the "V" where the number of ions and the conductance are at a minimum. As more titrant is added, the conductance rises again because OH^- ions are added to the system.

The analytical methods based on polarization measurements carried out under constant-current conditions are known as coulometric techniques, which are probably the oldest of electrochemical techniques (250). In this procedure, a constant current is applied to the system containing a given concentration of the ion to be determined, and the potential is recorded as a function of time. The potential adjusts itself to the value at which the electrode reaction can occur at the demanded rate (value of the constant current). As long as ions are available to support the current demand, the potential remains constant, but once the ions are used up by the electrochemical process, the potential drifts to a new value where some other electrochemical reaction can occur to support the demanded current. A curve containing a plateau similar to the right side of Fig. 2.1 is obtained. The length of the plateau region is called the transition time, τ. Since the number of coulombs, Q, passed in an electrochemical process is equal to It, where I is the value of the constant current and t is the time, the concentration, c, of the given ion solution may be determined from the plateau length by

$$c = I\tau/VnF \tag{6.2}$$

where V is the volume of solution, F the Faraday (charge on a mole of univalent ions, n the valence, and c the concentration in moles per unit volume.

This same method may be used to determine the amount of an adsorbed species or the thickness of an adsorbed film on a metal surface. Such a technique is identical with the charging curve techniques discussed in Chap. II.

When a coulometric method is carried out under conditions such that the process is controlled by diffusion in the presence of a supporting electrolyte, which is required to minimize the effects of migration (mass transport caused by electrostatic forces), the transition time is related to the concentration of the reacting species in solution by Sand's (221) equation,

$$\tau^{1/2} = \pi^{1/2}nFD^{1/2}c/2I \tag{6.3}$$

where D is the diffusion coefficient of the reacting species and c is its concentration. The method is known as chronopotentiometry as first suggested by Delahay and Mamantov (60), and the possibilities of using the method as an analytical tool were first pointed out by Gierst and Juliard (84). In this case, the product, $I\tau^{1/2}$, is independent of I. When the electrochemical process is not diffusion controlled, $I\tau$ varies in characteristic ways, depending on the nature of the controlling step of the electrode kinetics. Many of the problems associated with complications in the electrode kinetics such as the case where there is more than one reacting species in solution or where the electron transfer step takes place in more than one step have been worked out by Delahay and co-workers (20,59,61). As an analytical tool, however, the system must be carried out only under conditions of diffusion control.

Potentiostatic polarization methods are the basis of another group of electroanalytical techniques. In the electrogravimetric method, the potential of a weighed, inert cathode is held by a potentiostat at a point where the active species is electrodeposited on the electrode surface.

After the current has fallen to zero, the cathode is removed, dried, and weighed again. The difference in weight is the amount of the active species originally present in the sample. If the E_0 values of more than one active species present in the same solution are sufficiently far apart, the components may be separated and determined individually by this method.

Amperometric titrations are carried out ideally under potentiostatic polarization conditions. In general, the potential of the indicator electrode measured against a reference electrode (usually a saturated calomel electrode) is set at a value where the limiting current for the discharge of the active species is reached, and the current is recorded as the titrant is added to the sample. A plot of the current as a function of the volume of titrant added appears as two straight lines. The point at which the lines cross is the end point.

Probably the most useful and most reliable method of analysis in many cases employing electrochemical principles is the one termed *polarography* by Heyrovsky, who is credited (105,109) with the invention of this analytical technique. It is based on the fact that the potential at which the active species first begins to be discharged (the so-called decomposition potential) has a characteristic value for each kind of chemical species present in solution. In addition, as the controlled potential is increased (more cathodic for cations, more anodic for anions), the electrode kinetics become diffusion limited, and the magnitude of the diffusion current, I_d, is a measure of the concentration of the active species concerned.

The potential which is controlled by a potentiometer is scanned over the range of interest at a constant rate by a motor-driven contact on the potentiometer slide wire, and the current is recorded on a recording microammeter. Modern models are available in which a complete scan is obtained electronically during a single drop time.

A plot of the current as a function of the potential of the indicator electrode (usually a dropping mercury, platinum, or gold microelectrode) measured against a saturated calomel reference electrode appears as a series of steps or waves, depending on the number of oxidizable or reducible species present in solution. The potential measured at the current value determined half-way between the residual (prewave) and the diffusion (postwave) currents is called the half-wave potential, $E_{1/2}$, and has a characteristic value for each active species (ref. 149, p. 198), a fact which permits a qualitative analysis of the sample to be made. The $E_{1/2}$ value for a given species is independent of the magnitude of the concentration of that species.

The height of the wave is determined at a dropping mercury electrode (DME) as the difference between the residual and diffusion currents and is related to the concentration of the corresponding active species by the Ilkovic equation (127),

$$I_d = 607nD^{1/2}m^{2/3}t_d^{1/6}c \qquad (6.4)$$

where n is the number of electrons transferred, D the diffusion coefficient of the active species, m the mass of Hg emerging from the capillary per unit time, t_d the lifetime of Hg drop, and c the concentration of active species. For Eq. 6.4 to hold, the kinetics of the system must be under diffusion control. A desirable feature of the polarographic method is the fact that the currents drawn are very small (less than $50\ \mu A$) so that the composition of the sample is not sensibly affected. Most of the measured polarization appears at the indicator electrode because it is made small (Hg drop $= 0.5$ mm at maximum diameter) compared to the counterelectrode (large Hg pool). For a given current, the current density on the counterelectrode is very small so that polarization is negligible, but on the drop it is large so that nearly all of the measured polarization appears at the drop. To reduce the effects of migration currents, an indifferent supporting electrolyte, such as KCl or KNO_3, is added so that mass transport of the active species occurs only by pure diffusion.

Sign conventions normally employed in polarographic analyses are different from those found in Chaps. II–V but will be used here in Chap. VI whenever polarograms and polarographic data are discussed, because this is the form in which the reader will find these data presented in the literature. Polarograms are plotted with the currrent as the ordinate and the potential as the abscissa. A cathodic current is plotted in the positive direction and an anodic current in the negative direction.

A. THE DROPPING MERCURY ELECTRODE

In its simplest form, the dropping mercury electrode (DME) is composed of a vertically positioned capillary about 5–10 cm long and of about 0.05 mm bore diameter through which Hg is allowed to flow from a Hg reservoir under a pressure of about 30–60 cm at a rate of about 1–3 mg/sec so that the lifetime of a drop lies between 2 and 5 sec. Lingane and Laitinen (168) designed the arrangement shown in Fig. 6.1 in which the leveling bulb is connected to the capillary to provide a constant head of Hg to give a constant drop time. It has been pointed out (148,186)

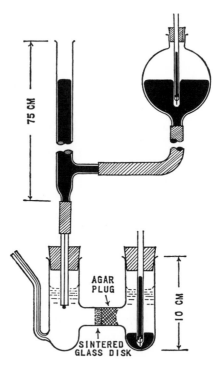

Fig. 6.1. A diagram of the dropping Hg electrode and the arrangement of the Hg reservoir (168). (By permission of the American Chemical Society.)

that it is essential that the capillary be cut at a 90° angle (within ± 5°) to the bore, otherwise Hg tends to adhere to the capillary walls, causing the drop size and the drop time to be erratic.

One of the great advantages of the DME is that the electrode surface is renewed every few seconds. Such an arrangement eliminates any long-term effects of adsorption of impurities and allows correction for the short-term effects. Another advantage is the use of Hg, which has a high hydrogen overvoltage. In fact, there is a virtual absence of adsorbed hydrogen atoms on the surface of cathodically polarized Hg (77), and in a deaerated solution of KCl, the Hg cathode acts as an ideally polarizable electrode (92), that is, the Hg–solution interface behaves like an electrical capacitor, up to potentials approaching a volt. In neutral or alkaline solutions, potentials more negative than −2V may be reached (ref. 56, p. 352) at a Hg cathode before water is reduced.

In the anodic direction, the maximum useful range is limited to about 0.65 V ($\sim$0.4 V vs. SCE) because Hg can be oxidized according to the reaction

$$Hg_2^{2+} + 2e \rightleftarrows 2Hg \qquad E_0 = 0.788 \tag{6.5}$$

Depending upon whether complexing agents are present, this useful anodic range may be greatly reduced. In such cases the anodic range is greatly increased by replacing the DME with a Pt or Au wire. Various modifications in the design of Hg electrodes may be found in Kolthoff and Lingane (ref. 149, p. 355).

1. The Electrochemistry of the Hg/O$_2$ System

Since this section is concerned with the determination of oxygen by means of the DME, it is important to understand the electrochemical behavior of Hg in oxygen-saturated electrolytes.

a. The Oxides of Hg. Mercury is stable in dilute acid or alkaline solutions such as H_2SO_4, $HClO_4$, and KOH and shows a strong reluctance to react with oxygen. Mercurous oxide, Hg_2O, does not exist (162,234) since the freshly precipitated oxide was shown by x-ray analysis (75) to be an intimate mixture of mercuric oxide, HgO, and metallic mercury.

The only stable oxide of mercury is HgO, which can exist in a yellow or red form. By precipitating a solution of a mercuric salt with alkali, the yellow form is made (ref. 234, p. 319), and by decomposing $Hg_2(NO_3)_2$ or $Hg(NO_3)_2$ with gentle heating, the red form is obtained. When metallic Hg is heated in air or O_2 above 300°C, red HgO is produced, but when HgO is heated above 360°C, the compound decomposes to the elements, indicating the low affinity Hg has for oxygen. It appears that the only difference between the two forms is the particle size (ref. 131, p. 335). Some evidence for this is obtained from solubility measurements made by Garrett and Hirschler (81), who found that the yellow form is slightly more soluble in H_2O than the red (0.0513 and 0.0487 g/liter at 25°C, respectively). For the yellow form, the particle size ranges between 1 and 10 μ, whereas for the red it ranges between 10 and 30 μ.

Brönsted (35) measured the potential of the Hg/HgO couple against a Pt/H$_2$ electrode in the same solution of NaOH and records a pH-independent value of 0.929 V for the potential of the cell. A value between 0.926 and 0.927 V has been reported by later investigators (2, 46,76,233a). By averaging all the available data in the literature, Hamer and Craig (100) arrive at a potential value of 0.9258 V at 25°C.

Since the potential of the hydrogen electrode in alkaline solution of pH 14 is -0.828 V, a potential of 0.098 V is obtained for the E_0 of the Hg/HgO reaction,

$$HgO + H_2O + 2e \rightleftharpoons Hg + 2OH^- \qquad E_0 = 0.098 \text{ V} \qquad (6.6)$$

in alkaline solution as recorded by Latimer (ref. 162, p. 179) and de Bethune and Loud (55).

It has been shown (2,65,220) that the potential of Eq. 6.6 is independent of the form of HgO used except that the yellow form was slower to settle down to a stable value than the red form (65). For this reason and for the fact that a highly reproducible standard state exists for metallic mercury, the Hg/HgO electrode qualifies as an excellent secondary reference electrode. This highly reproducible reference electrode (ref. 131, p. 336) is easily constructed (100,220). Since HgO is soluble in acids, the use of the Hg/HgO reference electrode is restricted to alkaline solutions.

b. Charging Curves on Hg Electrodes. Constant current charging curves were obtained (68) on Hg anodes and cathodes in both acid and alkaline solutions saturated with nitrogen. El Wakkad and Salem (68) plated Hg on Pt electrodes from acidified $Hg_2(NO_3)_2$ solutions. In $1N$ NaOH solution, the potential rose quickly from an open-circuit potential, which drifted with time (68a), to a plateau value of about 0.08 V (68). Jirsa and Lores (135) reported that a plateau value of about 0.09 V was observed when a Hg electrode was anodized galvanostatically. Eventually (68,135,194) O_2 is evolved at about 1.5 V. It was found (68) that the point at which O_2 is evolved is lowered as the temperature is increased, the plateau at 0.08 V is shortened as the current density is decreased, and the entire curve is shifted along the potential axis with changes in pH. At a current density of 0.04 A/cm^2, El Wakkad and Salem estimated that a film of HgO about 12 molecules thick was formed before O_2 was evolved. Because of the liquid nature of the metal substrate, a complete protective film cannot be built up on a Hg surface, and as noted by Newberry (194), bubbles of gaseous O_2 broke through the yellow film of HgO formed on a Hg anode in the NaOH solutions.

In N_2-saturated acid solutions, the only process occurring at a Hg anode in $0.1N$ $HClO_4$ solution is the dissolution of mercury which occurs at about 0.5 V (68). With anodic polarization, Hg also dissolves in sulfuric acid solution, but a gray film of $HgSO_4$ forms on the Hg surface (194). At higher polarization, O_2 is evolved, and bubbles of O_2 may be observed (194) breaking through the gray film.

Under galvanostatic conditions in N_2-saturated acid solutions, the potential of a Hg cathode drifts rapidly to the point where H^+ ions are discharged (68).

It is important to note the the charging curves of El Wakkad and Salem (68) show the presence of only one oxide in the Hg/O_2 system. No indication of the presence of a lower oxide than HgO could be found in their studies.

c. Oxygen Overvoltage Studies on Hg. The early studies (e.g., ref. 194) of oxygen overvoltage obtained on Hg electrodes are difficult to assess since control of the experimental conditions was not carried out rigorously enough. Steady-state polarization studies of the evolution of oxygen on stationary Hg anodes are virtually nonexistent in the published literature.

Some speculation with regard to the nature of the anodic process for the steady-state evoultion of O_2 on a stationary Hg electrode in alkaline solution may be made by taking into account the data obtained from charging curves (68,135,194). Since the potential at which O_2 is evolved on a Hg anode in NaOH becomes less noble with increasing temperature (68), the evolution process is activation controlled and probably occurs on oxide-free Hg sites along cracks in the HgO film. Oxygen is evolved on the Hg surface and not on the oxide surface because O_2 bubbles were observed (194) bubbling through the oxide film. A likely mechanism is the same as that offered for Pt, Eq. 3.8 followed by Eqs. 3.2–3.5. Because Hg forms only one oxide, the evolution of oxygen does not involve a higher oxide as in the case of Pd but probably occurs in parallel with oxide formation as suggested for Au anodes.

Steady-state polarization data were obtained for the reduction of O_2 in O_2-saturated acid and alkaline solutions on a hanging Hg drop (130) and on a stationary Hg pool (1,9). A Tafel region was found between 10^{-7} and 10^{-4} A/cm^2 in H_2SO_4 solutions and between 10^{-6} and 10^{-4} A/cm^2 in NaOH solutions (130). In acid solutions Iofa and co-workers (130) noted that the η depended of P_{O_2} when the P_{O_2} was varied between 1 and 0.03 atm. Both $d\eta/d \log i$ and $d\eta/d \log P_{O_2}$ were 0.116 in H_2SO_4 but 0.06 in NaOH. Bagotskii and Yablakova (9) reported Tafel slopes equal to 0.11 at pH 1, 0.08 at pH 6, and 0.03 at pH 10. It was concluded (41,42,44) that the rate-determining step in acid solutions is

$$(O_2)_{ads} + e \longrightarrow (O_2^-)_{ads} \qquad (6.7)$$

and that the reduction of O_2 on Hg proceeds by two two-electron steps with H_2O_2 as a stable intermediate. Apparently the kinetics for the reduction of O_2 in acid solutions on Hg cathodes are similar to those on Pt cathodes.

In alkaline solution, the mechanism is considered to be different from that in acid solutions. Bagotskii and Yablakova (9) came to the conclusion that the steady-state reduction of O_2 on Hg cathodes in alkaline solutions is diffusion controlled.

Iofa and co-workers (130) carried out overvoltage measurements on a hanging Hg drop in peroxide-stabilized acid solutions. At a H_2O_2 concentration of $0.6N$, a Tafel region was observed between 10^{-3} and 0.8 A/cm^2 with the very high Tafel slope of 0.300. In $0.1M$ H_2O_2-stabilized $2N$ H_2SO_4 solutions with O_2 stirring, a Tafel slope of high value (0.275) was also observed (121) on Rh cathodes. A clear explanation for the mechanism in this region is not presently available.

2. Polarographic Studies

In this section, the reduction of oxygen at a dropping mercury electrode (DME) is considered under polarographic conditions (diffusion-controlled kinetics carried out at controlled potential). Originally, the DME was used by Kucera (157) to determine the interfacial tension of polarized mercury. The polarographic method was first described by Heyrovsky in a series of papers (106) as an analytical tool. He showed (107) that the reduction of O_2 at a dropping mercury electrode proceeded through a stable intermediate, H_2O_2, by two two-electron steps because two waves of equal height were observed. These observations were confirmed by Kolthoff and Miller (150). The value for the diffusion coefficient, D_{O_2}, of O_2 in water solutions as calculated from the Ilkovic equation at 25°C is 2.6×10^{-5} cm^2 sec^{-1} (150). In methanol or ethanol solutions, D_{O_2} is larger (247) than in H_2O solutions.

The polarogram for the reduction of dissolved oxygen in air-saturated $0.1N$ KCl solution at a DME is shown in Fig. 6.2. In pure KCl solution, curve I is obtained, which exhibits a large current maximum beginning at about 0 V (vs. SCE) and returning to the limiting current just before the second wave begins. Such current maxima which are reproducible (110) are nearly always found in polargrams taken on a DME. Although maxima may occur on those taken at a Hg pool electrode (79,102), these maxima are not reversible. Kolthoff and Lingane (ref. 149, Chap. X) give an excellent account of the types of maxima observed and the theories presented to explain their existence. A complete theory is not available (ref. 149, p. 178) at this time.

It has been found (99,204,205,244) that the addition of traces of various capillary-active substances, such as gelatin, thymol, and methyl red,

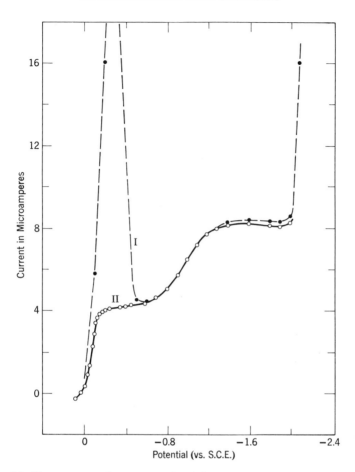

Fig. 6.2. The current–voltage curves (150) for the reduction of O_2 at a DME in air-saturated $0.1N$ KCl solution: I, in the absence of a maximum suppressor; II, in the presence of 0.01% thymol. (By permission of the American Chemical Society.)

suppresses the current maximum.   Curve II of Fig. 6.2 was taken in the presence of a trace of thymol.   In this case the two waves are clearly visible since the maximum is completely suppressed without affecting the shape of the second wave significantly.   The slope of the first wave —is very steep with $E_{1/2}$ occurring at -0.05 V (vs. SCE), but the slope of the second extends from -0.6 to -1.4 V (vs. SCE).   Both waves are of equal height, and the limiting currents are directly proportional to the concentration of dissolved oxygen in the solution.

As suggested by Heyrovsky (107) and Vitek (247), the first wave corresponds to the reduction of O_2 to H_2O_2,

$$O_2 + 2H^+ + 2e \longrightarrow H_2O_2 \qquad (6.8)$$

in acid and

$$O_2 + H_2O + 2e \longrightarrow HO_2^- + OH^- \qquad (6.9)$$

in alkaline solution. The second wave occurring at higher potentials corresponds to the reduction of H_2O_2 to H_2O,

$$H_2O_2 + 2H^+ + 2e \longrightarrow 2H_2O \qquad (6.10)$$

in acid and

$$HO_2^- + H_2O + 2e \longrightarrow 3OH^- \qquad (6.11)$$

in alkaline solution. Since Kolthoff and Miller (150) found no relationship between the pH and the $E_{1/2}$ of the first or second wave, they concluded that both reduction processes, $O_2 \rightarrow H_2O_2$ and $H_2O_2 \rightarrow H_2O$, occur irreversibly at Hg cathodes.

Because the H_2 overvoltage on Hg is very high, well-defined waves can be obtained for dissolved O_2 in solution in polarograms taken at a DME. Also, the Hg surface is continually renewed so that the effects of impurities are reduced to a negligible consideration. Therefore, the polarographic determination of dissolved O_2 with a DME is one of the most dependable methods available and in many cases the most convenient.

When H_2O_2 has accumulated in solution or when air-saturated solutions of H_2O_2 are polarized at a DME cathode, it is found that the diffusion current for the first wave is greater than the I_d for the second. This increase in I_d is referred to as an "exaltation" of the diffusion current. It was first observed by Heyrovsky and Bures (108) as the exaltation of the limiting currents of K^+ and Na^+ ions in very dilute air-saturated solutions of the alkali halides produced by the preceding discharge of dissolved O_2. In solutions of H_2O_2, the exaltation effect is explained (146) by a catalytic decomposition of H_2O_2 to produce additional O_2 which is reduced by Eq. 6.8 or 6.9, causing additional current to be added to the limiting current of the first wave. When the proper correction for H_2O_2 reduction is made to the I_d of the first wave, the first and second waves become equal.

With the discharge of the first electron, Eq. 6.7, $(O_2^-)_{ads}$ is formed on the electrode surface. The reduction of O_2 to water proceeds by the reaction of H^+ ions in acid or H_2O molecules in alkaline solutions, and the discharge of another electron as follows:

$$(O_2^-)_{ads} + H^+ \longrightarrow (H_2O)_{ads} \qquad (6.12)$$

$$(HO_2)_{ads} + e \longrightarrow (HO_2^-{}_{ads} \tag{6.13}$$

$$(HO_2^-)_{ads} + H^+ \longrightarrow H_2O_2 \tag{6.14}$$

When H_2O_2 has accumulated in solution, another process is possible. The $(O_2^-)_{ads}$ can react with an H_2O_2 molecule,

$$(O_2^-)_{ads} + H_2O_2 \longrightarrow OH^- + (OH)_{ads} + O_2 \tag{6.15}$$

to form adsorbed OH radicals. This is followed by a reaction chain proposed by Haber and Weiss (96) for the homogeneous catalysis of H_2O_2 decomposition by Fe^{2+} ion,

$$2(OH)_{ads} + H_2O_2 \longrightarrow 2(HO_2)_{ads} \tag{6.16}$$

and

$$(HO_2)_{ads} + H_2O_2 \longrightarrow H_2O + (OH)_{ads} + O_2 \tag{6.17}$$

Finally, the chain is broken by the step

$$(OH)_{ads} + e \longrightarrow OH^- \tag{6.18}$$

The additional O_2 producing the exaltation effect in the first wave is formed by Eqs. 6.15 and 6.17.

At a rotating Pt or Au cathode, Kolthoff and Jordan (146) added such OH-radical getters as acrolynitrile monomer to the system. Since no effect was observed on the exaltation of the first wave, they concluded that the chain mechanism, Eqs. 6.16 and 6.17, contributed negligibly to the kinetics, and the additional O_2 was produced by the mechanism, Eqs. 6.7, 6.15, and 6.18. The ratio of the I_d in the presence and absence of H_2O_2 is a measure of the exaltation effect and is directly proportional to the H_2O_2 concentration (146).

Another explanation for the presence of the additional oxygen may be given. Oxygen is reduced to H_2O_2 by a series of steps, Eqs. 9.7 and 6.12–6.14. Following this, the H_2O_2 is catalytically decomposed by a local cell process (117) originally suggested by Gerischer and Gerischer (83):

$$H_2O_2 + 2H^+ + 2e \longrightarrow 2H_2O \tag{6.19}$$

and

$$H_2O_2 \longrightarrow O_2 + 2H^+ + 2e \tag{6.20}$$

where the overall reaction is

$$2H_2O_2 \longrightarrow O_2 + 2H_2O \tag{6.21}$$

Since current is not contributed to the external circuit by the local cell mechanism, the local cell current does not affect the shape of the polarogram. The local cell mechanism is favored as the source (Eq. 6.21) of the additional O_2 responsible for the exaltation effect.

Recently, Cornelissen and Gierst (50) noted that in the presence of a suppressor polarograms which were taken at a DME in certain solutions of a particular composition showed an exaltation of the first wave in the absence of excess H_2O_2. Such behavior was first observed by Longmuir (172). Such solutions cited were $0.05N$ NaF, $0.1N$ $(C_2H_5)_4$-NCl, and $0.01N$ $(C_3H_7)_4$NCl in which the effect is explained in terms of the potential of the outer Helmholtz plane, ψ°. If ψ° is more positive in the system supporting electrolyte + salt than in the system supporting electrolyte alone, the rate of the catalytic reaction, Eq. 6.15, is increased while the rates of other reactions, such as Eq. 6.12, remain the same. They (50) suggest that the influence of ψ° may be great enough to make Eq. 6.15 detectable even though excess H_2O_2 is not present. The more positive value of ψ° occurs in NaF solutions because the exaltation takes place at potentials positive to the point of zero charge, -0.47 V vs. SCE (electrocapillary maximum), and in solutions of the tetraalkylammonium salts, because of cation adsorption.

Kolthoff and Izutzu (144) report that highly purified solutions of inorganic supporting electrolytes and of tetraalkylammonium salts do not exhibit an exaltation of the first wave at a DME in the presence of a suppressor. They propose that Cornelissen and Gierst's (50) observations were the result of effects produced by impurities present in the original substances. One of these impurities is the volatile tetraalkylammonium hydroxide which may be removed with N_2 stirring. Kolthoff and Izutzu condensed the impurities from the effluent N_2 flow in a cold trap. The purified salt did not exhibit an exaltation effect, but additions of the residue from the cold trap to the pure salt showed the same exaltation effect as found in the unpurified salt.

Impurities, apparently, may cause exaltation effects even though excess H_2O_2 may not be present. Possibly adsorption of these impurities inhibits the direct reduction of H_2O_2 to H_2O, Eq. 6.10, to a point where the catalytic decomposition of H_2O_2 by the local cell mechanism, Eqs. 6.19 and 6.20, adds enough O_2 to the system to be detected as an exaltation effect. Some indication of this inhibition by adsorption of impurities rests in the observation that the second wave obtained in the impure solutions (50) was shifted to more negative potentials. Solutions containing vinyl acetate are known (144) to give an exaltation of the first wave, and it was observed that the second wave was displaced to more negative potentials. This is to be expected if adsorbed vinyl acetate molecules inhibit the direct reduction of H_2O_2 to H_2O by Eq. 6.10.

A true exaltation of the first wave is found in solutions containing catalase (29) which is known (142) to accelerate the decomposition of H_2O_2 in a homogeneous catalytic process. As noted by Kolthoff and Izutzu (144), the presence of molybdate ion produces a genuine exaltation because MoO_5^{2-} ion is formed by reaction with H_2O_2,

$$H_2O_2 + MoO_4^{2-} \longrightarrow H_2O + MoO_5^{2-} \tag{6.22}$$

The discharge of MoO_5^{2-} ion

$$MoO_5^{2-} + 2H^+ + 2e \longrightarrow MoO_4^{2-} + 2H_2O \tag{6.23}$$

which occurs at potentials more positive than the discharge of O_2 (152), causes the observed exaltation of the first wave (144).   Kolthoff and Izutzu explain the exaltation produced by the presence of vinyl acetate in terms of the discharge of a complex formed by the interaction of vinyl acetate with peroxide at potentials positive to O_2 discharge.

A polarographic method was used by Delahay (58a) to obtain indirectly the polarization curve for the reduction of O_2 on bright Pt cathodes.   In this method, the test electrode was polarized at a constant potential in an air-saturated phosphate buffer of pH 6.9, while the oxygen content of the solution was determined with a DME as a function of time.   From the rate data obtained for the consumption of O_2, a value for the current flowing at the given value of potential was calculated. By repeating this experiment over a range of potential values, the complete polarization curve for the reduction of O_2 was determined.   The potentiostatic polarization of the Pt cathode and the stirring of the electrolyte were stopped during the polarographic measurements so that interference with the operation of the DME would not exist.   At low overvoltages this direct method gave the same results as the conventional method of determining O_2 overvoltage on Pt cathodes.   The advantage of the indirect method is the fact that only one single-electrode reaction is studied, whereas the direct method in many cases gives results which include contributions from several electrode reactions. From his investigations, Delahay concluded that the reduction of O_2 to H_2O_2 occurs over a wide range of potentials.   He also studied a number of other metal cathodes.

3. Use of DME in Analysis

A determination of the oxygen content of electrochemically inert gases may be made with a DME by saturating a solution of supporting electrolyte with the gas mixture.   After determining the residual current in N_2-saturated supporting electrode, the diffusion current is determined in the electrolyte saturated with air at atmospheric pressure.

Since the concentration of dissolved O_2 in solution is directly proportional to the partial pressure of the O_2 in the gas in equilibrium with the solution according to Henry's law, the oxygen content of the gas may be computed from the ratio of the diffusion current determined in the gas-saturated solution to that in the air-saturated solution and the known value of the partial pressure of oxygen in the water-saturated air. This procedure may be used to determine oxygen content from a few millimeters up to atmospheric pressure in inert gas samples. Because the diffusion of O_2 in anhydrous methanol or ethanol is several times larger than that in water (247), the use of these solvents may be used to extend the lower range of detection of oxygen.

It was pointed out by Vitek (247) that the complete polarogram need not be taken, since a single potential point in the limiting current region may be used. The ratio of the diffusion currents determined at this potential is a function of the oxygen content of the sample. Ordinarily a potential of about -0.7 V is chosen, but as cautioned by Wise (255), interfering waves due to the reduction of other substances cannot be detected in this single-point method. This disadvantage may be overcome by recording the first wave only.

Wise (255) describes a method in which the continuous determination of O_2 in a gas stream can be measured with a DME. The actual determination of the diffusion current at the usual form of DME must be made in unstirred solution, because if a polarogram is recorded while gas is bubbling through the solution, frequent irregular fluctuations in the current are obtained even if large damping condensers are used. Such irregular current fluctuations are damped out by using a shielded DME. A cylinder of metal gauze (about 100 mesh) or glass about 1 cm in diameter is placed around the tip of the capillary (255), ending just above the Hg pool anode. With a metal gauze of 100 mesh, Wise found that the system responded to changes in the P_{O_2} within about 200 sec. When a glass shield was used, the response time was increased by about 100 sec.

For determining traces of O_2 in gases or solution, Laitinen and co-workers (158) developed a rapid and sensitive method based on the measurement of the null potential of a DME. Ideally, if solution is completely free of capillary active substances and of electrochemically reducible and oxidizable substances, an isolated DME in this solution acquires a potential corresponding to the maximum on the electrocapillary curve (point of zero charge) which is -0.6 V (vs. SCE) in $0.1N$ KCl solutions. When the potential of the drop is shifted from the point of zero charge (PZC) by an externally applied potential, a double-layer charging current flows continuously through the circuit during the drop lifetime because the area of the Hg surface continually increases. Delahay (58) has suggested a method based on this principle to study the electrode kinetics of a given process under open-circuit conditions by observing the overvoltage developed as a result of the passage of the charging current as the area of an isolated electrode is changed. These charging currents are negative (corresponding to an anodic process according to polarographic convention) at potentials more positive than the PZC and are positive (cathodic process) at potentials more negative than the PZC. With the coulostatic method (58), Smith and Delahay (235a) have determined residual oxygen concentrations as low as 5×10^{-7} mole/liter.

If a substance which is reduced on the positive side of the PZC such as O_2 is present, a positive polarographic current will flow. Since the double-layer charging current and the O_2 polarographic are opposite in sign, a potential exists at which the two currents cancel, and the current in the external circuit is zero. A special flow-type cell was designed (158) for this system in which electrolyte was permitted to flow through the cell at a rate of 1.1 ml/min. Under steady-state conditions in N_2-saturated electrolyte, the potential corresponding to the electrocapillary maximum, E_{max} (0.5 V vs. Ag/AgCl in $0.1N$ KCl) was recorded. After the solution was saturated with the sample gas, the potential, E, was determined for zero current. It was shown (158) that the difference in potential is directly proportional to the concentration of O_2, c, according to

$$E - E_{max} = 17.2c \ t_d^{1/2} \tag{6.24}$$

where t_d is the drop time. Oxygen contents as small as 0.01% by volume in the gas mixture can be determined.

The DME is used for oxygen determinations in many biological systems which will be discussed in a later section.

The DME can be used to analyze for the O_2 content of a sample solution by amperometric titration with H^+ ion (ref. 166, p. 276). Kolthoff and Miller (150) have shown that the hydrogen-limiting current at a DME in $0.001M$ HCl solution occurs between -1.6 and -2.0 V (vs. SCE). If the potential of the system is held at -1.8 (vs. SCE), the total reduction current arises from reduction of H^+ ion to H_2. When O_2 is present, H^+ can be reduced in the reaction with O_2,

$$O_2 + 4H^+ + 4e \longrightarrow 2H_2O \tag{6.25}$$

As long as H^+ ion is in excess, the limiting current at -1.8 V is unaltered.

When the H^+ ion concentration falls below that required for the reduction of O_2 by Eq. 6.25, the electrode kinetics are controlled by the diffusion of O_2, and the limiting current is the O_2 diffusion current.

If one titrates an O_2-containing solution with a standard acid solution (e.g., HCl) using a DME indicator electrode potentiostated at -1.8 V (vs. SCE), the initially recorded current is the O_2 diffusion current as shown in Fig. 6.2 and remains constant until an excess of H^+ ion is added, at which time the recorded current increases. The end point is reached when the two diffusion rates are equal, giving the condition

$$C_H D_H^{1/2} = 4 C_{O_2} D_{O_2}^{1/2} \qquad (6.26)$$

Since D_H is 8.7×10^{-5} and D_{O_2} is 2.6×10^{-5} (ref. 1, p. 277), the end point occurs when

$$C_H = 2.17\, C_{O_2} \qquad (6.27)$$

Even though the polarographic wave for H_2O_2 is broad (second wave of Fig. 6.2), a well-defined diffusion current develops which may be used to determine the concentration of H_2O_2. It has been reported by Pellequer (198) that the determination of H_2O_2 by the polarographic method gives valid results for H_2O_2 concentrations as low as 1.5×10^{-5} mole/liter. At low concentrations of H_2O_2, the catalytic decomposition of H_2O_2 by Eq. 6.21 produces a negligible error because the reduction current of the O_2 produced by Eq. 6.21 is virtually equivalent to the reduction current, which would have been produced by the catalytically decomposed H_2O_2. However, at concentrations of H_2O_2 greater than $0.045M$ (85), significant error is introduced because the concentration of O_2 produced by Eq. 6.21 is so large that some of it escapes from solution and low results are obtained.

B. THE NOBLE METAL INDICATOR ELECTRODE

To determine the electrochemical properties of a given redox system, such as Fe^{2+}/Fe^{3+}, for which the reactants and products of the electrochemical reaction are soluble, an inert electrode is required which acts as a source or sink of electrons and a catalytic surface on which the electrode reaction takes place. The most commonly used material for constructing this inert electrode, called an indicator electrode because the potential responds to changes in the activity of the components of the redox system, is platinum. The potential of the indicator electrode is measured against an unpolarized reference electrode such as a SCE.

1. Properties of the Pt Indicator Electrode

Many investigators (6,8,10,54,151,161,164,167,173,187,213,223) have reported that a preanodized Pt electrode behaved differently toward various redox systems than one which had not been preanodized. In general, a freshly preanodized electrode behaved as a more reversible indicator electrode in reduction processes (54) such as the reduction of iodate (6) and reduction of iron (151) but inhibited oxidation processes such as the oxidation of iron and arsenic (10) and of iodide (8).

To explain the increased reversibility of preanodized Pt indicator electrodes, Kolthoff and Nightingale (151) and, later, Davis (54) suggested that preanodization produced an adsorbed film of oxygen which facilitated the transfer of electrons across what was called "oxygen bridges." From steady-state polarization measurements on a rotating Pt electrode, Veselovskii and co-workers (26,214) concluded that the electrochemical oxidation of a reducing agent in solution occurred mainly through chemical oxidation by the adsorbed oxygen film, and the anodic current was used to replace the consumed adsorbed oxygen. It would appear that the reducing agent of the investigated redox system could interact with the adsorbed oxygen layer on the freshly preanodized Pt indicator electrode used, for example, in a potentiometric titration. Such an interaction of the electrode with the redox system in solution would not only remove the oxygen bridges but also change the concentration ratio of the redox system, thus defeating the purpose of an indicator electrode in the electroanalytical determination. Since such interactions are not observed experimentally, it appears that the oxygen bridge mechanism is not the operating mechanism at a preanodized Pt indicator electrode.

Originally, Anson (6) subscribed to the oxygen-bridge explanation but later (7) rejected it in favor of an explanation in terms of platinized platinum. He reasoned that the anodized surface was reduced by the reducing agents in the system to be analyzed, producing a very thin layer of Pt black. This invisible roughening of the Pt surface is considered to be responsible for the increased reversibility of the preanodized Pt indicator electrode. A similar conclusion has been reached by a number of workers (25,64,73,233). According to this viewpoint, the decrease in activity of the Pt electrode after a given time was ascribed to the recrystallization of the Pt-black surface back to its original state. Nesterova and Frumkin (192) found that a loss of activity of Pt-black electrodes could be traced to a decrease in surface area due to the recrystallization of the Pt particles. X-ray diffraction studies carried out by Connolly and co-workers (49b) demonstrate that crystallites of supported Pt grow by a recrystallization process.   A suggested mechanism of crystallite growth in solutions at elevated temperatures (49c) involves a local cell mechanism set up between the large and small crystallites with the small ones acting as anodes.   Thacker (239a) has also observed the recrystallization of electrodeposited Pt blacks.

In partial support for the thesis that the reduction of the adsorbed film on a preanodized Pt electrode effectively produces a Pt-black surface, Anson and King (7) applied an alternating current to the Pt electrode such that it was anodized and cathodized just below the points of visible evolution of gas.   Within 10 min, the surface darkened and resembled a platinized platinum surface obtained in the absence of lead. This phenomenon was interpreted in terms of the formation of an oxide on the anodic half-cycle which was reduced to finely divided Pt on the cathodic half-cycle.

The data obtained from charging curves on Pt electrodes (24,228,245) indicate that once the monolayer of adsorbed oxygen is formed and evolution of molecular oxygen takes place the film ceases to grow with continued anodization.   Such an observation is in agreement with the concept that the adsorbed oxygen film on a Pt anode is a good electronic conductor.   Supporting evidence for this conclusion rests on the experimentally observed fact (13,115,170,228) that the presence of adsorbed oxygen does not lower the double-layer capacity of a Pt electrode in H_2SO_4 solutions.   If one accepts the viewpoint that the adsorbed oxygen film on Pt is an electronically conducting monolayer of adsorbed oxygen, it is difficult to see how mere formation and reduction of such a thin conducting film could produce a significant increase in the surface area. In fact, Bowden (27) reported that when Pt is alternately oxidized and reduced, changes in area are nil, but the catalytic activity of the electrode is increased.   Breiter (24,31) finds a negligible roughness effect.

Using a single-pulse technique similar to the one used by Hackerman and co-workers (34,224), it was found (116) that the surface area determined from double-layer capacity measurements did not increase significantly with the normal manual activation procedure of repeated anodization followed by cathodization.   However, the area did change with ac polarization.   It appears, then, that a more plausible explanation of the interaction of ac polarization with the Pt surface may be found by considering the diffusion of hydrogen into the body of the metal during the cathodic half-cycle.   On the anodic half-cycle, this dissolved hydrogen would be removed.   Since changes in the lattice parameters may accompany the penetration of hydrogen into the metal lattice (235), the repeated penetration and removal of hydrogen caused by the ac polarization with the resultant expansion and contraction of the lattice could produce a breaking up of the metal surface. In support of this conclusion, it was observed (116) that Pd, which is known to dissolve large quantities of hydrogen (e.g., ref. 78,122,155,235), darkened much more quickly than Pt, but Rh, which does not dissolve hydrogen in the massive state (ref. 234, p. 1515), did not darken even after a polarization time of an hour. When the ac polarization was biased with a constant dc source to the point where the potential did not drop below 0.05 V, it was found that neither Pt nor Pd became dark.

Recently, Gilman (88) has disagreed with the hydrogen dissolution explanation for the blackening of a Pt surface by ac polarization. He made studies of the roughening of a Pt surface produced by alternate anodic and cathodic polarization, using sequential pulse techniques and found that a darkening of the Pt surface was observed even though the potential was not permitted to fall below 0.4 V, the point above which hydrogen is no longer adsorbed. However, this potential limit, 0.4 V, for hydrogen adsorption was observed (24,210) at ambient temperatures ($\sim$25°C), whereas Gilman made his measurements at 120°C. At such a high temperature, the hydrogen adsorption limit has not been determined and may well be considerably more noble than 0.4 V. In fact, platinum is an endothermic occluder of hydrogen (ref. 235, p. 3), so that hydrogen is more soluble in platinum as the temperature is increased. Most likely, the darkening of the Pt surface observed by Gilman is caused by hydrogen dissolution in the Pt metal. Also, the solubility of hydrogen (123) and the resulting darkening of the Pt with ac polarization (116) can be increased by the presence of impurities.

As a point of interest, it was observed that with strong ac polarization dc biased above the hydrogen adsorption region the Pt surface became a bright golden yellow (116). Other workers (155,181,182,190,216) have observed this coloration, and it has been suggested (3,182,190,232) that the product of ac polarization is PtO_2. Similar observations were obtained in alkaline solutions. It may be mentioned here that alternating current superimposed on direct current will cause Pt to dissolve in acid or alkaline electrolytes (3,33,182,209,215,232). Strong dc polarization will produce brown oxides on Pt but will not dissolve the metal (182,190). Also, ac polarization alone dissolves Pt only if an oxidizing agent such as oxygen is present (215). Ruer (215) warns that if Pt electrodes are polarized with alternating current an inert atmosphere must be used to prevent loss of the Pt. Another interesting observation is the fact that the oxygen overvoltage is lowered by the superposition of alternating current (17,95,210,249), and since alternating current has no effect on the shape of the polarization curve (210), the lowering may be due to an increase in the surface area because of a roughening of the surface by the action of the alternating current.   Yet, this may not be the only cause, since the magnitude of the lowering on η depends on the frequency of the altenating current.

Since there may be some objections to the oxygen-bridge and platinized-platinum explanations for the higher reversibility of freshly anodized Pt indicator electrodes for certain electroanalytical systems, an explanation based on the concept of dissolved oxygen in the Pt metal is favored.   As pointed out earlier, oxygen may be dissolved in the Pt (139) metal; Schuldiner and Warner (227,229) obtained evidence that the presence of dissolved oxygen modified the catalytic activity of the Pt electrode surface toward electrode processes involving hydrogen or oxygen.   By treating Pt with HNO_3 (114), oxygen may be dissolved in the Pt to such an extent that an alloy of Pt and O atoms is considered (120) to be formed.   Such an electrode has been referred to as a Pt-O-alloy electrode, and from oxygen overvoltage measurements (118) it has been shown that the Pt-O-alloy/O_2 electrode is more reversible than the Pt/O_2 electrode.   Among other things, it was concluded that the Pt-O-alloy surface greatly facilitates the transfer of electrons.   From this point of view (119), it is the oxygen dissolved in the surface layers of the Pt metal (effectively a different electrode material than pure Pt) and not oxygen adsorbed on the metal surface which facilitates the electron transfer, thus generating a more reversible indicator electrode for electroanalytical use.

The presence of an adsorbed oxygen layer on a Pt surface inhibits the transfer of electrons (30,246), probably because the existence of the negative dipoles of the adsorbed oxygen atoms on the Pt surface increases the work function of the metal surface.   If a preanodized electrode is used as the cathode in an electrochemical reduction process, the inhibiting film is stripped off and a Pt surface containing dissolved oxygen atoms is exposed.   In this case, electron transfer is facilitated, and it is expected that the reduction of iron (151) and of iodate (6) would occur more reversibly at the preanodized Pt electrode.   On the other hand, if an anodic process is studied, the electrochemical process maintains the inhibiting film in most cases.   Consequently, preanodization which promotes film growth is expected to impede the oxidation of iron or arsenic (10) and of iodide (8).

A freshly preanodized Pt indicator electrode must be used (151) because with time the dissolved oxygen can escape.   The rate of escape of oxygen is rather slow (229) so that the Pt-O-alloy surface produced by preanodization provides an effective catalytic surface for many redox systems, since most investigations can be carried out well before the dissolved oxygen escapes.

Since the potential of the O_2 electrode depends on pH with a coefficient, dE/d pH, equal to 60 mV (245), Furman (80) pointed out that measurements of the rest potential of a Pt/O_2 electrode could be used to determine the end point of acid–base titrations. The potential of a platinized-platinum gauze over which O_2 gas was bubbled was measured against a SCE in a solution of the unknown sample. As acid is added to base or vice versa, the pH and hence the potential of the Pt/O_2 electrode changes along the familiar S-shaped curve of a typical potentiometric titration plot. The end point, of course, corresponds to the point where the change in potential for a given volume of added titrant is a maximum.

2. Polarography at Solid Electrodes

Most oxidation reactions cannot be studied at a DME because the Hg goes into solution and some other electrode material more inert must be used. In certain cases, particularly in biological studies, it is not advantageous to use a DME for O_2 determinations because of volume considerations, mercury poisoning, convenience, etc. The use of solid electrodes permits the construction of microelectrodes consisting of a short (4 or 5 mm) length of metal wire sealed in a glass tube. Schmidt (226) studied the shape of the waves and the diffusion currents obtained on solid electrodes of various materials as a function of the electrode material. The most clearly defined waves for both reduction and oxidation processes were obtained at Pt electrodes as reported by Lord and Rogers (173).

a. The Stationary Pt Wire Electrode. Diffusion currents were observed at Ag and Cu electrodes in the Ag/Ag^+ and Cu/Cu^{2+} systems, respectively, as early as 1897 by Salomon (217), but because of a lack of experimental detail an evaluation of the results is difficult to make. A proportionality between diffusion current and the concentration of various reducible substances, such as iodine and quinone, was found by Glasstone and Reynolds (89,90) at a stationary Pt wire in unstirred O_2-free solutions. From a study of Pt microelectrodes of various configurations (flat, wire, bead), Laitinen and Kolthoff (159) concluded that the Pt-wire microelectrode reached a steady-state diffusion current, which was least subject to irregular fluctuations, more quickly than any other of the electrode types.

Current–voltage curves were carried out for the reduction of O_2 in various buffer solutions over a pH range from 3 to 12 at a Pt microelectrode (160). As with a DME, it was found that the potential at which O_2 reduction begins is virtually unaffected by pH, which agrees with the suggestion that the rate-determining step in the kinetics of O_2 reduction is the discharge of the first electron to produce adsorbed O_2^- ions. The effects of specific adsorption of ions is shown by the change in shape of the *I–E* curve for solutions of the same pH, using different buffers. Hydrogen peroxide is the product of the polarographic reduction of oxygen at a Pt-wire microelectrode because the hydrogen η on Pt is so low that the potential at which hydrogen ion is reduced is reached before a potential at which H_2O_2 is reduced can be applied, as seen in Fig. 6.3. In fact, in the acid solutions, the first O_2-reduction wave is not completed before the hydrogen diffusion current appears. An additional advantage of a Pt electrode is the absence of the large maximum on the first wave for the reduction of O_2.

A flow system for the continuous determination of oxygen is described by Müller (187). The test solution flows through a glass tube containing a constriction in which the Pt-wire microelectrode is sealed. Limiting currents for the reduction of O_2 detected on the wire as the solution flows by is a function of both the oxygen concentration and the

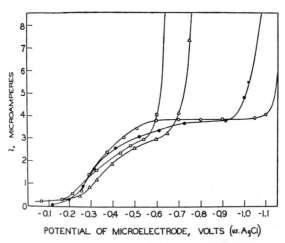

Fig. 6.3. *I–E* curves for the reduction of O_2 at a stationary Pt-wire microelectrode in air-saturated solutions (160) of various pH values: ($\bigcirc$) 0.01N NaOH + 0.01N KCl, pH 12; ($\bullet$) borate buffer, pH 9.0; ($\triangle$) acetate buffer, pH 3.7; ($\square$) phthalate buffer, pH 3.0. (By permission of the American Chemical Society.)

rate of flow.   Such a system may be used to determine the flow rate of a fluid containing a known amount of O_2.

b. The Rotated Pt Wire Electrode.   As pointed out by Laitinen and Kolthoff (159,160), reproducible diffusion currents, from which accuracies of $\pm 1\%$ in the analytical determination of a given sample may be achieved, are observed if time is allowed at each potential point for the steady-state current to be reached.   At a stationary Pt-wire electrode, at least 2 min is required to reach steady state.   This disadvantage in time may be eliminated at a rotating Pt wire electrode.

This electrode system is not to be confused with the rotating Pt disk electrode at which controlled hydrodynamic conditions may be maintained.   An excellent review of the construction, properties, and results of studies of the rotating disk has been written by Riddiford (208).

The rotating Pt-wire electrode may take the form of a small length (3–4 mm) of Pt wire (0.5 mm in diameter) sealed in glass tubing which is rotated by an electric motor.   Alignment of the shaft of the electrode holder in the chuck of the driving motor is desirable so that a minimum of wobble is produced.   Another form of the rotating wire electrode is the one in which a small Pt wire is welded perpendicular to the axis of rotation of an iron or steel rod which is mounted in the chuck of the driving motor as described by Laitinen and Kolthoff (161).   The steel rod is insulated from contact with solution.

Nernst and Merriam (191) were the first to describe the I–E curves which were obtained at a rotating Pt wire microelectrode.   Later, Fresenius (74) also reported observing diffusion currents at a rotating Pt-wire electrode.   The diffusion current observed at rotating electrodes is many times larger than those recorded at stationary electrodes of the same surface area.   By rotating the electrode at high enough speed, Hammett and co-workers (212) have shown that a point is reached where further increases in the speed of rotation do not affect the recorded current.   Under these conditions the electrode kinetics are no longer controlled by diffusion, and the laws of polarography do not apply.   Rotation speeds between 600 and 2000 rpm seem to be the most commonly used.

The I–E curves for the reduction of oxygen at a rotating Pt wire microelectrode were described by Laitinen and Kolthoff (161).   Curve I in Fig. 6.4 shows the first wave for the reduction of O_2 at a rotating Pt wire in $0.1N$ KCl and the greatly increased diffusion current produced by rotation of the electrode.   For electrodes of the same area, the diffusion current is increased from 3 μA at the stationary electrode to 61 μA at the rotating electrode.   Curves II and III show the results of reducing the O_2 content of the solution first by N_2 bubbling (curve II) and finally by removal of O_2 with Na_2SO_3 (curve III).   These data show that the large diffusion current is truly an O_2 diffusion current.   As in the stationary case, only the reduction of O_2 to H_2O_2 takes place before the hydrogen diffusion current occurs.

Polarograms obtained at solid electrodes in supporting electrolytes free of oxidizable or reducible materials frequently exhibit a small current wave in a certain region between the hydrogen reduction wave and the oxygen evolution wave (145,153,211).   When the polarogram was scanned from anodic evolution of O_2 to cathodic reduction of H^+ ion, a small current peak was observed, and the potential at which it appeared depended on pH, as seen in Fig. 6.5.   If, however, the polarogram was scanned in the reverse direction (negative to positive potentials), the peak did not appear.   The peak also appeared if the electrode was treated with oxidizing agents (Ce^{4+}, MnO_4^-) instead of preanodization before the polarogram was taken.   This small prewave is a reduction current produced by the reduction of the adsorbed oxygen film produced by anodization or the action of strong oxidizing agents.

The composition of this film is considered to be PtO by Kolthoff and Tanaka (153), but it is more likely to be a film of adsorbed oxygen atoms

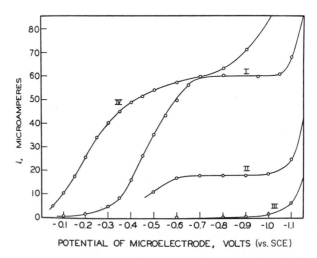

Fig. 6.4. *I–E* curves for the reduction of O_2 (161) at a rotating Pt-wire microelectrode: I in air-saturated $0.1N$ KCl solutions; II after bubbling N_2 through the solution for 10 min; III after removing dissolved O_2 with Na_2SO_3; IV in air saturated $0.1N$ KCl at a Ag-plated Pt-wire. (By permission of the American Chemical Society.)

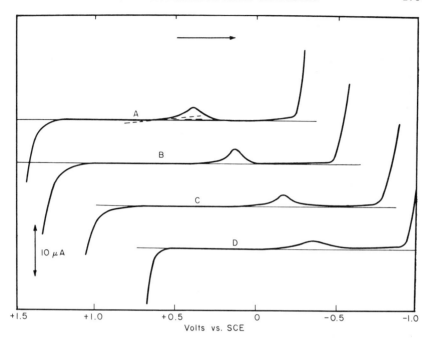

Fig. 6.5. $I–E$ curves for the reduction of O_2 (153) at Pt-wire microelectrodes measured from positive to negative potentials in O_2-free solutions: A, from 1.5 V in $0.1M$ $HClO_4$; B, from 1.35 V in acetate buffer, pH 5; C, from 1.1 V in sodium borate, pH 9; D, from 0.7 V in $0.1M$ NaOH. (By permission of the American Chemical Society.)

or dissolved oxygen which has diffused to the surface since it is unlikely (119) that a thin film of PtO_2 exists in the presence of Pt metal below 1.5 V (vs. NHE). Kolthoff and Tanaka conclude that the film of adsorbed oxygen accounts for the poorly defined anodic diffusion currents in automatic $I–E$ curves on rotating Pt wire electrodes. When these same measurements were carried out manually (153), well-defined and reproducible diffusion currents were obtained. To avoid irregular and unexpected waves, they stress the point that time must be allowed for the formation of the oxygen film in anodic studies and for the reduction of the film in cathodic studies. Since these abnormal residual currents fall rapidly with time to the normal value at a rotating electrode, steady-state reproducible data may be taken in about 1 min. Davis (54) warns that corrections to the observed currents for the formation or reduction of the adsorbed oxygen films, particularly in anodic studies, must be made to obtain meaningful results if automatic measurements are made.

c. **The Rotated Gold Electrode.**   Because Pt has such a low hydrogen overvoltage, the useful range of this electrode in reduction studies is limited because of the interference of H_2 evolution.   It is desirable to extend the range of usefulness of a solid electrode by finding another metal with higher overvoltage values than Pt.   Kolthoff and Jordan (146) carried out studies on electrodes made of Au, Ag, and amalgams of Au and Ag, and Lord and Rogers (173) investigated electrodes of graphite, Au, and PbO_2.

Graphite was found (173) to extend the useful range of the indicator electrode because of its high overvoltage for H_2 and O_2 evolution, but high resistance films were formed on C anodes which had to be removed before another polarogram could be taken.   At stationary graphite electrodes the data were less reproducible than on Pt or Au, and the rotated graphite electrodes gave poor results.   A detailed account of carbon indicator electrodes is given in Chapter VIII.   The amalgam and the Ag electrodes (146) gave results which were poorly reproducible. Measurements obtained with PbO_2 (173) were disappointing since the range of usefulness of the PbO_2 electrode was too low.

The gold electrode proved (146,173) to be a well-behaved indicator electrode with a much larger useful range than the Pt electrode.   Not only is the H_2 overvoltage larger on Au (about 0.4 V greater than on Pt at a given current density), but also the O_2 overvoltage is greater than on Pt (146).   The Au microelectrode is useful for determining O_2 diffusion currents in the pH range from 4 to 13.   In fact, a very poorly developed second wave for the reduction of H_2O_2 is found in $0.1M$ NaOH, as plotted in Fig. 6.6.   In $0.1M$ NaOH solution the oxygen diffusion current must be measured in the narrow range of -0.85 to -0.90 V (vs. SCE) at a rotating Pt electrode, but this range is increased at a rotating Au electrode from -0.8 V to -1.25 V (vs. SCE).

Another advantage in the use of a Au indicator electrode in preference to a Pt one is the fact that gold does not dissolve oxygen and only weakly adsorbs oxygen on the surface from solution (203).   Oxide formation does not begin until a potential more noble than 1.3 V is reached (14). Consequently, preanodization does not enhance the reversible behavior beyond the removal of adsorbed impurities from the surface, but on the other hand, aging effects similar to those found on Pt electrodes are not found on Au (193).   Many investigators prefer Au electrodes because the complications of dissolved oxygen and adsorbed oxides are not present.   To obtain reproducible residual currents, Kolthoff and Jordan (146) note that the Au microelectrode must be washed in $10M$ HNO_3 followed by precathodization at -2.0 V (vs. SCE) for 10 min.

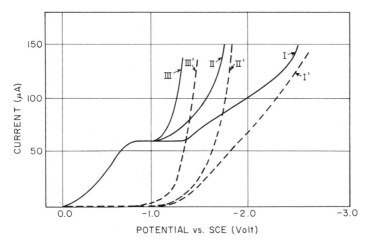

Fig. 6.6. *I–E* curves for the reduction of O_2 solutions (146) $2.62 \times 10^{-4}M$ oxygen at a rotating Au wire microelectrode: I $0.1M$ NaOH; II phosphate buffer, pH 7; III acetate buffer, pH 4; I′, II′, III′, corresponding residual current curves of O_2-free supporting electrolytes. (By permission of the American Chemical Society.)

A method of determining trace quantities of O_2 (less than $10^{-6}M$) based on the exaltation of the limiting current of O_2 by H_2O_2 at a rotating Au microelectrode provided the concentration of H_2O_2 is at least 200 times that of O_2 has been demonstrated by Kolthoff and Jordan (147). However, if the H_2O_2 concentration is greater than $10^{-3}M$, the H_2O_2 yields an *I–E* curve which interferes with the oxygen wave. The limiting current is measured at -0.8 V (vs. SCE), and correction for the residual current is obtained from the measurement in O_2-free supporting electrolyte. The oxygen-diffusion currents in air-saturated supporting electrolyte, I_d^{sat}, and in the supporting electrolyte saturated with the unknown gas, I_d^g, are determined. Afterward the required amount of H_2O_2 is added to the gas-saturated electrolyte, and the limiting current of the exalted O_2 wave, I_l, is recorded. The exaltation, J is defined as I_l/I_d^g. Assuming that air contains 20.7% O_2, the oxygen content of the sample in volume per cent, X, is given by the relationship

$$X = 20.7 I_l/J I_d^{sat} \tag{6.28}$$

Reported values of J (47) range from 90 in solutions of pH 4–30 in $1M$ NaOH. Such determinations must be made in a thermostat since the exaltation increases with temperature (146). There is a linear relationship between the exalted limiting-current density and oxygen concentration determined at a rotating gold microelectrode in solutions ($0.1M$

NaOH and $10^{-3}M$ H_2O_2) from 1×10^{-7} to $5 \times 10^{-6}M$ in O_2. It is also noted that I_l is proportional to the H_2O_2 concentration at Au electrodes.

In a private communication, Professor Kolthoff informs me that later workers in his laboratory have not been able to confirm the work by Kolthoff and Jordan. No exaltation has been found either on Pt electrodes that were treated in various ways or on Au electrodes.

d. Membrane-Covered Electrodes. One of the first investigations reporting the use of a Pt microelectrode in studies of the oxygen content of biological tissue was made by Davies and Brink (53). When the Pt cathode was constructed by sealing the Pt wire in a soft glass tube and by grinding the tip of the glass flush with the Pt metal, the observed diffusion was dependent on the relative motion of the fluid to be tested. To minimize the effects of solution flow and convection currents, Davies and Brink used a recessed electrode which was made by sealing a Pt wire in a soft glass tube which extended a few millimeters beyond the tip of the Pt wire. By using a microflame or a heated wire carefully, a perfectly cylindrical recess was formed, and electrodes as small as 25 μ were reported to have been made.

Such recessed electrodes obey the laws of linear diffusion, and consequently, are free from the mass transfer effects in the system to be analyzed. Continuous determination of dissolved oxygen could be made, but it required from 5 to 20 min to reach a steady-state value of the diffusion current. This electrode could not follow rapid changes in O_2 content of a given system, and in such cases, the flush electrode must be used. Whether the flush or recessed Pt microelectrode was used, the sensitivity of the indicator electrode fell slowly with time. This aging effect appears to be caused by the presence of surface active agents in the biological fluids which poison the catalytic surface of the Pt electrode for the reduction of O_2.

One of the solutions to the loss of sensitivity or poisoning of the indicator electrode by the adsorption of catalytic poisons present in the test solution was obtained by Olsen et al. (195), who applied a 5- to 10-cycle square wave to the solid electrode. The diffusion current was recorded on the negative half-cycle, and during the positive half-cycle the adsorbed impurities were removed.

To minimize the aging effect, Davies and Brink (53) covered the end of the recessed electrode with a membrane made of collodion which was permeable to the dissolved O_2 in solution but prevented the larger surface active molecules, such as protein molecules, from reaching the Pt surface. The recess was filled with water.

According to Mancy and co-workers (179), Kamienski (140) was one of the first to attempt to isolate the surface of a Pt-indicator electrode by a semipermeable membrane. He covered his electrode with a protective film of silica gel. Bowers and co-workers (27a) have developed the mathematics governing the reduction of cations such as Tl^+ and Ni^{2+} at a membrane-covered Hg-indicator electrode. They showed that the experimental results agreed with the findings calculated from a model in which the rate of the electrode process was determined by diffusion of ions through a cellophane membrane across which a linear concentration gradient was assumed.

Clark (47) described a Pt microelectrode covered with cellophane to study the oxygen content or oxygen tension, P_{O_2}, of blood. With this type of indicator electrode, which bears his name in the biological literature, the protective membrane of cellophane prevented the presence of the red blood cells from interfering with the measurements. The diffusion current was independent of the rate of stirring of the blood sample. The most notable improvement in the Clark type of electrode was made by enclosing both the indicator electrode and the reference electrode by a semipermeable membrane (48) in an electrode housing filled with a supporting electrolyte (usually $1M$ KCl solution). Since O_2, which diffuses through the membrane and dissolves in the supporting electrolyte is reduced at the Pt cathode in a system entirely isolated from the system to be analyzed, dissolved O_2 in a nonconducting medium may be investigated (97,138,222). With this electrode system, the O_2 content of a gaseous sample may be determined directly by bringing the sample of gas in contact with the membrane without the intervening process of saturating a given supporting electrolyte with the gaseous sample when using a bare wire electrode or a DME.

A study of the various types of membrane one may use in constructing a Clark type of electrode was carried out by Sawyer and co-workers (222). They covered a Pt disk (diameter of $\frac{3}{16}$ in.) imbedded in a glass rod through which a sealed Pt wire acted as an electrical contact with films of Saran, Mylar, Teflon, polyethylene, natural rubber, silicone rubber, and polyvinyl chloride. Polyethylene films gave the best performance, but in highly corrosive media for high-temperature applications, Teflon has the advantage. Since Teflon is about 2.5 times more permeable to oxygen than polyethylene, the use of Teflon membranes lowers the limits of detection of dissolved oxygen in the test sample.

Since the response time of this electrode system depends on the length of the diffusion path and the thickness of the membrane, the solution layer between the membrane and the cathode surface should be kept to a minimum.   In line with these requirements, the membrane is stretched tightly against the surface of the cathode and held in place by an O-ring or plastic ring (97,138,222).   As noted by Sawyer et al. (222), the sensitivity is inversely proportional to the membrane thickness, which is limited by the physical strength of the plastic film.   It appears that film thicknesses between 1 and 0.5 mil are in common usage.   The mathematics for the analysis of the diffusion properties of a system in which the diffusion path is comprised of two different media has been discussed (32,179).

Sawyer and co-workers (222) used a SCE for a reference electrode, but Halpert and Foley (97) used a Ag/AgCl reference electrode, which took the form of a silver wire coated with AgCl.   The AgCl film was formed anodically in a KCl solution.   The Ag wire was wound as a helix around the glass or plastic rod used as a holder for the Pt cathode.   Such a system can be built in a very compact form.   Oxygen diffusion currents are usually determined by holding the cathode at -0.6 to -0.8 V (vs. SCE).   Before the membrane-covered electrode can be used for O_2 determinations, the internal supporting electrolyte must be purged of any dissolved oxygen which may be accomplished by saturating the solution with an inert gas or by cathodizing the Pt electrode until zero current is recorded with the membrane in contact with an O_2-free medium. Usually the electrode is calibrated against a sample containing a known amount of O_2, such as an air sample.

Gold may be substituted for Pt as the cathode and in some respects it gives a better performance.   As noted before, the potential range available for observing O_2 diffusion currents is larger on Au, and interference by the presence of adsorbed oxygen films is greatly reduced.

Unfortunately, the membrane-covered electrodes have a large dependency on temperature being about $2\%/°C$ (37,43) which requires these measurements to be made under thermostatic conditions.   This, however, over long periods of time imposes an impossible limitation to the automatic, continuous recording of dissolved O_2 in natural settings, such as lakes and rivers.   By using thermistors (devices which change in resistance with changes in temperature) in series with the load circuit, Carrit and Kanwisher (37) were able to construct a cell composed of a Pt cathode and a Ag/AgO_2 reference electrode enclosed in a solution of KOH behind a polyethylene membrane with which temperature-independent determinations of dissolved O_2 in deep sea water were made. Briggs and Viney (32) demonstrated that a bridge circuit of thermistors makes a more reliable temperature-compensating device.

There is a need for a rugged O_2 detector which could be sterilized for biological studies. Pittman (200) has described a galvanic detector suitable for such studies. A Pt-wire cathode is sealed in a glass capillary rod at the top of which a helix of Ag wire is wound as the anode. This assembly is mounted in a glass tube which has a hole in its side near the bottom end. Enough wire extends beyond the end of the glass capillary to thread through the hole in the glass tube and to wind five complete turns on the outside of the tube, which is imbedded in a cement except for the strands directly across the hole. This Pt-wire helix is covered with a silicone rubber tube as the porous membrane. Because solutions of KCl interacted with the silicone rubber, the cell was filled with $0.5M$ NH_4Cl. Oxygen was removed from the NH_4Cl solution by polarizing the cell until zero current was recorded, and the system was calibrated with air. This detector can be used for gaseous and liquid samples. Gold cathodes were also used, and the diffusion current was determined at 0.5 V.

Under impure conditions (long-term use in bacterial cultures, for example), the electrode becomes poisoned and the sensitivity reduced. By superimposing an ac polarization of about 1 V on the direct current with a 60-cycle transformer connected in series with the dc circuit similar to the method used by Olson (195), the adsorbed impurities may be desorbed on the positive half-cycle. However, rectification currents are added to the diffusion currents by the reduction of O_2 on the negative half-cycle in clean systems and by the oxidation of impurities on the positive half-cycle of impure systems. To get useful results, superimpose the alternating current on the dc polarization from the beginning of the experiment so that the electrode never becomes poisoned, and then remove the alternating current during the time when actual measurements are taken. It is reported (200) that about 2 min was required to reach the steady state.

A similar detector which uses a Au cathode and a Ag anode in $0.1N$ KCl solution mounted behind Teflon, polypropylene, and polyethylene membranes and has good stability and response characteristics has been reported (143). Beckman Instruments have developed commercially (4) a flexible microelectrode which fits in the tip of a catheter and is a Clark-type polarographic detector. This electrode may be worked into the heart chamber where it may remain to record continuous readings of dissolved oxygen in biological fluids. The Beckman Oxygen Analyzer, Model 77700, commercially available, consists of a Au cathode and Ag anode mounted in an epoxy casting and connected electrically by a KCl gel. A Teflon membrane is stretched tightly over the Au cathode, and the diffusion current is determined at 0.8 V (vs. SCE). This O_2 detector may be used for gaseous or liquid samples.

For studies at low temperatures, Halpert and co-workers (98) replaced the KCl solution in a Teflon-covered Pt-Ag/AgCl O_2 detector with a H_2O–CH_3OH–KCl mixture as the electrolyte and increased the effective area of the Pt cathode.

3. Reduction of O_2 in Nonaqueous Solutions

A review of polarographic techniques in nonaqueous solutions has been presented by Gutman and Schöber (95a). One of the important concerns with such systems is the conductivity of the electrolyte. The polarographic reduction of a number of systems including O_2 in acetonitrile (with a dielectric constant of about 38) was studied by Coetzee and Kolthoff (49a). They chose $KClO_4$ or KCNS as supporting electrolytes.

Two waves were obtained for the reduction of O_2 at a DME, but the half-wave potentials were shifted to much more negative values than in the case with water solutions (from -0.05 V vs SCE in H_2O to -0.75 V in acetonitrile for the first wave). They (49a) concluded that the first wave was a two-electron process. Kolthoff and his co-workers have reported the results of the reduction of O_2 at a DME in dimethylsulfoxide (DMSO) (152a) with a dielectric constant of about 46 (35a,125a) and in N-methyl-acetamide (NMA) (143a) with a dielectric constant of about 165 (54a). In these cases, two waves of unequal height were observed, and the half-wave potentials were shifted to more negative potentials as in the case of acetonitrile. The heights of the two waves were in a ratio of about 3 : 2. It may be mentioned that repurification of these materials is required. Sufficient data were not available to interpret their observations (143a,152a).

Maricle and Hodgson (181a) concluded that a two-electron process was unlikely in a solvent with such a low availability of protons since the reduction of O_2 consumes protons. A more likely process is a one-electron reduction producing the superoxide ion O_2^-. The structure and paramagnetic analysis of the O_2^- ion has been given by Bennett et al. (16a), and although the only superoxides known are the superoxides of the alkali or alkaline earth metals, McElroy and Hashman (174a) have prepared tetramethylammonium superoxide (pale yellow solid) in liquid ammonia solutions. Russell (216a) has also postulated the existence of the O_2^- ion in the reactions of O_2 with organic anions in DMSO.

Potential sweep studies on Pt electrodes and polarographic reduction of O_2 at a DME in dimethylformamide (DMFA) and DMSO with tetrabutylammonium perchlorate as a supporting electrolyte were carried out by Maricle and Hodgson (181a). Oxygen is more soluble in DMFA than in DMSO. The SCE reference electrode was separated from the cell by a double salt bridge to prevent contamination of the electrolyte with moisture from the SCE. Two waves of unequal height (in ratio of 3 : 2) were obtained for the reduction of O_2 and the half-wave potential of the first wave was -0.73 V. They interpreted their data in terms of a one-electron process, the formation of the O_2^- ion. Electron spin resonance (ESR) measurements did not permit the conclusive identification of the paramagnetic species present but did give support to the conclusion that the O_2^- is involved.

The two-electron change reported by Coetzee and Kolthoff (49a) was explained (181a) by the presence in the acetonitrile of H_2O which had diffused across the salt bridge from the SCE to the electrolyte. Water interacts with the O_2^- ion to produce O_2, HO_2^- ions, and OH^- ions. After O_2 had been reduced at a Pt electrode in tetrabutylammonium perchlorate-supported DMSO, H_2O was added (181a), and vigorous evolution of O_2 occurred. When a source of protons, phenol, was added to the solution, a two-electron wave was obtained at the DME.

The reduction of O_2 was studied by Peover and Smith (198a) in pyridine, acetonitrile, acetone, DMFA, DMSO, and methylene chloride containing tetraalkylammonium perchlorates as supporting electrolytes. From ESR spectrum analysis, it is deduced that the O_2^- ions has a half-life of at least several minutes (up to $\frac{1}{2}$ hr in pyridine) in these solvents.

Although the difference in heights of the two polarographic waves for the reduction of O_2 at a DME in aprotic solutions such as DMSO is not explained, it appears that the characteristic process in such a system is the reduction of O_2 to the O_2^- ion by a one-electron reaction. The O_2^- ion can then react with the supporting electrolyte to produce superoxides. If proton donors are present, a complex with the donor is made (198a) before the electron transfer takes place, resulting in the formation of peroxides.

C. GALVANIC OXYGEN DETECTORS

For many applications such as the continuous determination of dissolved oxygen in lakes and rivers (37), polarographic determination with a membrane-coated cathode requires the use of an external source of current for the controlled potential polarization of the indicator cathode. Because the cathode and reference electrode are separated by a membrane, specially designed, expensive measuring equipment is needed. This high impedance problem may be reduced by enclosing the cathode and reference electrode in the same cell envelope behind the membrane, but the external source of current is still demanded which is not convenient for the continuous recording of dissolved O_2 in natural waters in the field.

By using a galvanic cell in which the electrochemical reaction at an inert cathode is the reduction of O_2 and the reaction at a suitable anode involves the consumption of OH^- ions, an alkaline electrolyte such as KOH would remain invariant, and the cell could be designed so that the current delivered by the cell would be proportional to the concentration of O_2 in the medium in contact with the cathode. Several galvanic systems have been used as oxygen detectors.

1. Fuel Cells

Because a discussion of O_2-diffusion electrodes and fuel cells is presented in Chap. VIII, a detailed description of these electrodes will not be given here. Suffice it to say that fuel cells employ diffusion electrodes which are composed of a porous material (graphite, nickel plaque, etc.) The oxidant and reductant are supplied to the back side of the porous electrodes mounted so that the electrolyte separates the anode and cathode. Usually a catalyst is applied to the solution side of the diffusion electrode. The reactant (in gaseous or vapor form) diffuses through the pores of the electrode and dissolves in the electrolyte on the front or solution side where the electrochemical process takes place on the catalyst surface.

One of the first applications of a fuel cell to the detection of oxygen was disclosed by Paris (197) who used a Zn–air cell. The oxygen content of a gas sample was determined by replacing the air stream to a porous carbon cathode by the gas sample. Oxygen is reduced to OH^- ions at the cathode, and Zn^{2+} ions produced at the anode form a complex hydroxide with the electrolyte (20% NH_4Cl). The current developed is proportional to the concentration of O_2 in the gas sample. Moiseev and Brikman (183) used an electrolyte of NH_4Cl paste (moistened NH_4Cl powder) to separate the porous carbon cathode from a Zn anode and noted that the cell had a high temperature coefficient. The cell current was independent of the gas flow over a wide range, but response of the device to changes in P_{O_2} was about 1 min. To make this measuring system more efficient, Kordesch and Marko (154) applied metal catalysts to the porous carbon cathode. By using an acidic electrolyte, Jacobson (132) described a Zn–air cell which could tolerate large amounts of CO_2 in the gas sample.

In a recent news release from NASA (5), a low-temperature H_2–O_2 fuel cell is described as an oxygen detector. A sulfuric acid electrolyte is contained in a chamber made of a special glass membrane. Porous Pt disks are pressed against each side of the glass chamber and H_2 is fed to the anode. When an oxygen-containing gas is fed to the cathode, a current proportional to the O_2 concentration of the gas flows through the external circuit. Although a linear dependence was not found, the system could be calibrated with O_2 samples of known concentration.

A more rugged and convenient O_2 detector may be made by replacing the liquid electrolyte with an ion-exchange membrane (248). A very sensitive O_2 detector based on a high-temperature fuel cell with a solid electrolyte has been developed at Westinghouse Electric Corp. (112). The electrolyte takes the form of a ceramic tube (8 in. long, 0.5 in. diameter) which is made of zirconium oxide stabilized with calcium oxide and becomes conducting at elevated temperatures ($\sim$1000°C). Electrodes are formed by depositing a metal film (probably Pt or Ag) on the surfaces of the tube. With the external surface of the tube sealed to the atmosphere, the test gas is passed through the tube. When oxygen is present in the gas, the potential is found to depend logarithmically on the P_{O_2} in the gas. The device may be used in a vacuum or at pressures above atmospheric without any interference. It is to be noted that the cell acts as an oxygen concentration cell and consequently is ultra-sensitive to changes in P_{O_2}. Similar cells using ZrO_2–CaO solid electrolytes with Pt electrodes coated with FeO, NiO, or Cu_2O and operating between 800 and 1200°C have been reported (91) to detect the P_{O_2} in CO–CO_2 and argon–oxygen mixtures from 10^{-1} to 10^{-20} atm. Conduction occurs in these electrolytes by oxide ions through lattice vacancies.

As a matter of fact, any fuel cell could be adapted as an O_2 detector. If a fuel such as H_2 or Zn is supplied at the anode, the current developed by the cell is a function of the P_{O_2} in the gas sample fed to the cathode. On the other hand, if the fuel is not supplied to the anode and it is isolated from the external environment, the potential of the cell acting as an O_2 concentration cell is a function of the P_{O_2} of the gas sample. These devices must be calibrated before use.

2. Membrane-Covered Galvanic Devices

Hersch (103) pointed out that although oxygen could be detected with galvanic cells (154,183,197), by the proper choice of materials and by proper design, a much more sensitive and convenient O_2 detector can be made. He showed that a cell composed of a Pt-wire spiral cathode placed in a 24% solution of KOH with a lead amalgam anode could be used as an efficient detector of dissolved O_2 in gas samples which were bubbled through the cell over the Pt cathode. When the Pt cathode was only partially submerged, the current produced by the cell for a given gas flow was increased. This phenomenon will be discussed in Chap. VIII.

Later, Hersch (104) designed a cell in which he replaced the Pt cathode with Ag. In this cell, Pb foil was wrapped around a steel tube, and over the lead was wrapped a Ag gauze separated by a sheet of porous polyvinyl chloride (PVC) film. The entire assembly was held in a glass tube, and the PVC film was saturated with 24% KOH solution ($\sim5M$). The gaseous test sample passed down the center of the steel tube and returned up the glass tube over the Ag gauze. Oxygen is reduced to OH^- ions at the cathode, and Pb ions produced at the anode react with the OH^- ions to form plumbate ions,

$$Pb + 3OH^- \rightarrow PbO_2H^- + H_2O + 2e \qquad (6.29)$$

The current is reported to be linear with O_2 concentrations below 0.01%, and the response to changes in P_{O_2} is 30 sec.

This cell is known as the Hersch cell. Baker et al. (11) describe a Hersch cell in which the KOH solution is held in No. 50 filter paper and kept from drying out by wicking in a reservoir of 24% KOH solution. Recently, Bozzan and Bordonali (28) reported that oxygen concentrations of 2–100 ppm gave linear currents using 25% KOH solution.

The Hersch cell may be used to detect O_2 only in gaseous samples. To analyze other samples, such galvanic detectors must be isolated from the sample medium by a semipermeable membrane. Mancy (179) was the first to describe such an O_2 detector.

a. O_2 Detectors with Pb Anodes. For a compact, self-contained detector of dissolved oxygen in natural waters which is relatively maintenance-free, Mancy and co-workers (179) described a galvanic detector composed of Pb and Ag anodes in a KOH electrolyte. Because the behavior of a Pt cathode depends on the previous history of electrode preparation, they chose Ag as the cathode material. The Ag cathode which was a disk 0.6 cm in diameter and the Pb anode which was a ring placed around the Ag disk were embedded in polystyrene. The tip of the assembly was polished to expose the two electrodes. After a disk of lens paper was saturated with $1M$ KOH solution, it was pressed against the tip of the electrode assembly. A piece of polyethylene film was placed over the cell tip and held in place with a polyethylene ring. This type of cell is called the Mancy cell.

In the absence of oxygen, the current output of the cell is virtually zero, but in the presence of oxygen the current developed is proportional to the P_{O_2} of the O_2 present. With this detector, the sample to be analyzed may be in the gaseous or liquid form. If the liquid sample is anhydrous, there is some drying of the membrane, which has some permeability to water vapor, and the KOH solution must be replaced periodically. A 0–15 microammeter must be used, since the currents produced are small. The presence of CO_2 in the sample can cause carbonate formation, which lowers the sensitivity of the device. This detector has a fast response to changes in P_{O_2}; to reach 99% of the steady-state current, it requires only 8 sec. Although the cell has a high temperature coefficient which requires a calibration curve, this inconvenience can be removed by the use of temperature-compensating devices employing thermistors (32,37). The temperature effects for the galvanic devices are much larger than for the Clark-type cell being at least 5%/°C.

The mathematics for the diffusion currents in this double film system has been discussed (32,179). At constant temperature T, the steady-state current I is given by

$$I = nFAPc/b \qquad (6.30)$$

where A is the effective mambrane area, P the membrane permeability, b the membrane thickness, and c the concentration of O_2 in the sample. At constant c, I varies with T according to

$$I = Ke^{-J/T} \qquad (6.31)$$

where K and J are constants for a given system and are determined by the cell geometry and the nature of the membrane material, respectively. With proper design, the characteristics of the cell operation may be varied to accomodate a variety of experimental conditions.

A number of designs based on the Mancy cell have been reported in the literature (32,137,169,175,257). To lower the effects of interference by CO_2, saturated solutions of $KHCO_3$ have been used (32,175). A highly sensitive O_2 detector is described (169) in which the potential developed across a standard resistor in the load circuit by the cell current is opposed through a voltage divider by the potential of a 1.3-V Hg cell. By using this bucking voltage, a change in P_{O_2} of 7×10^{-5} μl of dissolved O_2 per milliliter of sample per minute may be detected; a 1-mV change in load voltage is equivalent to 0.066 μl/ml/min of dissolved O_2. Acetate buffers and Teflon membranes are used (137) and dissolved O_2 in organic liquids are measured (257).

b. O$_2$ Detectors with Cd Anodes.   Neville (136,193) found that Au gave more reproducible results than Pt for long-term usage, since the poisoning and aging of Pt was not found for Au cathodes.   For the anode material he tried Zn, Fe, and Cd.   Zinc was too reactive, iron had poor oxide characteristics, but cadmium fulfilled the requirements.   In a construction similar to the Mancy cell, a Au rod surrounded by a Cd ring was mounted in a nylon cell filled with KCl solution.   A 1-mil thick polyethylene membrane was stretched over the Au cathode.   The electrolyte remained invariant because for every OH$^-$ ion produced by the reduction of O$_2$ at the cathode an OH$^-$ ion was removed in the formation of hydroxides of cadmium at the anode.   With time it was noted that the nylon swelled with the absorption of water, but other construction materials can be used.   Apparently the oxide films produced on the anode surface had no effect on the operating characteristics of the detector.   Temperature compensation was achieved through the use of thermistors.   Continuous operation was reported for lifetimes of 3–6 months.

Another design (67) featured a Ag cathode composed of a porous ceramic tube packed with small Ag balls (0.028–0.033 in. in diameter). Around the outside of the ceramic tube was wound a helix of Cd wire to serve as the anode.   This electrode assembly is mounted in a borosilicate glass tube with Kel-F end pieces.   The region between the ceramic and glass tubes is filled with $2M$ KCl solution, the pH of which is maintained by injecting K$_2$CO$_3$ into the cell from an external reservoir.   Nitrogen gas is bubbled through the KCl solution. The test solution flows through the 30-cm long column packed with silver balls against gravity, and it is reported that most of the O$_2$ dissolved in the sample is removed in the first 5 cm of the column.   The current generated by the cell depends on the concentration of dissolved O$_2$ and on the rate of flow of the sample as given by the relationship

$$c(\text{mg/l}) = 4.97 I \ (\text{mA})/f \ (\text{ml/min}) \tag{6.32}$$

For samples rich in dissolved O$_2$, this detector gives the same results as others, but for small O$_2$ concentrations the results are too low (at 0.15 ppm, 5% error; at 0.005 ppm, 15% error) which is explained by the loss of dissolved O$_2$ through adsorption processes on glass surfaces, on grease films, and in the Kel-F end pieces.

D. APPLICATIONS OF O$_2$ ELECTRODES TO BIOLOGICAL SYSTEMS

Besides the short review of O$_2$ electrode applications to biological studies found in Kolthoff and Lingane (ref. 149, p. 555), a detailed account of the oxygen cathode has been written by Davies (52). An excellent description and discussion of reference electrodes and O_2 cathodes used in biological studies has also been presented by Cater and Silver (ref. 131, p. 465 *et seq.*)

1. Determination of O_2 in Natural Waters

Continuous determinations of dissolved O_2 in lakes, reservoirs, rivers, and streams are important to pollution control engineers since algae and water plant life add oxygen to the water through photosynthesis processes. Such measurements may be used to determine the rate of stagnation of lakes and ponds. In oceanographic investigations, data concerning the distribution of dissolved O_2 in both space and time are required as a basis for the understanding of the behavior and interaction of parts of large bodies of water. As pointed out by Carritt and Kanwisher (37), Worthington (256) evaluated the oxygen content data obtained in the North Atlantic Ocean over a period of 25 years and came to the conclusion that the deep water in this region had not exchanged water with the surface for about 140 years. Such information is important to those who suggest the use of deep oceans as a dumping ground for radioactive wastes.

Early determinations of the O_2 content of water had to be carried out by the lengthy and relatively clumsy Winkler titration method. The more convenient method of determining the O_2 diffusion current at a DME was used to measure the oxygen content of lake water (180,185, 237) and sea water (89,93). The current was recorded in most cases at -0.8 V (vs. SCE). To measure the O_2 content of rivers, Spoor (237) allowed the water to flow through the cell, but to get useful information the water flow and the head of Hg (about 60 cm) had to be kept constant. Blaedel and Todd (22) reported that diffusion currents could be reproduced within 1% for steady water flow and at high flow rates the current reached 99% of its steady-state value in 1 or 2 min.

The oxygen content of solutions (or gases) may be determined indirectly with a DME. Japanese investigators (239) have described a process in which the solution is passed through a Jones reductor and the Zn^{2+} ion liberated is determined polarographically at a DME at -1.5 V against the Hg pool.

It has been reported (37,71,72) that the diffusion current obtained at a streaming Hg electrode with a Zn reference electrode has been used to measure the dissolved O_2 in subsurface waters of Norwegian fjords.

A Pt-wire microelectrode makes a mechanically simpler indicator electrode than the DME. Such a system has been used to determine the dissolved oxygen concentration of lake and sea water (86,87,174).

For rapid determination of O_2 content of natural waters, a rotating Pt cathode may be used (86,174,196).

Metals other than Pt have been used. To improve the sensitivity of the indicator electrode, Husmann and Stracke (126) used high-over-voltage alloys containing Fe, Cr, Ni, Sn, Sb, and Pb; and Wilson and Smith (254) used a rotating amalgamated Ag wire. Eckfeldt (66) used a gold disk (3 in. in diameter and $\frac{1}{16}$ in. thick) with a labyrinth of channels cut in one face through which the solution flowed as the indicator electrode. The channel-containing face was pressed against a porous ceramic membrane 0.25 in. thick which separated the Au-indicator electrode from a chamber filled with $2N$ KCl solution. In the chamber was mounted a Ag/AgCl reference electrode. The diffusion current at -0.82 V (vs. SCE) depended on the rate of flow of water through the channels and may be used as a continuous determination of the O_2 content of the flowing water.

In all cases, with the use of the DME and wire electrodes, adsorption of impurities from solution poisoned the electrode surface, producing aging effects and loss of sensitivity. By using a membrane-covered wire electrode (the Clark electrode), some of these difficulties may be overcome. Carritt and Kanwisher (37) embedded a Pt electrode in an epoxy resin in a Ag tube. Over this electrode assembly was stretched a polyethylene sack containing the $0.5M$ KOH electrolyte. The Ag was anodized to produce a Ag/Ag_2O reference electrode. The presence of the polyethylene membrane protected the electrode surface from contamination by impurities in the system to be measured, and temperature effects were compensated for with thermistors. KOH solution was chosen over KCl as used by Solov'ev (236) because the reduction of O_2 at the Pt cathode produces OH^- ions, and the KOH electrolyte remains invariant with usage, whereas the KCl does not.

Because all of these polarographic methods require an external polarizing source, such devices are inconvenient to use for the continuous monitoring of dissolved oxygen in lakes and streams in remote locations. A self-contained system such as the Ag-Pb cell described by Mancy and co-workers (179) is the most desirable. By using thermistors to provide temperature compensation (32), this O_2 detector gives reliable measures of the vertical profile of the O_2 concentration in lakes and seas and in the O_2 content of lakes and rivers during changes of the seasons. With an electrolyte of saturated $KHCO_3$ (32,175), the device is made free of the interference of CO_2 present in the water sample. Such cells may be used to measure O_2 concentrations in lakes and rivers continuously from 2 to 6 months (32,175); but in certain waters, a bacterial film may form on the membrane, which lowers the sensitivity and which may have to be removed periodically by wiping the membrane.

2. Determination of O_2 in Sewage Wastes

A knowledge of the oxygen content of sewage wastes and activated sludges is highly important to the sewage disposal and sanitation engineer as an indication of the content and control of undesirable waste products in disposal tanks. It is very desirable to obtain a continuous record of the dissolved O_2 present in the sewage.

Ingols (129) reported that stationary and rotating Pt electrodes cannot be used for the detection of O_2 in sewage and sludges because the electrode surface becomes filmed with impurities, rendering them inoperative as O_2 detectors. Lynn and Okun (174) arrived at similar conclusions. Even covering the electrode with a membrane (129) or using a superimposed square wave similar to that used by Olson and co-workers (195) to remove impurities was not effective enough to give reliable data.

The advantage of the DME, continouus renewal of the electrode surface, is an attractive one for this application. Ingols (128) has described the use of a DME in the continuous detection of O_2 in sewage and activated sludges, but Moore and co-workers (185) noted that industrial wastes contain interfering substances. To use polarographic techniques for the determination of dissolved O_2 in sewage, NaCl must be added to increase the conductivity of sewage and activated sludges. It is interesting that maxima suppressors are not required in the detection of O_2 with a DME in raw sewage (185) since naturally occurring suppressors are always present.

The effects of the presence of surface active agents on the properties of the DME when used for the detection of O_2 have been investigated (178,225). A flowing electrolyte (0.01M KCl) was circulated through the cell containing a DME, and the I–E curves and electrocapillary curves for various concentrations of surface active agents were determined (178). The sodium salt of dioctyl sulfosuccinic acid (Aerosol OT) and alkyl benzene sulfonate (a synthetic detergent which according to McKinney (176) accounts for at least 50% of the surface active material found in sewage wastes) were used as anionic agents and alkylbenzyldimethylammonium chloride (Roccal) as cationic agents. A calibration curve of the dependence of the diffusion current on flow rate was prepared, and a specially designed cell was required to prevent distortion of the drop by the solution flow.

Since surface active agents can alter the energy relationships at the Hg–electrolyte interface (ref. 77, p. 65 *et seq.*), the adsorbed layer can modify the electrode kinetics (57). Anions are adsorbed on the positive branch of the electrocapillary curve and are almost completely desorbed at potentials more negative than -1.4 V (vs. SCE) (178). The presence of surface-active anions decreases the diffusion current and shifts the first wave to more negative potential values. However, at potentials more negative than -1.4 V, the curves for the cases in which anionic surface-active agents are present or absent coincide. Mancy and Okun (178) suggest that the adsorbed film inhibits the electron-transfer step.

When cationic surface-active agents are present, the first wave is unaffected, but the hydrogen discharge potential is shifted to more positive potentials. As a result, the second wave is masked by hydrogen currents. If nonionic surface-active agents are present, it is found (178) that the diffusion currents are lowered and both waves are shifted to more negative potentials.

Consequently, Mancy and Okun conclude that the polarographic determination of O_2 with a DME in solutions which contain anionic surface agents is valid only if the diffusion current is measured at large negative values which correspond to the current value of the second wave. Determinations obtained from currents corresponding to the first wave give reproducible results when cationic agents are present, but if nonionic surface agents are present, O_2 determinations with a DME are not feasible.

Since the surface-active agents present in most waste waters are of the anionic type, the diffusion current corresponding to the second oxygen wave should be used in detecting the oxygen content of sewage and waste waters. As cautioned by Mancy and Okun, for the determination of dissolved O_2 in sewage wastes of unknown composition, some precaution and experimentation should be taken to find the most reliable conditions for measuring the diffusion current. Recently, Malz and Bortlisz (177) found that the best results for the continuous detection of dissolved O_2 in sewage with a DME were obtained with short drop times.

3. Determination of O_2 in Blood

Since the blood circulatory system of a creature is the body's highway over which life-sustaining oxygen is transported to the cells and tissues, a knowledge of the distribution of O_2 in blood from various parts of the body is essential to many biological investigations. Samples of body fluids must be transferred from the body to the cell without permitting

any contact with the oxygen of the air.   This was done (16) by collecting
the sample under mineral oil equilibrated with O_2 at a pressure near that
estimated for the O_2 tension of the body fluid.   A DME and a SCE reference were used (16,253).   Wiesinger (253) noted that irreproducible
results were obtained with whole blood but those with the blood plasma
were satisfactory.

Oxygen determinations using a DME are inconvenient, and Hill (113)
has described a continuous method of recording the O_2 concentration in
body fluids with a Pt-wire microelectrode.   As noted by Clark (47), the
red blood cells of the blood interfere with the behavior of the Pt-wire
microelectrode, producing aging effects and loss of sensitivity.   In addition, the recorded current depended on the rate of stirring the blood
sample.   To improve the properties of the microelectrode, Clark and co-
workers (47,49) covered the Pt electrode with membranes of cellophane
and polyethylene.   This construction made the behavior of the Pt-
indicator electrode virtually independent of stirring effects and the
presence of the red blood cells.   A more compact unit was obtained
when both the indicator and reference electrodes were enclosed (48)
behind the polyethylene membrane.

Various modifications of the Clark cell have been described in the
literature (36,45,51,70,156,231,241).   Torres (241) evaluated the use
of a Clark electrode for measuring the O_2 tension of blood and concluded
that reliable results were obtained only if the measurement was made
as soon as possible after the sample was drawn.   He found the response
time was not linear with the P_{O_2}.   Qualitatively, the response time increased as the P_{O_2} increased.

Krog and Johansen (156) describe an electrode assembly based on the
Clark electrode (48) which may be used on the tip of a heart catheter to
measure the O_2 concentration inside the heart.   A Pt wire (0.1–0.05 mm
in diameter) is sealed in a glass capillary by melting the glass.   Around
this is slipped a Ag tube 1 mm in diameter and sealed to the glass with
a cement.   The tip of the assembly is ground smooth so that a disk of
filter paper soaked in physiological saline solution is in contact with the
Pt and Ag electrodes.   A Teflon membrane is stretched over this electrode assembly, holding the filter paper in place, and is secured by a
polyethylene O-ring.   The entire assembly is placed in the catheter so
that about 2.5 mm extends beyond the catheter tip.   They report that
the system is very stable with little aging by contact with body fluids.
Not only does the device have high mechanical strength, but also it may
be heat sterilized.   The device marketed by Beckman Instruments (4)
is based on this principle.   It is a flexible microelectrode of the Clark

type which may be fitted in the tip of a catheter.   It is reported that the catheter may be worked into the heart chamber, pulmonary artery, bone marrow, or urinary tract to remain in place and to deliver continuous readings of the concentration of dissolved O_2

When a membrane-covered Pt electrode of 1–2 mm diameter is used to record the O_2 tension of blood, the reduction of O_2 causes a depletion of O_2 in the vicinity of the Pt cathode, and vigorous stirring of the blood sample is required (231).   Such a procedure required a relatively large blood sample, and in fact a sample so large that duplicate determinations cannot be made in the case of an infant or a small animal.   If a small cathode about 40 μ in diameter is used, the diffusion current is low enough so that O_2 depletion in the neighborhood of the electrode is not experienced.   However, the current is so low, expensive measuring equipment is required.

Butler and co-workers (36) described a modification of the Clark electrode, which combines the advantages of both the large and small electrodes.   A spiral of 40-μ thick Pt ribbon was embedded in an epoxy resin along with a Ag wire to one side.   By grinding the end of this assembly properly, only the 40-μ edge of the Pt ribbon spiral was exposed. Over these exposed electrodes was placed a disk of porous film, such as filter paper, saturated with KCl solution which was held in place with a polyethylene film.   Anodization of the Ag wire produced a Ag/AgCl reference electrode.   With such an arrangement, the greater area and resulting greater current of a large electrode are sufficiently dispersed along the edge of the Pt ribbon to avoid regions of O_2 depletion without stirring.   Since stirring is not required with this electrode, blood samples of the order of 0.2 ml are sufficient for reliable determinations of dissolved O_2.

Dewey and Gray (63) describe a method in which the concentration of O_2 in biological fluids is measured with a Hersch cell.   After a carrier gas is bubbled through the sample, it is led to the Ag electrode of the Hersch cell.   The current developed by the cell is proportional to the O_2 content of the gas which, in turn, is proportional to the O_2 content of the sample.

4. Determination of O_2 in Cells and Tissues

Many biological investigations involving such studies as the behavior of tumors or the functioning of various parts of the brain require a knowledge of the concentration of O_2 or the oxygen tension of the biological tissue under consideration.   For experimental considerations, only a probe type of oxygen electrode is expedient for these studies.   The microelectrode is inserted into the tissue to be studied and may be guided into place with a modified hypodermic needle. (39). A reference electrode may be placed in any tissue. To obtain interference-free electrical measurements, the laboratory bench must be well insulated and the measuring instruments well grounded since the specimen to be examined must be isolated electrically.

Carter and Silver (ref. 131, p. 516) point out that differences in potential exist between different points on the skin of animals and humans and are called bioelectric potentials. Between injured and normal skin, a potential difference may also be recorded. In making polarographic determinations of the dissolved O_2 in various tissues, one may have to take into account, under certain circumstances, the current due to the presence of bioelectric potentials through measurements of the residual current. If this is not possible (studies *in vivo*), several electrodes may be placed (43) at various points on the skin to obtain an average reading, or the reference electrode may be kept as close to the indicator electrode as possible. In general, local anesthetics such as procaine (41) are used.

The first to report the use of Pt microelectrodes in studies of animal tissue were Davies and Brink (53), who described the use of two types of electrodes, the recessed and the flush. With a flush electrode they were able to detect the consumption of O_2 when muscle fiber contracts, which occurred over a period of time of about 0.1 sec. Poisoning effects produced by the adsorption of impurities from body fluids (240) may be reduced by the use of membrane-covered electrodes (43,44,62).

The use of microelectrodes for the study of the oxygen content of living tissue has been reported by many investigators (18,19,39–41,43,69,134, 184,243). When an animal is placed in an atmosphere box, it is possible to detect the changes in the O_2 tension of the skin as the concentration of O_2 in the gas which the animal breathes is varied (39,40). If the blood supply to a tissue is cut by pressure, the O_2 tension of the tissue falls (243), and the rate of fall may be used to estimate the consumption of O_2 by the tissue. In fact, Montgomery and Horwitz (184) report that measurements of the O_2 tension of the skin of a limb may be used to aid in the determination of the need of amputation in the case of arterial embolism. When an animal died, the O_2 tension of the skin tissue fell to zero (40).

Carter and co-workers (39–43) have made detailed studies of the oxygen tension in tumor tissue of animals and breast carcinoma of women. When the patient breathes medical O_2 in place of air, not much effect is noticed in the O_2 tension of a tumor which has outgrown the blood supply (42), but if the tumor becomes ulcerated, large effects are observed. It was also observed (42,43) that little effect in the O_2 tension of bone marrow is found when the animal breathes medical O_2.

The O_2 tension of the interstitial brain fluids of various animals has been studied (38,41,238) with membrane-covered microelectrodes as a function of the injection of various drugs in the animals' systems. When the blood supply to the brain is cut by drugs (41), the O_2 tension of the brain tissues goes to zero. The O_2 tension of other tissues, such as those of the mammary gland and testes of rats, were also studied in the presence of various drugs.

Metals other than Pt have been investigated as materials suitable for microelectrodes. Gold (40,43,69), Pd (40), Rh (40), In (69), Sb (69), 80–20 Pt-Ir alloy (40), and Au-plated Cu (18) have been examined as indicator electrode material and Fe (18) as reference electrode material instead of the usual Ag wire (Ag/AgCl reference). Rhodium has poor mechanical properties, and In and Sb are not suitable. Platinum and palladium interact with O_2, and although Pt gives a high current, Au reaches the steady-state value of current more quickly (40). Evans (69) reports that only Au and Pt are suitable for tissue measurements.

An important consideration in these studies, as pointed out by Jamieson and Van den Brenk (134), is the dimensions of the microelectrode. Usually, bare or collodion-covered Au or Pt wires about 200–300 μ in diameter are used, but physical damage to the tissue by their insertion can cause large errors in the value of the O_2 tension (133) since they are large compared to the cell dimensions. They (134) found that Au electrodes about 60 μ in diameter caused much less damage and gave more accurate determination of O_2 in the tissues. Results of O_2 studies on tissue from brain cortex, liver, kidney, and spleen of animals were compared for 60- and 330-μ Au-wire microelectrodes when the air breathed by the animals was varied from ambient pressure up to 5 atm. The 330-μ electrode caused macroscopic bruising of the tissue which, in turn, produced pooling of blood. As a result, the O_2 readings are much too high, and the effect of pressure is exaggerated. When an Au electrode 60 μ in diameter was used, macroscopic bruising was not observed, and more reliable O_2 tension data were obtained. They tried a 20-μ Pt wire, but such an electrode was not rigid and was too fragile to use.

5. Determination of O_2 in Photosynthesis Studies

A continuous determination of O_2 which is exchanged between plants and the surrounding media requires a short response time of the indicator electrode to follow the rapid changes in O_2 concentration when the system is illuminated.

Petering and Daniels (199) used a DME to measure the O_2 content of a suspension of algae in nutrient solution with illumination and in the dark. They standardized their solutions by using the Winkler titration method for dissolved O_2. However, Blinks and Skow (23) found that the response time of the DME is too slow because too much solution is required to accomodate the drops of Hg and the time lag between the beginning of illumination and the response time of the electrode is too long. Besides, current oscillations may obscure the rapid changes in O_2 concentration.

At first, Blinks and Skow used a stationary Hg pool to study the O_2 exchange of leaves of plants by placing them directly on the Hg surface. Because of the danger of poisoning the plant tissues with Hg at the Hg cathode surface, they replaced the Hg pool with a Pt-foil cathode. The leaves were pressed against the electrode surface and held in an agar gel. So that the current could be measured only from the front side of the electrode, the back of the Pt foil was coated with wax. A SCE was used as the reference electrode. If the electrode was in contact with the plant tissue, an instant response to illumination was detected as the plant cells evolved O_2.

Platinum-wire microelectrodes have been used to determine photosynthetic O_2 production in suspension of algae (23) and in leaves, stems, and roots of plants (189). In general, the results observed show that when the plant tissue is illuminated with light, a rapid increase in O_2 concentration is detected, which is produced by the O_2 evolved by the illuminated Pt cells. In the dark, the O_2 content of the system falls because O_2 is consumed by the respiration of algae or plant cells.

The photosynthetic production of O_2 by algae at low partial pressures of O_2 (10^{-4} to 10^{-1} torr) has been investigated by Whittingham (251). Whittingham and Brown (252) measured the photosynthetic evolution of O_2 by illuminated algae with a Hersch-type cell suspended in alkaline solution as a function of the length of flashes of light. With short flashes of light (less than 5 msec), no production of O_2 was detected, but with long flashes (35 msec), O_2 was produced, corresponding to about 1 mole of oxygen for 800 moles of chlorophyll. As an interesting note, a long flash preceded by a short flash yielded twice as much O_2 as the long flash alone. When the order of flashes was reversed, less O_2 was produced than before but was more than with the long flash alone. Even with a background of illumination, the yield of O_2 was increased with a short flash.

6. Determination of O_2 in Biological Cultures

As a measure of the behavior of biological cultures, a knowledge of the concentration of dissolved O_2 in the system is desirable. One of the best methods for determining the O_2 content of these systems is the polarographic method.

A number of reports of such investigations appear in the literature (12,94,124,125,163,171,218,219). Hospodka (125) and Le Petit (163) used a Pt cathode and Ag/Ag_2O reference electrode covered with a polyethylene membrane. The KOH solution was adsorbed in a disk of filter paper placed between the electrodes and the polyethylene film. This electrode responded to changes in O_2 concentration of yeast cultures (125) in 15 sec. A rotating Pt electrode in the form of a Pt bead melted at the end of a Pt wire attached to the rim of a rotating glass disk (600 rpm) was used by Grinyuk (94) and Somokhvalova and coworkers (218) in studies of the respiration of microorganisms such as growing *Penicillium* cultures. For the determination of O_2 content of of microbal cultures, Horn and Jacob (124) chose a rotating (1300) rpm Au electrode coated with polystyrene. Continuous monitoring of the dissolved O_2 in fermentation cultures was carried out by Balatti et al. (12) with amalgamated Ag electrodes covered with collodion. Longmuir (171) has reported the temperature coefficients of the respiration of bacteria.

7. Determination of O_2 in Other Media

The polarographic determination of dissolved O_2 in a variety of media has been carried out.

To evaluate the role of dissolved O_2 in reactions which cause a deterioration of flavor in canned orange juice, Lewis and McKenzie (165) elected to determine the dissolved O_2 in orange juice polarographically with a DME since the results of chemical methods were vitiated by the deep color of the juice. Neither a maximum nor a second O_2 wave was observed. It was suggested that this behavior may be traced to the presence of citric acid in the juice sample. The diffusion current was determined at -0.4 V (vs. SCE), and the residual current was small. After the system was calibrated with a manometric method, the diffusion current was found to be directly proportional to the O_2 content of the juice sample.

Hartman and Garrett (101) needed a rapid and accurate measurement of the dissolved O_2 in milk for studies of oxidation reactions in dairy products. They chose the polarographic method using a DME. To minimize losses of O_2 produced by bacteria, the measurements were made at 0°C by cooling the sample in ice water. Diffusion currents could be determined at -0.8 or -1.2 V (vs. SCE), and the polarograms exhibited no maxima. It is reported that the solid content of the milk did not interfere with the readings.

In studies of the exchange of oxygen between the blood and the urine of humans, Reeves and co-workers (206,207) used a polyethylene-covered Pt microelectrode as a cathode but did not make any compensation for changes in temperature. They observed that the O_2 tension is lower in urine than in the renal venous blood, and concluded that since the O_2 content of human urine is chiefly determined by a gaseous equilibrium with the uteral walls, these results cannot be used as a measure of the internal O_2 tension.

Measurements of the O_2 content of soils have been made (141,199) by allowing the sample of soil to equilibrate with an added amount of $0.1N$ KCl solution. The O_2 diffusion current was determined with a DME. Although the O_2 content varies with the type of soil, it is less than expected because of the respiration of microorganisms (141). Additions of nitrogen or organic materials increase the activity and growth of microorganisms which cause a decrease of the O_2 present in the soil. If the O_2 content is decreased sufficiently, root activity decreases and the plant is affected.

The O_2 dissolved in oil field water was detected by Garst and McSpadden (82) with Pt-disk cathode in an electrode housing made of Lucite-filled NH_4Cl solution buffered at pH 8.2. Contact with the sample was made through the polyethylene membrane covering the Pt disk. A carbon electrode served as the counterelectrode.

Beckmann (15) reported a method of analyzing the retort gases obtained from the carburetion of shale. The test cell was so designed that the gas bubbled through a $0.1M$ KCL solution. After saturation of the solution with the gas, the diffusion current was recorded with a DME. Although the O_2 content depends on the gas flow rate, it reached a constant value for a constant flow rate. Since the pressure of H_2S and NH_3 interferes with the readings, the gas was led through a purification train before entering the HCl saturator. The system was calibrated with air.

Because of the high concentration of soaps and detergents in laundry wastes, Rand and Heukelekian (202) reported that the $I–E$ curves obtained in laundry wastes were distorted as compared to the conventional curves. Cater (43) reported that Pt and Au microelectrodes were used to detect the concentration of O_2 in a large variety of studies, including those concerned with the O_2 content of butchered meat and semen.

References

1. Akopyan, A. U., *Zh. Fiz. Khim.*, **33**, 82 (1959).
2. Allmand, A. J., *Z. Elektrochem.*, **16**, 254 (1910).
3. Altman, S., and R. H. Busch, *Trans. Faraday Soc.*, **45**, 720 (1949).
4. Anon., *Ind. Res.*, **1965**, 73, Dec.
5. Anon., NASA Tech. Brief 65–10066, March, 1965.
6. Anson, F. C., *J. Am. Chem. Soc.*, **81**, 1554 (1959).
7. Anson, F. C., and D. M. King, *Anal. Chem.*, **34**, 362 (1962).
8. Anson, F. C., and J. J. Lingane, *J. Am. Chem. Soc.*, **79**, 1015, 4901 (1957).
9. Bagotskii, V. S., and I. E. Yablakova, *Zh. Fiz. Khim.*, **27**, 1663 (1953); *Dokl. Akad. Nauk SSSR*, **85**, 599 (1952).
10. Baker, B. B., and W. M. MacNevin, *J. Am. Chem. Soc.*, **75**, 1473, 1476 (1953).
11. Baker, W. J., et al., *Ind. Eng. Chem.*, **51**, 727 (1959).
12. Balatti, A. P., L. A. Mazza, and R. J. Ertola, *Ind. Quim. (Buenos Aires)*, **23**, 123 (1963); *Chem. Abstr.*, **61**, 7652 (1964).
13. Banta, M. C., and N. Hackerman, *J. Electrochem. Soc.*, **111**, 114 (1964).
14. Barnartt, S., *J. Electrochem. Soc.*, **106**, 722 (1959).
15. Beckmann, P., *Chem. Ind. (London)*, **1948**, 791.
16. Beecher, H. K., R. Follansbee, A. J. Murphy, and F. N. Craig, *J. Biol. Chem.*, **146**, 197 (1942).
16a. Bennett, J. E., et al., *Phil. Mag.*, **46**, 443 (1955).
17. Bennewitz, K., *Z. Phys. Chem. (Leipzig)*, **72**, 202 (1910).
18. Berezin, I. P., I. M. Epshtein, and L. A. Kashchevskaya, *Eksperim. Khirurg. i Anesteziol.*, **9**, 18 (1964); *Chem. Abstr.*, **61**, 10988 (1964).
19. Berezovs'kii, V. A., *Fiziol. Zh. Akad. Nauk Ukr. RSR*, **10**, 825 (1964); *Chem. Abstr.*, **62**, 10796 (1965).
20. Berzins, T., and P. Delahay, *J. Am. Chem. Soc.*, **75**, 4205 (1953).
21. Bewick, A., M. Fleischmann, and M. Liler, *Electrochim. Acta*, **1**, 83 (1959).
22. Blaedel, W. J., and J. W. Todd, *Anal. Chem.*, **30**, 1821 (1958).
23. Blinks, L. R., and R. K. Skow, *Proc. Natl. Acad. Sci. U. S.*, **24**, 420 (1938).
24. Böld, W., and M. Breiter, *Electrochim. Acta*, **5**, 145 (1961).
25. Bonnemay, M., *Z. Elektrochem.*, **59**, 798 (1955).
26. Borisova, T. I., and V. I. Veselovskii, *Zh. Fiz. Khim.*, **27**, 1195 (1953).
27. Bowden, F. P., *Proc. Roy. Soc. (London)*, **A125**, 446 (1929).
27a. Bowers, R. C., A. M. Wilson et al., *J. Am. Chem. Soc.*, **80**, 2968 (1958); **81**, 1840 (1959); *J. Phys. Chem.*, **65**, 672 (1961).
28. Bozzan, T., and C. Bordonali, *Comit. Nazl. Energia Nucl.*, RT/CHI-17 (1964); *Chem. Abstr.*, **62**, 8381 (1965).
29. Brdicka, R., K. Wiesner, and K. Schäferna, *Naturwiss.*, **31**, 390 (1943).
30. Breiter, M. W., *Electrochim. Acta*, **9**, 441 (1964).
31. Breiter, M. W., *Electrochim. Acta*, **11**, 905 (1966).
32. Briggs, R., and M. Viney, *J. Sci. Instr.*, **41**, 78 (1964).
33. Brochet, A., and J. Petit, *Z. Elektrochem.*, **11**, 448 (1905); *Ann. Chim. Phys.*, **5**, 328 (1905).
34. Brodd, R. J., and N. Hackerman, *J. Electrochem. Soc.*, **104**, 704 (1957).
35. Brönsted, J. N., *Z. Physik. Chem. (Leipzig)*, **65**, 84 (1909).
35a. Butler, J. N., *J. Electroanal. Chem.*, **14**, 89 (1967).
36. Butler, R. A., J. F. Nunn, and S. Askill, *Nature*, **196**, 781 (1962).

37. Carritt, D. E., and J. W. Kanwisher, *Anal. Chem.*, **31**, 5 (1959).
38. Cater, D. B., S. Garattini, F. Marina, and I. A. Silver, *Proc. Roy. Soc. (London)*, **B155**, 958 (1961).
39. Cater, D. B., and A. F. Phillips, *Nature*, **174**, 121 (1954).
40. Cater, D. B., A. F. Phillips, and I. A. Silver, *Proc. Roy. Soc. (London)*, **B146**, 289 (1957).
41. Cater, D. B., A. F. Phillips, and I. A. Silver, *Proc. Roy. Soc. (London)*, **B146**, 382, 400 (1957).
42. Cater, D. B., and I. A. Silver, *Acta Radiol.*, **53**, 233 (1960).
43. Cater, D. B., I. A. Silver, and G. M. Wilson, *Proc. Roy. Soc. (London)*, **B151**, 256 (1959).
44. Charlton, G., *J. Appl. Physiol.*, **16**, 729 (1961).
45. Charlton, G., D. Read, and J. Read, *J. Appl. Physiol.*, **18**, 1247 (1963).
46. Chow, M., *J. Am. Chem. Soc.*, **42**, 488 (1920.).
47. Clark, L. C., *J. Appl. Physiol.*, **6**, 189 (1953).
48. Clark, L. C. *Trans. Am. Soc. Artificial Internal Organs*, **2**, 41 (1956).
49. Clark, L. C., R. Wold, D. Granger, and Z. Taylor, *J. Appl. Physiol.*, **6**, 186 (1953).
49a. Coetzee, J. F., and I. M. Kolthoff, *J. Am. Chem. Soc.*, **79**, 870, 1852, 6110 (1957).
49b. Connolly, J. F., R. J. Flannery, and G. Aronowitz, *J. Electrochem. Soc.*, **113**, 377 (1966).
49c. Connolly, J. F., R. J. Flannery, and B. L. Meyers, *J. Electrochem. Soc.*, **114**, 241 (1967).
50. Cornelissen, R., and L. Gierst, *J. Electroanal. Chem.*, **3**, 219 (1962).
51. Daly, J. J., M. White, and J. Bamforth, *Clin. Sci.*, **24**, 413 (1963).
52. Davies, P. W., *Phys. Tech. Biol. Res.*, **4**, 137 (1962).
53. Davies, P. W., and F. Brink, *Rev. Sci. Instr.*, **13**, 524 (1942).
54. Davies, D. G., *J. Electroanal. Chem.*, **1**, 73 (1959); *Talanta*, **3**, 335 (1960).
54a. Dawson, L. R., P. G. Sears, and R. H. Graves, *J. Am. Chem. Soc.*, **77**, 1986 (1955).
55. de Bethune, A. J., and N. A. S. Loud, *Standard Aqueous Electrode Potentials and Temperature Coefficients*, Clifford A. Hampel, Skokie, Ill., 1964.
56. Delahay, P., *New Instrumental Methods in Electrochemistry*, Interscience, New York, 1954, p. 396.
57. Delahay, P., *Double Layer and Electrode Kinetics*, Interscience, New York, 1965, Chap. 11.
58. Delahay, P., *J. Electroanal. Chem.*, **10**, 1 (1965).
58a. Delahay, P., *J. Electrochem. Soc.*, **97**, 198, 205 (1950).
59. Delahay, P., and T. Berzins, *J. Am. Chem. Soc.*, **75**, 2486 (1953).
60. Delahay, P., and G. Mamantov, *Anal. Chem.*, **27**, 478, (1955).
61. Delahay, P., and C. C. Mattax, *J. Am. Chem. Soc.*, **76**, 874 (1954).
62. Del Monte, U., I. Bose, and G. Mascherpa, *Ric. Sci. Rend. Sez.*, **B2**, 238 (1962); *Chem. Abstr.* **60**, 2030 (1964).
63. Dewey, D. L., and L. H. Gray, *J. Polarog. Soc.*, **7**, 15 (1961).
64. Dietz, H., and H. Göhr, *Electrochim. Acta*, **8**, 343 (1963); *Z. Physik. Chem. (Leipzig)*, **223**, 113 (1963).
65. Donnan, F. G., and A. J. Allmand, *J. Chem. Soc.*, **99**, 845 (1911).
66. Eckfeldt, E. L., *Anal. Chem.*, **31**, 1453 (1959).

67. Eckfeldt, E. L., and E. W. Schaffer, *Anal. Chem.*, **36**, 2008 (1964).
69. El Wakkad, S. E. S., and T. M. Salem, *J. Phys. Chem.*, **56**, 621 (1952).
68a. El Wakkad, S. E. S., and T. M. Salem, *J. Phys. Colloid Chem.*, **54**, 1371 (1950).
69. Evans, N. T. S., and P. F. D. Naylor, *J. Polarog. Soc.*, **1960**, 40, 46.
70. Fabel, H., D. W. Luebbers, and B. Rybak, *Bull. Soc. Chem. Biol.*, **46**, 811 (1964).
71. Føyn, E., *Rept. Norwegian Fish Market Invest.*, **11**, 3 (1955).
72. Føyn, E., *Chem. Abstr.* **55**, 12715 (1961).
73. French, W. G., and T. Kuwana, *J. Phys. Chem.*, **68**, 1279 (1964).
74. Fresenius, L. R., *Z. Physik. Chem. (Leipzig)*, **80**, 481 (1912).
75. Fricke, R., and P. Ackermann, *Z. Anorg. Allgem. Chem.*, **211**, 233 (1933).
76. Fried, F., *Z. Physik. Chem. (Leipzig)*, **123**, 406 (1926).
77. Frumkin, A. N., in *Advances in Electrochemistry and Electrochemical Engineering*, Vol. 1, P. Delahay, Ed., Interscience, New York, 1961, p. 66.
78. Frumkin, A. N., and N. Aladjalova, *Acta Physicochim. URSS*, **14**, 1 (1944).
79. Frumkin, A. N., and B. Burns, *Acta Physicochim. URSS*, **1**, 232 (1934).
80. Furman, N. H., *J. Am. Chem. Soc.*, **44**, 2685 (1922).
81. Garrett, A. B., and A. E. Hirschler, *J. Am. Chem. Soc.*, **60**, 299 (1938).
82. Garst, A. W., and T. W. McSpadden, *Chem. Eng. News*, **1963**, 65, Apr. 8; *Chem. Abstr.*, **62**, 377 (1965).
83. Gerischer, R., and H. Gerischer, *Z. Physik. Chem. (Frankfurt)*, **6**, 178 (1956).
84. Gierst, L., and A. Juliard, *J. Phys. Chem.*, **57**, 701 (1953).
85. Giguere, P. A., and J. B. Jaillet, *Can. J. Res.*, **B26**, 767 (1948).
86. Giguere, P. A., and L. Lauzier, *Can. J. Res.*, **B23**, 76 (1945).
87. Giguere, P. A., and L. Lauzier, *Can. J. Res.*, **B23**, 223 (1945).
88. Gilman, S., *J. Electroanal. Chem.*, **9**, 276 (1965).
89. Glasstone, S., *Trans. Am. Electrochem. Soc.*, **59**, 277 (1931).
90. Glasstone, S., and G. D. Reynolds, *Trans. Faraday Soc.*, **29**, 399 (1933).
91. Goto, K., and G. R., St. Pierre, *Tetsu To Hagane*, **49**, 1760 (1963); *Chem. Abstr.*, **62**, 8381 (1965).
92. Grahame, D. C., *Chem. Rev.*, **41**, 441 (1947).
93. Grasshoff, K., *Kiel Meersforch.*, **91**, 8 (1963); *Chem. Abstr.*, **60**, 7792 (1964).
94. Grinyuk, T. I., *Lab. Delo*, **5**, 31 (1959); *Chem. Abstr.*, **54**, 10038 (1960).
95. Grube, G., and B. Dulk, *Z. Elektrochem.*, **24**, 237 (1918).
95a. Gutman, V., and G. Schöber, *Angew. Chem.*, **70**, 98 (1958).
96. Harper, F., and J. Weiss, *Naturwiss.*, **20**, 948 (1932); *Proc. Roy. Soc. (London)*, **A147**, 332 (1934).
97. Halpert, G., and R. T. Foley, *J. Electroanal. Chem.*, **6**, 426 (1963).
98. Halpert, G., A. C. Madsen, and R. T. Foley, *Rev. Sci. Instr.*, **35**, 950 (1964).
99. Hamamoto, E., *Collection Czech. Chem. Commun.*, **5**, 427 (1933).
100. Hamer, W. J., and D. N. Craig, *J. Electrochem. Soc.*, **104**, 206 (1957).
101. Hartman, G. H., and O. F. Garrett, *Ind. Eng. Chem. Anal. Ed.*, **14**, 641 (1942).
102. Herasymenko, P., *Trans. Faraday Soc.*, **24**, 257 (1928).
103. Hersch, P., *Nature*, **169**, 792 (1952).
104. Hersch, P., *Anal. Chem.*, **32**, 1030 (1960).
105. Heyrovsky, J., *Chem. Listy*, **16**, 256 (1922).
106. Heyrovsky, J., *Compt. Rend.*, **179**, 1044, 1267 (1924); *Trans. Faraday Soc.*, **19**, 785 (1924); *Chem. Listy*, **20**, 122 (1926); *Mikrochemie*, **12**, 25 (1932).
107. Heyrovsky, J., *Arkiv. Hem. Ferm.*, **5**, 162 (1931); *Chem. Abstr.*, **26**, 1542 (1932).

108. Heyrovsky, J., and M. Bures, *Collection Czech. Chem. Commun.*, **8**, 446 (1936).
109. Heyrovsky, J., and M. Shikata, *Rec. Trav. Chem. Pay-Bas*, **44**, 496 (1925).
110. Heyrovsky, J., and E. Vascautzanu, *Collection Czech. Chem. Commun.*, **3**, 418 (1931).
111. Hickling, A., *Trans. Faraday Soc.*, **38**, 27 (1942).
112. Hickman, W. M., *Chem. Eng. News*, **1964**, 39, Mar. 9; *Ind. Res.*, **1964**, 89, Dec.
113. Hill, D. K., *J. Physiol.*, **105**, 24 (1946).
114. Hoare, J. P., *J. Electrochem. Soc.*, **110**, 1019 (1963).
115. Hoare, J. P., *Nature*, **204**, 71 (1964).
116. Hoare, J. P., *Electrochim. Acta*, **9**, 599 (1964).
117. Hoare, J. P., *J. Electrochem. Soc.*, **112**, 608 (1965).
118. Hoare, J. P., *J. Electrochem. Soc.*, **112**, 849 (1965).
119. Hoare, J. P., *J. Electroanal. Chem.*, **12**, 260 (1966).
120. Hoare, J. P., S. G. Meibuhr, and R. Thacker, *J. Electrochem. Soc.*, **113** (1966).
121. Hoare, J. P., *Electrochim. Acta*, in press.
122. Hoare, J. P., and S. Schuldiner, *J. Electrochem. Soc.*, **102**, 485 (1955).
123. Hoare, J. P., and S. Schuldiner, *J. Electrochem. Soc.*, **103**, 237 (1956).
124. Horn, G., and H. E. Jacob, *Chem. Tech.*, **16**, 237 (1964).
125. Hospodka, J., *Chem. Abstr.*, **62**, 4311 (1965).
125a. Hovermale, R. A., and P. G. Sears, *J. Phys. Chem.*, **60**, 1433, 1579 (1956).
126. Husmann, W., and G. Stracke, *Wasserwirtschaft*, **48**, 13 (1957).
127. Ilkovic, D., *Collection Czech. Chem. Commun.*, **6**, 498 (1934); *J. Chim. Phys.*, **35**, 129 (1938).
128. Ingols, R. S., *Sewage Works J.*, **13**, 1097 (1941); *Ind. Eng. Chem. Anal. Ed.*, **14**, 256 (1942).
129. Ingols, R. S., *Sewage Ind. Wastes*, **27**, 7 (1955).
130. Iofa, Z. A., Ya. B. Shimshelevich, and E. P. Andreeva, *Zh. Fiz. Khim.*, **23**, 829 (1949).
131. Ives, D. J. G., and G. J. Janz, *Reference Electrodes*, Academic Press, New York, 1961, Chap. V, p 231.
132. Jacobson, M. G., *Anal. Chem.*, **25**, 587 (1953).
133. Jamieson, D., *J. Colloid Radiol. Australia*, **6**, 94 (1962).
134. Jamieson, D., and H. A. S. van den Brenk, *Nature*, **201**, 1227 (1964).
135. Jirsa, F., and K. Lores, *Z. Physik. Chem. (Leipzig)*, **113**, 235 (1924).
136. Johnson, L. F., J. R. Neville, R. W. Bancroft, and T. H. Allen, *Chem. Abstr.*, **61**, 1513 (1964).
137. Johnson, M. J., J. Borkowski, and C. Engblom, *Biotechnol. Bioeng.*, **6**, 457 (1964).
138. Kahn, J. S., *Anal. Biochem.*, **9**, 389 (1964).
139. Kalish, T. V., and R. Kh. Burshtein, *Dokl. Akad. Nauk SSSR*, **81**, 1093 (1951); **88**, 863 (1953).
140. Kamienski, B., *Bull. Acad. Polon. Sci.*, **5** [6A], 85 (1949).
141. Karsten, K. S., *Am. J. Botany*, **26**, 855 (1939).
142. Keilin, D., and E. F. Hartree, *Biochem. J.*, **39**, 293 (1945).
143. Kensey, D. W., and R. A. Bottomley, *Inst. Brewing Proc. Conv.*, **7**, 30 (1962); *Chem. Abstr.*, **61**, 3940 (1964); *J. Inst. Brewing*, **69**, 164 (1963); *Chem. Abstr.*, **59**, 8114 (1963).
143a. Knect, L. A., and I. M. Kolthoff, *J. Inorg. Chem.*, **1**, 195 (1962).
144. Kolthoff, I. M., and K. Izutzu, *J. Electroanal. Chem.*, **7**, 85 (1964).

145. Kolthoff, I. M., and J. Jordan, *J. Am. Chem. Soc.*, **74**, 382 (1952).
146. Kolthoff, I. M., and J. Jordan, *J. Am. Chem. Soc.*, **74**, 570, 4801 (1952).
147. Kolthoff, I. M., and J. Jordan, *Anal. Chem.*, **24**, 1071 (1952).
148. Kolthoff, I. M., and G. J. Kahan, *J. Am. Chem. Soc.*, **64**, 2553 (1942).
149. Kolthoff, I. M., and J. J. Lingane, *Polarography*, Vols. 1, 2, 2nd ed., Interscience, New York, 1952.
150. Kolthoff, I. M., and C. S. Miller, *J. Am. Chem. Soc.*, **63**, 1013 (1941).
151. Kolthoff, I. M., and E. R. Nightingale, *Anal. Chim. Acta*, **17**, 329 (1957).
152. Kolthoff, I. M., and E. P. Perry, *J. Am. Chem. Soc.*, **73**, 5315 (1951).
152a. Kolthoff, I. M., and T. B. Reddy, *J. Electrochem. Soc.*, **108**, 980 (1961).
153. Kolthoff, I. M., and N. Tanaka, *Anal. Chem.*, **26**, 632 (1954).
154. Kordesch, K., and A. Marko, *Mikrochem. ver. Microchim. Acta*, **36**, 620 (1950).
155. Kozawa, A., *J. Electroanal. Chem.*, **8**, 20 (1964).
156. Krog, J., and K. Johansen, *Rev. Sci. Instr.*, **30**, 108 (1959).
157. Kucera, G., *Ann. Physik*, **11**, 529, 698 (1905).
158. Laitinen, H. A., T. Higuchi, and M. Czuha, *J. Am. Chem. Soc.*, **70**, 561 (1948).
159. Laitinen, H. A., and I. M. Kolthoff, *J. Am. Chem. Soc.*, **61**, 3344 (1939).
160. Laitinen, H. A., and I. M. Kolthoff, *J. Phys. Chem.*, **45**, 1061 (1941).
161. Laitinen, H. A., and I. M. Kolthoff, *J. Phys. Chem.*, **45**, 1079 (1941).
162. Latimer, W. M., *Oxidation Potentials*, 2nd ed., Prentice-Hall, Englewood Cliffs, New Jersey, 1952, p. 175.
163. Le Petit, G., *Chem. Abstr.*, **62**, 10803 (1965).
164. Lewartowicz, E., *Compt. Rend.*, **253**, 1260 (1961); *J. Electroanal. Chem.*, **6**, 11 (1963).
165. Lewis, V. M., and H. A. McKenzie, *Anal. Chem.*, **19**, 643 (1947).
166. Lingane, J. J., *Electroanalytical Chemistry*, 2nd ed., Interscience, New York, 1958, p. 4.
167. Lingane, J. J., *J. Electroanal. Chem.*, **2**, 296 (1961).
168. Lingane, J. J., and H. A. Laitinen, *Ind. Eng. Chem. Anal. Ed.*, **11**, 504 (1939).
169. Lipner, H., L. R. Witherspoon, and V. C. Champeaux, *Anal. Chem.*, **36**, 204 (1964).
170. Llopis, J., and F. Colom, *Proc. CITCE*, **8**, 414 (1958).
171. Longmuir, I. S., *Biochem. J.*, **57**, 81 (1954).
172. Longmuir, I. S., *J. Polarog. Soc.*, **1**, 11 (1957).
173. Lord, S. S., and L. B. Rogers, *Anal. Chem.*, **26**, 284 (1954).
174. Lynn, W. R., and D. A. Okun, *Sewage Ind. Wastes*, **27**, 4 (1955).
174a. McElroy, A. D., and J. S. Hashman, *J. Inorg. Chem.*, **3**, 1798 (1964).
175. Mackereth, F. J. H., *J. Sci. Instr.*, **41**, 38 (1964).
176. McKinney, R. E., *Sewage Ind. Wastes*, **29**, 654 (1957).
177. Malz, F., and H. Bortlisz, *Z. Anal. Chem.*, **206**, 409 (1964).
178. Mancy, K. H., and D. A. Okun, *Anal. Chem.*, **32**, 108 (1960).
179. Mancy, K. H., D. A. Okun, and N. Reilley, *J. Electroanal. Chem.*, **4**, 65 (1962).
180. Manning, W. M., *Ecology*, **21**, 509 (1940).
181. Margules, M., *Wied. Ann.*, **65**, 629 (1899); **66**, 540 (1899).
181a. Maricle, D. L., and W. G. Hodgson, *Anal. Chem.*, **12**, 1256 (1965).
182. Marie, C., *Compt. Rend.*, **145**, 117 (1907); **146**, 475 (1908); *J. Chim. Phys.*, **6**, 475 (1908).
183. Moiseev, A. S., and N. M. Brikman, *J. Appl. Chem. USSR*, **12**, 620 (1939).
184. Montgomery, H., and O. Horwitz, *J. Clin. Invest.*, **29**, 1120 (1950).

185. Moore, E. W., J. C. Morris, and D. A. Okun, *Sewage Works J.*, **20**, 1041 (1948).
186. Müller, O. H., *J. Chem. Soc.*, **66**, 1019 (1944).
187. Müller, O. H., *J. Am. Chem. Soc.*, **69**, 2992 (1947).
188. Müller, O. H., *Trans. Electrochem. Soc.*, **87**, 441 (1965).
189. Mukhamedzhanov, M. V., A. D. Snezhko, and A. P. Yazykov, *Uzbeksk. Biol. Zh.*, **8**, 7 (1964); *Chem. Abstr.*, **61**, 13633 (1964).
190. Nagel, K., and H. Dietz, *Electrochim. Acta*, **4**, 1 (1961).
191. Nernst, W., and E. S. Merriam, *Z. Physik. Chem. (Leipzig)*, **53**, 235 (1905).
192. Nesterova, V. I., and A. N. Frumkin, *Zh. Fiz. Khim.*, **26**, 1178 (1952).
193. Neville, J. R., *Rev. Sci. Instr.*, **33**, 51 (1962).
194. Newberry, E., *J. Chem. Soc.*, **109**, 1051, 1066 (1916).
195. Olson, R. A., F. S. Brackett, and R. G. Crickard, *J. Gen. Physiol.*, **32**, 681 (1949).
196. Ozaki, T., J. Suzuki, and K. Izawa, *Bunseki Kagaku*, **13**, 107 (1964); *Chem. Abstr.*, **61**, 1262 (1964).
197. Paris, A., *Ind. Chim.*, **20**, 807 (1933).
198. Pellequer, H., *Compt. Rend.*, **222**, 1220 (1946); **225**, 116 (1947).
198a. Peover, M. E., and B. S. White, *Electrochim. Acta*, **11**, 1061 (1967).
199. Petering, H. G., and F. Daniels, *J. Am. Chem. Soc.*, **60**, 2796 (1938).
200. Pittman, R. W., *Nature*, **195**, 445 (1962).
201. Presbrey, C. H., and S. Schuldiner, *J. Electrochem. Soc.*, **108**, 895 (1961).
202. Rand, M. C., and H. Heukelekian, *Sewage Ind. Wastes*, **23**, 1141 (1941).
203. Rao, M. L. B., A. Damjanovic, and J. O'M. Bockris, *J. Phys. Chem.*, **67**, 2508 (1963).
204. Rasch, J., *Collection Czech. Chem. Commun.*, **1**, 560 (1929).
205. Rayman, B., *Collection Czech. Chem. Commun.*, **3**, 314 (1931).
206. Reeves, R. B., D. W. Rennie, and J. R. Pappenheimer, *Federation Proc.*, **16**, 693 (1957).
207. Rennie, D. W., R. B. Reeves, and J. R. Pappenheimer, *Am. J. Physiol.*, **195**, 120 (1958).
208. Riddiford, A. C., in *Advances in Electrochemistry and Electrochemical Engineering*, Vol. 4, C. Tobias, Ed., Interscience, 1966, p. 47.
209. Ruis, A., J. Llopis, and F. Colom, *Anales Real Soc. Espan. Fis. Quim.*, **B51**, 11 (1955).
210. Ruis, A., J. Llopis, and F. Colom, *Anales Real Soc. Espan. Fis. Quim.*, **B51**, 21 (1955); *Proc. CITCE*, **6**, 280 (1954).
211. Rogers, L. B., M. M. Miller, R. B. Goodrich, and H. F. Stehney, *Anal. Chem.*, **21**, 777 (1949).
212. Rosenthal, R., A. E. Lorch, and L. P. Hammett, *J. Am. Chem. Soc.*, **59**, 1795 (1937).
213. Ross, J. W., and I. Shain, *Anal. Chem.*, **28**, 548 (1956).
214. Rozental', K. I., and V. I. Veselovskii, *Zh. Fiz. Khim.*, **27**, 1163 (1953).
215. Ruer, R., *Z. Phys. Chem.*, **44**, 80 (1903); *Z. Elektrochem.*, **9**, 235 (1903); **11**, 661 (1905).
216. Ruer, R., *Z. Elektrochem.*, **14**, 309, 633 (1908).
216a. Russell, G. A., *Chem. Eng. News*, **42**, 49 (Apr. 20, 1964).
217. Salomon, E., *Z. Physik. Chem. (Leipzig)* **24**, 55 (1897); **25**, 365 (1898).

218. Samokhvalov, L. A., L. V. Churnikova, and N. M. Smolenskaya, *Lab. Delo*, **1964**, 494; *Chem. Abstr.*, **61**, 13620 (1964).
219. Samokhvalov, L. A., and M. S. Shul'man, *Mikrobiologiya*, **32**, 896 (1963).
220. Samuelson, G. J., and D. J. Brown, *J. Am. Chem. Soc.*, **57**, 2711 (1935).
221. Sand, H. J. S., *Phil. Mag.*, **1**, 45 (1901).
222. Sawyer, D. T., R. S. George, and R. C. Rhodes, *Anal. Chem.*, **31**, 2 (1959).
223. Sawyer, D. T., and L. V. Interrante, *J. Electroanal. Chem.*, **2**, 310 (1961).
224. Schmid, G. M., and N. Hackerman, *J. Electrochem. Soc.*, **109**, 243 (1962).
225. Schmid, R. W., and C. N. Reilley, *J. Am. Chem. Soc.*, **80**, 2087 (1958).
226. Schmidt, A., *Z. Elektrochem*, **44**, 699 (1938).
227. Schuldiner, S., B. J. Piersma, and T. B. Warner, *J. Electrochem. Soc.*, **113**, 573 (1966).
228. Schuldiner, S., and R. M. Roe, *J. Electrochem. Soc.*, **110**, 332 (1963).
229. Schuldiner, S., and T. B. Warner, *J. Electrochem. Soc.*, **112**, 212 (1965).
230. Schwartz, W. M., and I. Shain, *Anal. Chem.*, **35**, 1770 (1963).
231. Severinghaus, J. W., and A. F. Bradley, *J. Appl. Physiol.*, **13**, 515 (1958).
232. Shibata, S., *Nippon Kagaku Zasshi*, **79**, 239 (1958); **80**, 453 (1959); *Bull. Chem. Soc. Japan*, **33**, 1635 (1960).
233. Shibata, S., *Bull. Chem. Soc. Japan*, **36**, 525 (1964).
233a. Shibata, F. L. E., et al., *J. Chem. Soc. Japan*, **52**, 399, 404 (1931).
234. Sidgwick, N. V., *The Chemical Elements and Their Compounds*, Oxford University Press, London, 1950, p. 292.
235. Smith, D. P., *Hydrogen in Metals*, Univ. of Chicago Press, Chicago, 1948.
235a. Smith, F. R., and P. Delahay, *J. Electroanal. Chem.*, **10**, 435 (1965).
236. Solov'ev, L. G., *Okeanologiya*, **4**, 149 (1964); *Chem. Abstr.*, **60**, 15592 (1964).
237. Spoor, W. A., *Science*, **108**, 421 (1948).
238. Sugioka, K., and D. A. Davies, *Anestesiology*, **21**, 135 (1960).
239. Takahashi, T., H. Sakurai, and T. Sakamoto, *Buseki Kagaku*, **13**, 627 (1964); *Chem. Abstr.*, **61**, 10037 (1964).
239a.   Thacker, R., *Nature*, **212**, 182 (1966).
240. Tobias, J. M., and R. Holms, *Federation Proc.*, **6**, 215 (1947).
241. Torres, G. E., *J. Appl. Physiol.*, **18**, 1008 (1963).
242. Underkofler, W. L., and I Shain, *Anal. Chem.*, **35**, 1778 (1963).
243. Urbach, F., and G. Peirce, *Science*, **112**, 785 (1950).
244. Varasova, E., *Collection Czech. Chem. Commun.*, **2**, 8 (1930).
245. Vetter, K. J., and D. Berndt, *Z. Elektrochem.*, **62**, 378 (1958).
246. Vielstich, W., *Z. Instrumentenk.*, **71**, 29 (1963).
247. Vitek, V., *Collection Czech. Chem. Commun.*, **7**, 537 (1935).
248. Warner, H., U. S. Pat. 3,149,921, (Sept. 22, 1964).
249. Wartenburg, H. V., and E. H. Archibald, *Z. Elektrochem.*, **17**, 812 (1911).
250. Weber, H. F., *Wied. Ann.*, **7**, 536 (1879).
251. Wittingham, C. P., *J. Exptl. Botany*, **7**, 273 (1956).
252. Whittingham, C. P., and A. H. Brown, *J. Exptl. Botany*, **9**, 311 (1958).
253. Wiesinger, K., *Helv. Physiol. Pharmicol. Acta*, **6**, 13, 34, 71 (1948); *Chem. Abstr.*, **42**, 7359 (1948).
254. Wilson, L. D., and R. J. Smith, *Anal. Chem.*, **25**, 218 (1953).
255. Wise, W. S., *Chem. Ind.* (*London*), **1948**, No. 3, 37.
256. Worthington, L. V., *Deep-Sea Research*, **1**, 244 (1954).
257. Wotring, A. W., and W. W. Keeney, *Proc. Natl. Anal. Instr. Symp.*, **8**, 207 (1962).

chapter VII

The Oxygen Electrode on Some Active Metals

The first five chapters have dealt with the behavior of the oxygen electrode on the noble metals. In such cases, the thickness of the adsorbed oxygen films formed on these electrodes is of the order of a monolayer, unless the electrode is subjected to unusual conditions of severe polarization. In this chapter, the behavior of the oxygen electrode on a number of active metals, particularly those of interest in commercial battery systems, will be considered. As pointed out by Glicksman and Morehouse (206a), the desirable characteristics of battery cathodes such as a high reversible potential, large coulomb capacity, good stability, and compatibility with other components of the battery system are found in the metal oxides. A review of the properties of many of the battery systems in common use was presented by Morehouse et al. (360).

Such metals as silver, nickel, and lead form thick layers of oxides on the electrode surface. Because such an oxide film may act as a non-conducting insulator, an electronically conducting semiconductor, or an ionically conducting defect structure, the electrochemical properties of such a system are more complicated than those encountered in the noble metal systems. In acid and neutral solutions, one must be concerned with the process of the dissolution of the metal which injects metal ions into the electrolyte.

Since a variety of reference electrodes were used in the investigations reported in the literature, it is important to note that all potentials are recorded in this chapter with respect to the NHE unless stated otherwise.

A. SILVER

Interest in the electrochemistry of the silver–oxygen system has been generated by the development of the silver–zinc and silver–cadmium alkaline battery systems. The advantages of the Ag/Zn battery are the high energy production per unit weight and per unit volume and the ability to deliver electrical energy at a high rate. As early as 1910, a battery employing a zinc anode and a silver oxide cathode was described in a patent by Morrison (361) and was studied by Zimmerman (537). Other silver oxide systems using Cd anodes (272) and Fe anodes (265) have been considered, but all of these systems suffered because of the solubility of silver oxide in the alkaline electrolyte. Consequently, these batteries had a high rate of self-discharge.

To immobilize the electrolyte and to minimize the migration of soluble components from one electrode to the other, Andre (8) described the use of a cellophane separator placed between the anode and the cathode by enclosing the silver electrode in a cellophane bag. White and co-workers (520) studied the Ag/Zn system as a primary battery. With the use of improved battery plate separators, the Ag/Zn secondary battery has been designed to a practicable state (154). Recently, Schaffer (447) analyzed the use of ion-exchange membranes as battery plate separators.

1. The Oxides of Silver

A number of workers (20,35,55,153,181,290,303,388,389,440,463,475) have studied the nature of the adsorbed oxygen layer on a Ag surface when Ag is exposed to an oxygen atmosphere. Initially, the adsorption of oxygen is rapid, and Bagg and Bruce (20) calculated that a monolayer of adsorbed O atoms was adsorbed at oxygen pressures as low as 10^{-2} torr. Further adsorption takes place slowly. At 0.2 torr, Temkin and co-workers (475) found that oxygen was adsorbed even after 185 hr. They (303) found that a curve obtained from a plot of the amount of oxygen adsorbed as a function of $\log P_{O_2}$ has a break in it, indicating two types of adsorption. The first rapid adsorption appears to be a chemisorption process where the heat of adsorption falls from 47 kcal/mole at a coverage θ of 0.05–17 kcal/mole at a θ of 0.4 (388). Over the complete range of coverages, $0 < \theta < 1$, the heat of adsorption fell from $\sim$130 to $\sim$10 kcal/mole with increasing coverage (389).

Fogel et al. (181) bombarded a Ag ribbon with a beam of positive argon ions at 10^{-8} A/cm² and 20 kV in an oxygen atmosphere of 10^{-5} to 10^{-4} torr. An analysis of the secondary positive and negative ions was made with a mass spectrometer. As the P_{O_2} increased, the number of negative ions decreased, and the number of positive ions increased. From these studies they concluded that the adsorbed oxygen existed partly as O atoms and partly as O_2 molecules. In addition, some of the adsorbed O_2 molecules may exist as adsorbed O_2^- ions.

Sandler and Durigon (440) concluded from $^{18}O_2$ exchange studies that no evidence for the presence of undissociated adsorbed oxygen could be found. There are two forms of adsorbed oxygen atoms on silver, a weakly adsorbed form having a heat of adsorption of 16 kcal/mole (35) and a strongly bonded form the heat of adsorption of which is unknown (440). Earlier work gave poorly reproducible and conflicting results because of a lack of control of impurities (440).   From high purity studies of the adsorbed oxygen on Ag, Sandler and Durigon suggested that on the Ag surface two types of areas exist which differ in the arrangement of the Ag atoms and in the way in which oxygen atoms are imbedded in the surface.   Also, the more weakly bonded O atoms are mobile, and diffusing over the outer surface of the Ag can exchange with the strongly bonded oxygen.   Ostrovskii and Temkin (389) also observed two types of adsorbed oxygen, one associated with two Ag atoms and the other with one Ag atom.

Using the data from a similar series of $^{18}O_2$ exchange studies, Boreskov and co-workers (54,55) determined that the thickness of the steady-state oxygen layer at 20°C was two to three monolayers.

During the slow adsorption region, silver may also dissolve oxygen in the bulk metal according to evidence obtained by Bagg and Bruce (20). This dissolved oxygen probably corresponds to the "deep chemisorbed adsorbed" oxygen of Temkin (475).

It has been well known for some time that silver can dissolve large quantities of oxygen under certain conditions.   In 1820, Chevillot (104) found that molten silver saturated with air gave up O_2 when it solidified, and Gay-Lussac (190) recorded that one volume of Ag gave up 22 volumes of O_2 when the metal solidified.   Several investigators (63,149, 420) reported that some of the oxygen, which was dissolved in the Ag at elevated temperatures, was retained in the metal when the Ag was cooled.   Eichenauer and Mueller (153) have shown that the presence of adsorbed oxygen on the Ag surface can distort the measurement of oxygen solubility in the metal, and as a result, some caution should be exercised in the interpretation of such measurements.

At elevated temperatures below the melting point of Ag, 960°C, O_2 can diffuse through Ag, which was reported as early as 1884 by Troost (483). Coles (113) gives the results of the permeability of Ag to O_2 over a temperature range from 500 to 850°C and a pressure range from 120 to 2050 torr.

In what form the dissolved oxygen exists in Ag is a question which has been the subject of several investigations (146,147,457,460,464,465). In all cases, it was found that the amount of dissolved oxygen is proportional to the square root of the partial pressure of oxygen.   From a study of the solubility of O_2 in molten Ag at 1075°C, Sieverts and Hagensacker (457) concluded that the dissolved oxygen existed in the atomic form.   Donnan and Shaw (147) observed that the freezing point of molten silver determined in an inert atmosphere was 962°C but in air was lowered to 955°C.   From these studies, it was concluded that the dissolved oxygen could be in the atomic form either as O atoms or as Ag_2O. They favored the Ag_2O hypothesis. If the molten silver were cooled quickly by pouring it rapidly into water at 0°C, the dissolved oxygen might be retained in the form existing at high temperatures, according to Simons (460). The solid dry sample of Ag was sealed in tubes at 400°C, and pure O_2 was found in the tubes. In an attempt to determine if Ag_2O was present, Simons dissolved the Ag with Hg, but no Ag_2O was found, although an increase in the pressure of the system had built up. He also suggested that the dissolved oxygen is present in the form of Ag_2O. Recently, Domanski (146) proposed that the dissolved oxygen exists as Ag_2O_3, but this seems less likely, since Ag_2O_3 is so much more unstable than Ag_2O.

It has been found (460,464) that oxygen diffuses very slowly through Ag at room temperature, but above 400°C the diffusion rate becomes appreciable and increases with temperature. The amount of oxygen present in solid Ag at ambient temperatures is very low (465). According to Steacie and Johnson (465), the solubility of N_2 in Ag is negligible. When one tries to diffuse air through a silver membrane at temperatures above 400°C, as Spencer (464) found, only the oxygen passes through. Possibly the actual entity that migrates through the Ag lattice is the oxide ion, O^{2-}.

a. Preparation of Silver Oxides. From a study of the silver–oxygen system, Lewis (333) concluded that the only stable components are O_2, Ag, and Ag_2O. Evidence for the existence of a suboxide, Ag_4O, is lacking, and so, if it exists at all, it must be unstable. The oxide Ag_2O may be made by precipitating a solution of a soluble silver salt with alkali (422). When heated, Ag_2O begins to evolve O_2 at 160°C (456), and although Sidgwick (456) reports that the nearly black oxide is sensitive to light, Riley and Baker (422) observed that Ag_2O was stable in the light if placed in a sealed container. Changes in Ag_2O exposed to air while being irradiated were considered (422) to be caused by the reducing action of impurities in the air. However, Veselovskii (494) and Blocher and Garrett (44) observed definite photoeffects when silver oxide electrodes were irradiated. Veselovskii found that the photocurrent (difference in observed current when a polarized silver oxide electrode is placed in the dark and in the light) increased with potential up to 0.3 V. Long-term experiments involving the oxides of silver should be carried out in the dark (134,140).

The higher oxide, silver dioxide AgO, is obtained by the oxidation of a silver salt in solution with HNO_3 or $K_2S_2O_8$ to give an impure product or by the anodic oxidation of Ag metal to give pure AgO. When concentrated solutions of $AgNO_3$ are mixed with $K_2S_2O_8$, a black precipitate is formed (18,261,269,531) which Austin (18) analyzed as 77% AgO, which contains a sulfate impurity, according to Yost (53), and which is called silver peroxysulfate, $Ag_7O_8SO_4$, by Wales and Burbank (512). This black product evolves O_2 when heated and when in contact with acids. If the black precipitate is washed with boiling water, a very pure sample of AgO is obtained, as reported by Jirsa (261).

By anodizing a silver electrode in solutions of nitric acid, a black precipitate is formed at the electrode (19,22,23,74,260,350,373,514,515) with an empirical formula of Ag_7NO_{11}, which Weber (515) suggests may be written as $2Ag(AgO_2)_2 \cdot AgNO_3$. The black Ag_7NO_{11} decomposes slowly in H_2O (514) but rapidly in boiling water to give nearly pure AgO (260,273,514).

There is some question whether silver dioxide is really a peroxide, Ag_2O_2. Brow (74) gives a survey of the extensive early literature. When the dioxide is dissolved in acid, a test for peroxide is negative (373,531). Although Klemm (289) found AgO to be slightly paramagnetic, Sugden (468) and Neiding and Kazarnovskii (373) reported it to be diamagnetic. In a series of investigations, Noyes and co-workers (377–381) studied the oxidation of silver ion in solution and the properties of complex silver ions. They, too, found that solutions of Ag_7NO_{11} did not contain peroxides and that their magnetic measurements indicated the dissolved silver is in the divalent state instead of the trivalent state (Ag_2O_3), as proposed by others (19,22,74,260,531). In reactions with O_3, they (377–381) found only Ag^{2+}.

Jirsa and Jelinik (263), in an attempt to make Ag_2O_3, oxidized Ag with dry ozone, but only obtained Ag_2O. When they stirred powdered Ag with O_3 for 12 hr, they got a black product which they concluded to be a mixture of AgO and Ag_2O instead of Ag_2O_3. As an interesting note, Strutt (467) reported that in the interaction of O_3 with Ag_2O every collison resulted in a reaction. All attempts to isolate Ag_2O_3 have ended in failure.

b. Standard Potentials of Silver Oxide Couples. Before the results of charging curves are discussed, a consideration of the potentials of various silver electrodes should be considered. The potential of the Ag/Ag^+ couple,

$$Ag^+ + e \rightleftarrows Ag \qquad E_0 = 0.799 \qquad (7.1)$$

was measured by Jellinek and Gordon (258) and reported as 0.808 V.

The accepted value (126,315) is 0.799 V. The average of a number of determinations of the Ag^+/Ag^{2+} couple,

$$Ag^{2+} + e \rightleftarrows Ag^+ \qquad E_0 = 1.98 \text{ V} \qquad (7.2)$$

made by Noyes and co-workers (379–381) was 1.914 V in only fair agreement with the accepted (126,315) values of 1.98 V.

A number of workers (43,96,184,332,340,374,424) determined the potential of the Ag/Ag_2O couple,

$$Ag_2O + H_2O + 2e \rightleftarrows 2Ag + 2OH^- \qquad E_0 = 0.345 \text{ V} \qquad (7.3)$$

and recorded potentials in NaOH solutions from 1.150 to 1.172 V (vs. Pt/H_2 in the same solution). However, these potentials were not steady and drifted to less noble values with time. One explanation for the drifting potentials was given by Rørdam (424) in terms of two forms of Ag_2O, although Bitton (43) considered such a possibility improbable.

Heat capacity data obtained for Ag_2O (153,195,213,214,262,264,290, 390,404) have been used to investigate the possibility of two modifications of Ag_2O and to determine the thermodynamic properties of the $Ag–Ag_2O–O_2$ system. Pitzer and Smith (404) concluded from their measurements that only one modification of Ag_2O exists, but they did record that an anomalous peak in the specific heat occurs between 20 and 45°K. According to Kobayashi (290), a physical change, which is associated with relief of strains in the crystal, takes place between 100° and 200°C. The potential change accompanying this physical change was estimated to be 6 or 7 mV and could explain the drifting potentials observed by Luther and Pokorny (340) and Fried (184). From more recent work, Pitzer and co-workers (195,214) concluded that the Ag_2O lattice can exist in one of two independent sublattices. In very small crystallites, there is a transition from a symmetrical, high temperature-stable structure to a slightly distorted, low temperature-stable structure between 20 and 30°K, where the anomalous peak in the heat capacity is observed. Such a transition is not observed for large crystals. Well-behaved heat capacity data may be obtained with large crystal size Ag_2O which has been annealed at about 150°C to remove the low temperature transition observed by Pitzer and the high temperature physical change observed by Kobayashi.

At first, in an attempt to make a stable Ag/Ag_2O electrode, Hamer and Craig (225) plated a Pt-wire helix with Ag by heating the Pt wire covered with a paste of Ag_2O at 450°C according to the method described by Harned (228) and measured the potential against the Hg/HgO electrode in a NaOH solution. The results agreed with those reported by Luther and Pokorny and by Fried.   The potentials drifted with time (lost 10 mV in 17 days), but if the old Pt wire was replaced with a freshly plated one, the potential returned to the original value (1.17 V vs. Pt/H_2).   There was a surface effect, since electrodes made of foils decreased in potential faster than those made of wires.   Consequently, Hamer and Craig concluded that the potential drift was not the result of changes in the Ag_2O but depended on the role which the Ag surface plays in the potential-determining mechanism.

In a second series of experiments, Hamer and Craig made a small Pt gauze cup by folding a piece of gauze in the form of a cone which was filled with moist Ag_2O paste.   After the cone had been completely covered with Ag_2O, it was partially reduced (appearance of a gray color) in hydrogen at 60°C.   Such a prepared electrode was placed in the cell immediately after being removed from the H_2 atmosphere.   In this case a stable potential (a drift of less than 1 mV in 38 days) was observed, and the average value reported is 0.342 V for the Ag/Ag_2O potential.

Since HgO can exist in a red or yellow modification, Hamer and Craig suggested that a more reliable value for the potential of the Ag/Ag_2O couple could be obtained if the potential of the Ag/Ag_2O electrode was measured against the hydrogen electrode.   Dirkse (136) measured the emf of a cell composed of a Ag/Ag_2O and a Pt/H_2 electrode in KOH solutions.   A pH-independent value of 0.338 V was recorded.

Because Hamer and Craig (225) heated the Ag/Ag_2O electrode in its preparation and because Gregor and Pitzer (214) had noted anomalies in the heat capacity of unannealed, finely divided Ag_2O, Gregor and Pitzer (215) studied methods of making a stable Ag/Ag_2O electrode without heating the system during its preparation.   Three experiments are recorded in which the electrode was made by mixing Ag powder with Ag_2O.   Electrical contact was made with a Pt wire, and the counter-electrode was a Hg/HgO electrode.   The Ag_2O had to be partially reduced electrolytically or else the potential of the Ag/Hg cell would drift.   The experiments differed by the heat treatment given the Ag_2O: (a) unannealed, being heated only to 110°C to facilitate drying, (b) annealed by heating in a bomb at 180°C and 130 atm of O_2 for 2 days, and (c) macrocrystalline by culturing in a bomb under water at 325°C and 200 atm of O_2 for 20 days.   From studying the samples by differential thermal analysis, it was determined that (a) had not undergone a Kobayashi annealing (290) but (b) and (c) had.   Both (a) and (b) looked alike, whereas (c) contained large crystallites.   Electrodes made from samples (a) and (b) gave the same results, and a potential of 0.343 V was recorded, in good agreement with Hamer and Craig's values for the potential of the Ag/Ag_2O electrode. These electrodes were very stable since the potential decreased by only 0.2% in 45 days. The potential of electrodes fabricated from (c) gave lower potentials (0.341 V), and Gregor and Pitzer suggested that the large Ag_2O crystals provided a less intimate contact between the Ag and Ag_2O than in the cases of (a) and (b). The results support the conclusion arrived at by Hamer and Craig that changes in the potential of the Ag/Ag_2O electrode do not depend on changes in the Ag_2O but on the nature of the contact between the surfaces of the silver and silver oxide. Newton (375) prepared Ag_2O by various methods and reported that electrodes made from these oxides behaved in the same way.

Johnston and co-workers (266) have measured the solubility of Ag_2O in alkaline solutions and reported that the solubility increased with increasing alkali concentration. Identical results were obtained with KOH, NaOH, and BaOH. The solubility of Ag_2O at 25°C in $1N$ alkali was 3×10^{-4} g equiv. Consequently, if the solubility of Ag_2O can be tolerated in a given alkaline system, the Ag/Ag_2O electrode serves as an excellent reference electrode (256,475a) if properly constructed (215, 225). Latimer (ref. 315, p. 191) lists a value of 0.344 V, and more recently, de Bethune and Loud (126) give values of 0.345 V (calculated) and 0.342 V (experimental) for the potential of the Ag/Ag_2O electrode.

The potential of the Ag_2O/AgO couple,

$$2AgO + H_2O + 2e \rightleftharpoons Ag_2O + 2OH^- \qquad E_0 = 0.607 \text{ V} \qquad (7.4)$$

has been measured (52,237,263,340). The earliest recorded value is 0.57 V, obtained by Luther and Pokorny (340), and is the value accepted by Latimer (ref. 315, p. 192). Jirsa and Jelinek (263) observed a potential of about 0.6 V, and Hickling and Taylor (237) found a value of 0.63 V. Recently, Bonk and Garrett (52) measured the potential of the cell,

$$Pt/Ag, Ag_2O, NaOH (1M), AgO, Ag_2O/Pt$$

By using a potential of 0.342 V for the Ag/Ag_2O electrode, a potential of 0.604 V is obtained for the E_0 of Eq. 7.4. They found that Pt could not be replaced with Ag because reduction of AgO takes place in the presence of Ag, according to

$$AgO + Ag \rightarrow Ag_2O \qquad (7.5)$$

and the potential fell to zero in 30 min. The latest recorded value for the potential of the Ag_2O/AgO electrode is given as 0.607 V by de Bethune and Loud (126).

If a Ag electrode is anodized in HNO_3 solutions, a black precipitate is formed, and the potential of this system is about 1.57 V (96,260,340) against Pt/H_2 in the same solution. These authors considered 0.74 V as the potential of the AgO/Ag_2O_3 couple as listed by Latimer (ref. 315, p. 192) and de Bethune and Loud (126). As will be seen later on, x-ray diffraction studies provide no evidence for the presence of Ag_2O_3 on Ag anodes, and Dirkse (133) has suggested that Ag_2O_3 might be a solid solution of AgO and dissolved oxygen.

A summary of the potentials and electrode reactions possible in the $Ag-O_2$ system may be found in the potential–pH diagram constructed by Delahay et al. (130). From the corrosion diagram (130), it is seen that Ag goes into solution at potentials more noble than 0.4 V in strongly acid solutions. This corrosion region becomes narrower as the pH increases until the solution becomes so alkaline (pH > 12) that Ag is passive over the entire potential range. Delahay et al. point out that values involving Ag_2O_3 are considered to be provisional, and if Ag_2O_3 exists, it quickly decomposes when the polarizing circuit is broken.

c. Charging Curves on Silver Electrodes. A large number of investigations (90,103,133–135,140,207,237,269,270,382,406,407,413,485, 493,498,510,530,537) have been concerned with the constant current charging curves obtained on Ag anodes and cathodes in alkaline solutions. In general, the charging curve obtained on a Ag anode in KOH solution is similar to the curve plotted in Fig. 7.1. As soon as the current is applied, the potential rises from an open-circuit value to a short plateau at about 0.44 V (0.1 V vs. Ag/Ag_2O) after which it rises rapidly to a peak value of about 0.75 V before decaying to a long second plateau at about 0.69 V (0.35 V vs. Ag/Ag_2O). Finally, at the end of the second plateau, the potential rises to a steady value (about 1.0 V), where O_2 is evolved.

It is generally agreed that the first plateau is associated with the formation of Ag_2O and is short because a layer of Ag_2O only about three molecules thick (133,237) is formed before the potential rises to the second plateau. As soon as the surface of the Ag is covered with Ag_2O, it appears, the potential rises to the second plateau. The resistance of Ag_2O is very high, 7×10^8 ohm-cm (90,133), and apparently the film of Ag_2O passivates the Ag surface. Consequently, the potential rises under galvanostatic conditions until a new process begins.

The new electrode process occurring in the second plateau is the formation of AgO, but this second plateau is preceded by a potential peak. Several explanations for the origin of this peak have been offered. Hickling and Taylor (237) proposed that the peak resulted from the

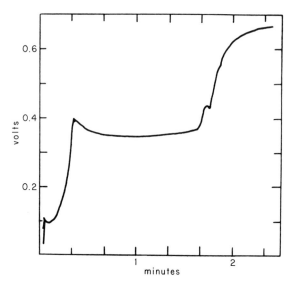

Fig. 7.1. Anodic charging curve (134) for a Ag electrode in 15% KOH solution at 5 mA/cm² (potential vs. Ag/Ag₂O). (By permission of The Electrochemical Society.)

initial formation of Ag_2O_3 which subsequently decomposed to form AgO and O_2. According to this scheme, AgO is not the primary product of the anodization process. This explanation offered by Hickling and Taylor has been severely criticized by later investigators who point out that the peak potential never reaches (133) the potential (0.74 V) assigned to the AgO/Ag_2O_3 couple, and x-ray diffraction studies (513) give no indication of the presence of Ag_2O_3.

Jones and co-workers (270) studied the anodization of Ag wires in $1N$ KOH solutions and concluded that AgO was the primary product of the electrode process. However, they reported that the formation of AgO was inefficient since only about 30% of the current is used to form AgO and the rest is used in the parallel process of the evolution of O_2. Although complete conversion of all the silver to AgO was not observed, both Dirkse (134) and Wales (513) obtained 100% current efficiency in the AgO formation. Jones and co-workers favored the process in which the difficulty in nucleating the AgO in the film of Ag_2O as the cause of the potential peak at the beginning of the second plateau.

Because Ag_2O possesses such a high specific resistance, Dirkse (134) suggested that the potential peak was produced from a passivation of the Ag surface by the protective film of Ag_2O. When the potential rises to a value high enough, Ag_2O is converted to the better conducting (5×10^3 ohm-cm) AgO with a resulting drop in potential. LeBlanc and Sachse (325) measured the conductivity of the silver oxides and found that the conductivity of Ag_2O is much lower than that of AgO. From resistance measurements, Neiding and Kazarnovskii (373) found AgO to be an electronic semiconductor.

From an estimate of the resistance of a thin Ag_2O film, Cahan and co-workers (90) assumed that the resistance of such a film could not account entirely for the peak. Instead, the film has thin spots at which the current is concentrated locally, producing a local high current density and a resultant high overvoltage. Once the higher conducting AgO is formed, the polarization is lowered to the plateau value. Yoshizawa and Takehara (530) and Vidovich et al. (498) believe that the overvoltage producing the peak is generated by the difficulty with which Ag^+ ions can diffuse across the film of Ag_2O.

It seems, then, that the peak observed at the beginning of the second plateau of the anodic charging curve on a Ag electrode in alkaline solutions is produced by an overvoltage caused by the difficulty with which charge is transferred across the poorly conducting film of Ag_2O. In this case, the charge carriers may be Ag^+ ions.

Evidence of the magnitude of the overvoltage present along various parts of the anodic charging curve may be obtained from the intermittent charging techniques employed by Rädlein (413) and Lange and co-workers (207,369,382). In this technique, the charging current is interrupted every 3 sec for 3 sec so that the open-circuit potential may be observed on an oscilloscope throughout the entire charging curve as seen in Fig. 7.2. In the first plateau, the open-circuit potential is

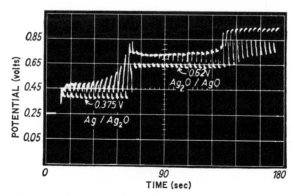

Fig. 7.2. Anodic, intermittent charging curve (369) on a Ag electrode in $1N$ KOH solution at 5.84 mA/cm². (By permission of Bunsengesellschaft.)

constant at the Ag/Ag$_2$O potential, but the polarization increases with time. At the point where the potential peak is observed, the polarization is at a maximum value, and the open-circuit potential begins to shift to more noble values. The long second plateau is characterized by a constant low polarization, and the open-circuit potential is constant at the Ag$_2$O/AgO value. Once O$_2$ evolution sets in, the open-circuit value drifts with time to more noble values, a behavior in contrast to the flat potential plateaus of the Ag/Ag$_2$O and Ag$_2$O/AgO regions.

Additional evidence for the nature of the processes taking place during the constant-current charging of a Ag anode were obtained by Cahan et al. (90) using a method of Cahan and Rüetschi (91) in which an ac square wave is superimposed on the dc constant current bias. With this technique, the impedance of the electrode may be observed along with the potential as a function of time. The results of such an investigation are shown in Fig. 7.3 as a plot of potential, resistance, and double layer capacity as a function of time. Of particular interest is the large spike in the resistance–time curve at the beginning of the second plateau corresponding to the potential peak. As the Ag surface finally becomes covered with a layer of Ag$_2$O, the resistance rises dramatically

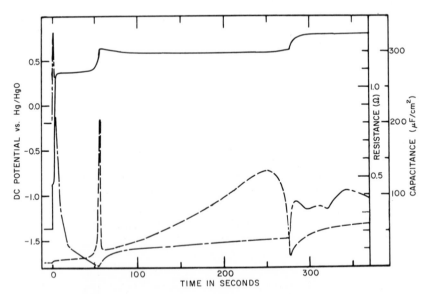

Fig. 7.3. The electrode potential (solid line), ohmic resistance (dashed line), and double layer capacity (dot-dashed line) of Ag during anodization (90) at 3 mA/cm^2 in KOH solution as a function of time. (By permission of The Electrochemical Society.)

but falls just as quickly as the AgO is formed. With a thickening of the AgO film, the resistance increases slowly until O_2 evolution is reached. During the second plateau, AgO is considered (134) to be formed from Ag without any appreciable amount of Ag_2O accumulating. The drop in resistance with O_2 evolution is interpreted to mean that oxygen is added to the film and has been offered as evidence of the existence of a higher oxide than AgO.

During the first plateau where the surface is being covered with the poorly conducting layer of Ag_2O, the double layer capacity decreases while the resistance increases with increasing thickness of the Ag_2O layer. On the other hand, along the second plateau where the surface is covered with the highly conducting layer of AgO, both the double layer capacity and the resistance increase as the AgO layer increases in thickness. This same kind of behavior was noted for the highly conducting layer of adsorbed oxygen on Pt (241) and the poorly conducting layer of adsorbed oxygen on Au (240). The double-layer capacity has been measured at Ag electrodes by other workers (57,237,328,344,414, 493,506,528). Yakovleva et al. (528) obtained the same behavior as reported by Cahan et al. (90). According to Borisova and Veselovskii (57), the potential of the capacity minimum at 0.6 V (vs. Pt/H_2) is the point of zero charge on oxidized Ag in alkaline solution.

After the Ag electrode had been anodized to the point of O_2 evolution, the current was reversed, and the constant-current discharge curve was obtained for the silver oxide cathode. A typical discharge curve is the solid line in Fig. 7.4. As soon as the current is applied, the potential falls to a short plateau at the Ag_2O/AgO potential. The length of this Ag_2O/AgO plateau becomes shorter as the current density is increased until a point is reached beyond which it disappears. With time, the potential falls to a value corresponding to the Ag/Ag_2O potential, where it remains until all of the oxide is reduced, after which the system shifts quickly to the evolution of H_2 at very negative potential values.

An interesting observation is the fact that most of the oxide is reduced at the Ag/Ag_2O potential, whereas most of the oxide was deposited at the Ag_2O/AgO potential. When the current is first applied, some Ag_2O is formed, and since the Ag_2O has a much higher resistance, the potential falls to the Ag/Ag_2O value at the time when the surface of the AgO becomes covered with Ag_2O. At this point some metallic Ag can be formed. The resistance is low at the beginning of the upper plateau in Fig. 7.4, but as the film of Ag_2O forms, the resistance rises rapidly at the transition from the upper to the lower plateau. Once the lower plateau is reached where some Ag may be deposited as filaments in the

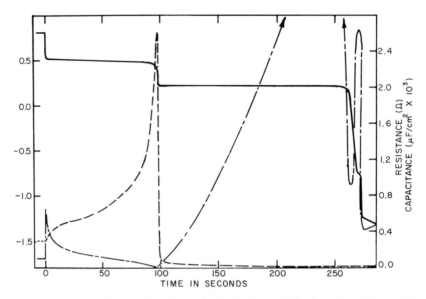

Fig. 7.4. Same as Fig. 7.3 for the cathode discharge (90) of a silver/silver oxide electrode. (By permission of The Electrochemical Society.)

oxide film, as suggested by Cahan and co-workers (90), the resistance falls to very low values. The initial decrease in the double layer capacity in Fig. 7.4 parallels the formation of the poorly conducting film of Ag_2O. The large increase in the double layer capacity after the plateau transition is likely associated with the pseudocapacitance of the reduction of the oxides to Ag.

Oscillograph traces of the intermittent discharge of a silver oxide cathode are shown in Fig. 7.5. Along the upper plateau, the open-circuit potential corresponds to the Ag_2O/AgO value while the polarization continuously increases with time as the Ag_2O film is formed. When the AgO is covered with Ag_2O, the system shifts to the lower plateau where the open-circuit potential corresponds to the Ag/Ag_2O value. This indicates that the surface of the electrode is a layer composed of an intimate mixture of Ag_2O and finely divided Ag. Throughout the lower plateau the polarization is very low. Ohse (382) observed that, although the corresponding plateaus during charge and discharge are of unequal length, the sum of the lengths of the plateaus during charge is equal to the sum during discharge. The current efficiency for both processes is virtually 100%, according to Dirkse (133–135) and Burbank and Wales (82,513).

The decrease in resistance at the end of the upper plateau of the changing curve where evolution of O_2 begins led Cahan and co-workers (90) to suggest that a higher oxide than AgO is present, but since it is very unstable, it is impossible to detect it analytically. Pleskov (406) interpreted charging data obtained on a rotating Ag disk in terms of the presence of Ag_2O_3. Using the intermittent charging technique, Rädlein (413) found that the potential of a Ag anode at which O_2 had been evolved for several minutes decayed slowly when the polarizing circuit was broken, as shown in Fig. 7.6. He offers this observation as possible evidence for the presence of the unstable Ag_2O_3.

Recently, Casey and Moroz (103) studied the anodic polarization of Ag foil electrodes in $KOH-H_2O$ eutectic at $-40°C$. Below a critical current density (15 $\mu A/cm^2$ for bright foil), the usual charging curve (as in Fig. 7.1) was observed, but above it, the upper plateau did not appear. Instead, the potential rose directly from the end of the Ag_2O plateau to the evolution of O_2. When the polarizing circuit was broken, a decay in potential similar to the one observed by Rädlein was found. If the anodized Ag was polarized cathodically, a small kink was observed at about 0.6 V (vs. Hg/HgO) which is near the potential assigned to the AgO/Ag_2O_3 couple (130,260,266,340). After the kink, the potential fell to the Ag/Ag_2O plateau without any evidence of an AgO plateau. At such low temperatures, it may be possible that Ag_2O_3 is stable enough to be observed.

The Volta potential has been measured by Göhr and Reinhard (208) and by Rädlein (413) on Ag-O, Ag_2O, and AgO surfaces using the condenser method (538). Although the Volta potential depends on the nature of the surface layer present, it is independent of how the given layer was produced.

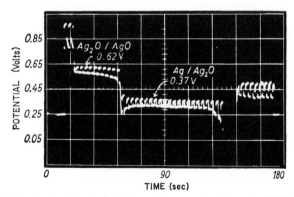

Fig. 7.5. Cathode, intermittent charging curve on a Ag electrode (369) in $1N$ KOH solution at 5.84 mA/cm^2. (By permission of the Bunsengesellschaft.)

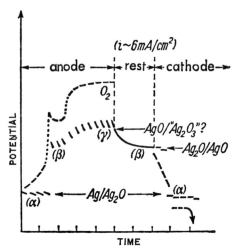

Fig. 7.6. Intermittent charging curves obtained on a Ag electrode (413) in $1N$ NaOH at 6 mA/cm². (By permission of the Bunsengesellschaft.)

d. X-ray Studies. The x-ray diffraction patterns for Ag and for Ag_2O are well known and may be found in the *ASTM Powder Diffraction File*. Both Ag and Ag_2O have a face-centered cubic lattice. X-ray diffraction studies an AgO have been made by a number of investigators (62,211,341,445,471,539), and the diffraction pattern for AgO has been established by Graff and Stadelmair (211) and Scatturin et al. (445). The AgO lattice is monoclinic. Although other crystal forms of silver oxide have been reported, only the presence of face-centered cubic Ag_2O and monoclinic AgO has been established on a Ag electrode in alkaline solutions (513). Both McMillan (341) and Graff and Stadelmair (211) concluded that there is only one crystal modification of AgO. However, Scatturin and co-workers (445) reported evidence for two Ag—O distances in the AgO lattice. From later work using neutron diffraction techniques, Scatturin and co-workers (446) concluded that all the Ag atoms are not equivalent and two Ag—O distances are present in the monoclinic lattice of diamagnetic AgO. They have interpreted these findings in terms of covalent bonds of mono- and trivalent silver. In recent work, McMillan (342,343) considers the paramagnetic monoclinic AgO to be a double oxide of mono- and trivalent silver, and that when AgO is dissolved to form diamagnetic solutions, the dissolved form to be not $Ag(OH)_3^-$ as proposed by Dirkse (141) but really a combination of $Ag(OH)_2^-$ and $Ag(OH)_4^-$.

Zvonkova and Zhdanov (539) took x-ray patterns of the black oxide obtained from the anodic oxidation of Ag in $AgNO_3$ solutions. The structure deduced from an analysis of their data agrees with the formula, Ag_7NO_{11}. Although the exact structure could not be determined, Braekken (62) suggested Ag_2O_3 was present in the black Ag_7NO_{11}, but that if the Ag_7NO_{11} were boiled, only AgO was found (471). Schwabe and Hartmann (471) made AgO by several methods but always obtained the same pattern. In HNO_3 and HF baths, the product of the anodization of Ag gave some evidence of a higher oxide than AgO, such as Ag_2O_3, according to Graff and Stadelmair (211), since lines other than those for AgO were present. These lines belonged to a face-centered cubic (fcc) lattice and could not be identified with Ag_7NO_{11} because the same pattern was observed in the fluoride bath. At room temperature, the anodic product slowly decomposed to AgO, but in boiling water the decomposition took place within 2 hr. They (211) suggested that Ag exists in the $3+$ state in the product obtained from the anodization of Ag in HNO_3 or HF. The x-ray pattern assigned to AgO by Jones and Thirsk (269) has been shown (512,513) to belong to $Ag_7O_8SO_4$.

Several investigators (53,131,134,135,512) have made an x-ray analysis of the products formed on a Ag anode in alkaline solutions as a function of potential. There is one difficulty with these studies in the fact that the polarization processes must be stopped and the electrode transferred from the cell to the x-ray system for analysis. This procedure, of course, requires time, and changes in the system may take place over the time interval between anodic formation and x-ray analysis. Several attempts to overcome this handicap have been made (67,68,163,439). Briggs and co-workers (67,68) used an oscillating wire partly submerged in the solution, but when unstable products were formed, self-discharge occurred on the upper part of the wire where the current density was low. By placing the electrochemical system in a polyethylene bag, Falk (163) attempted to make x-ray analyses of the polarized electrode *in situ*, but the presence of the polyethylene film interfered with the x-ray pattern. A horizontal electrode with a vertical x-ray goniometer was used by Salkind and Bruins (439), but this system was inconvenient because corrections had to be made for the position of the electrode.

Recently, Burbank and Wales (82,513) have designed an experimental cell in which x-ray analysis of the polarized surface of the electrode may be carried out without interrupting the polarizing circuit.

In this Teflon cell (82), the test electrode is in the form of a disk (about 57 mm in diameter and up to 6 mm thick) which is submerged half-way in the electrolyte. A horizontal x-ray beam is directed above the cell and is reflected from the top half of the disk electrode. By rotating the disk from 10 to 60 rpm continuously during the experiment the electrode is always covered with a film of electrolyte, and as a result, the electrochemical parameters can be measured simultaneously with the determination of the x-ray diffraction pattern. Higher speeds of rotation are desirable to minimize any unevenness in the diffraction pattern and to detect the presence of unstable products before they decompose, but the rotation speed is limited by splashing or the development of thick films of electrolyte. Some thinning of the electrolyte film may be obtained by directing a jet of air against the disk.

With this cell, Burbank and Wales (82,513) studied the anodic formation and the cathodic reduction of oxides on silver. In agreement with other investigators (53,58,134,135,512), face-centered cubic Ag_2O was formed during the first plateau of the anodic charging curve obtained on a Ag anode in alkaline solutions. The Ag_2O lines developed gradually over the region of the first plateau without any broadening or shifting of the Ag lines which indicated that Ag_2O was not formed by oxygen penetrating the face-centered cubic lattice of the Ag which would produce a slight expansion of the lattice parameters. The Ag_2O formed on the surface of the Ag in its own crystal form. With time the Ag_2O lines became stronger and the Ag lines weaker.

No evidence of a suboxide suggested by Jones and co-workers (270) was found by Wales and Burbank (513), in agreement with the findings of Levi and Quilico (331), who prepared what was supposed to be Ag_4O. Levi and Quilico did not find any lines other than those for Ag and Ag_2O and concluded that a silver suboxide does not exist. It is possible that a layer of adsorbed oxygen, Ag-O, may exist on a Ag surface under certain conditions as assumed by Göhr and Reinhard (208) since x-rays could not detect a monolayer of Ag-O.

During the anodic oxidation of a Ag electrode, the pattern for AgO did not appear until the beginning of the second plateau was reached when about 0.5% of the active material was converted. Also, in this case the lines of the Ag_2O pattern did not broaden nor shift as the AgO lines grew in intensity. Therefore, the AgO formed as a distinct crystal at the expense of the Ag_2O at the electrode–solution interface. Initially, most of the current went into the oxidation of Ag_2O to AgO, but later on, additional silver from the bulk metal was oxidized. In agreement with Yoshizawa and Takehara (503), Wales and Burbank (513) found the thickness of the AgO layer was much greater than that of the Ag_2O layer at the end of the Ag_2O/AgO plateau.

Because of the effect of the rotation of the disk, the overvoltage for O_2 evolution was about 0.1 V lower than for stationary electrodes.   In spite of extensive polarization at the potentials of O_2 evolution, an x-ray pattern which could be associated with Ag_2O_3 was not observed. Under certain conditions (low charging rate), the potential decayed relatively slowly when the gassing electrode was open-circuited as noted by others (103,413).   Extended periods of gassing produced a more perfect crystalline AgO as noted (513) from the sharper diffraction pattern. With rapid charging, the surface of the electrode was smoother, the crystal size of the AgO was smaller, and the thickness of the oxide layer was decreased.   With slow charging, a thicker layer of large AgO crystals was formed producing a rougher surface.

When a fully charged Ag electrode was left under open-circuit conditions for 60 hr, no pattern for Ag_2O was visible (513) although the AgO pattern was slightly weaker, indicating that the self-discharge rate according to Eq. 7.5 was low in solution.   Even in the dry state, Eq. 7.5 can take place but, of course, extremely slowly.   Dirkse (135) stored charged Ag electrodes dry in the dark for 2 yr and found that complete conversion to Ag_2O had not taken place.

On discharge, Ag_2O was detected early (513) and appeared to be formed first in the areas where the AgO layer was thin.   The x-ray diffraction patterns showed that the Ag_2O formed as distinct crystals and not by the gradual shrinking of the AgO lattice.   The Ag/Ag_2O plateau was reached when Ag_2O had completely covered the electrode surface. From the rate of change of intensity of the lines with time of discharge, it was concluded that both AgO and Ag_2O were being discharged simultaneously and the layer of Ag_2O which always covered the AgO remained at approximately the same thickness until nearly the end of the Ag/Ag_2O plateau was reached.   Apparently, the discharged Ag did not trap any pockets of oxide during the discharge process as proposed by Cahan et al. (90).   At the end of the discharge, only Ag lines were observed, even though small spots of Ag_2O could be seen visually.

A point of interest is the observation that the Ag_2O/AgO plateau is shorter after a slow charge than after a fast charge even though the thickness of the oxide layer is greater for a slow than for a fast charge. At first (509), it was thought that during a slow charge AgO could react slowly with the Ag base to form Ag_2O (Eq. 7.5).   Such a charged electrode would contain much more Ag_2O and require a shorter time to acquire a layer of Ag_2O, thus accounting for the short plateau.   Since no

Ag_2O was observed after a slow charge (513), a more favorable explanation rests in terms of the crystal size. Although the thickness of the oxide formed with a fast charge is relatively thin, the crystal size is small, representing a large surface area. Consequently, a greater time for a given discharge rate is required to cover the surface with Ag_2O; hence, a longer Ag_2O/AgO plateau is observed. Under slow charging conditions, the thick oxide layer is composed of larger crystals representing a smaller surface area to be covered with Ag_2O. A short plateau is observed in this case.

If, as suggested by McMillan (342) and Scatturn (445,446), AgO is a double oxide of Ag(I) and Ag(III), one would not expect to detect Ag_2O_3 at the high anodic potentials because such lines are already present in the pattern assigned to AgO. As noted by Dirkse (135), appreciable amounts of oxygen may be added to the oxide layer during O_2 evolution but are released on discharge without contributing to the coulombic capacity of the electrode. It is also released on open circuit (103,413). This additional oxygen could increase the ratio of Ag(III) to Ag(I) in AgO but not enough to give a different pattern. According to Wales and Burbank (512,513), O_2 evolution may modify the crystal size. Under more stringent conditions of oxidation, such as a high rate of charge at $-40°C$ (103) or oxidation of Ag in NHO_3 and HF (211), the ratio of Ag(III) to Ag(I) in the AgO could be increased to the point where an oxide approximating Ag_2O_3 could be formed. Apparently, however, any oxide with a ratio of Ag(III) to Ag(I) greater than the one which is stable in the double oxide known as AgO is highly unstable and decomposes spontaneously to AgO. To obtain a clearer and more conclusive picture of the nature of the higher oxides of Ag, one must await results of further research.

Some information about the charging curves obtained on Ag electrodes in acid solutions is available (148,269,512). In this case, as in the alkaline case, two plateaus separated by a peak are observed in the constant-current charging curve, and two plateaus are observed in the discharge curve. During the first plateau, the Ag surface becomes covered with a layer of white Ag_2SO_4. After passing over the high potential peak, black spots of AgO, according to Jones and Thirsk (269), appeared randomly throughout the white layer and expanded with time until the entire layer was converted to black AgO at the end of the second plateau. A study of the nucleation of AgO in the layer of $AgSO_4$ was made by Dugdale and co-workers (148). AgO was formed by the reaction of Ag^+ ions with adsorbed OH radicals produced by the discharge of water. Once the nuclei were formed, they grow in two dimensions in the Ag_2SO_4.

Oxygen evolution takes place at the end of the second plateau on the oxide surface, but AgO is also formed while O_2 is evolved. Jones and Thirsk (269) observed that the self-discharge curve of a Ag/AgO electrode in H_2SO_4 solution was similar to the constant-current discharge curve. On discharge, the Ag/Ag_2SO_4 plateau is longer than the Ag_2SO_4/AgO plateau.

From x-ray diffraction studies, Jones and Thirsk found lines for Ag and Ag_2SO_4 in the first plateau, and for Ag and AgO in the second. Faint lines for Ag_2O were detected in the second plateau. However, Wales and Burbank (512) examined a Ag electrode charged to the middle of the second plateau. The x-ray diffraction pattern showed lines for Ag, Ag_2SO_4, and $Ag_7O_8SO_4$ with a trace of AgO. After allowing such a charged Ag electrode to stand in air for 8 days, the diffraction pattern showed lines only for Ag, Ag_2SO_4, and AgO. With time, the peroxysulfate decomposes to AgO. Therefore, in H_2SO_4 solutions with increasing time of charging, Ag is oxidized to Ag_2SO_4 and further to $Ag_7O_8SO_4$ before evolution of O_2 takes place. On standing, the peroxysulfate can decompose to AgO. Wales and Burbank were not able to observe either Ag_2O or the suboxide reported by Jones and Thirsk and concluded that the pattern reported by Jones and Thirsk for AgO was really the one for $Ag_7O_8SO_4$, since it did not agree with the accepted pattern for AgO reported by Scatturin (445) and McMillan (341).

The peak in the charging curve at the beginning of the second plateau was explained by Jones and Thirsk in terms of the difficulty with which the oxide could nucleate in the layer of Ag_2SO_4.

e. Potential Sweep Traces on Silver Electrodes. Experiments have been performed on silver-wire electrodes, in KOH solution, which were polarized with a continuously increasing potential, by Dirkse and de Vries (138) using a standard model polarograph. The current–potential plot exhibited four peaks; the second was identified with Ag_2O, the third with AgO, and the fourth with O_2 evolution. The first was unidentified and disappeared in strong KOH solutions. When the potential of the electrode was held constant at the potential of the first peak for 30 min in $5M$ KOH, a rest potential of 0.028 V vs. SCE which was independent of the partial pressure of oxygen was obtained. X-ray examination of the electrode showed only the lines for Ag. Since the potential was independent of P_{O_2}, the presence of a layer of adsorbed oxygen, Ag-O, seems unlikely. They suggest that this 0.028-V potential may be determined by a $Ag/Ag(OH)_2^-$ couple. In strong KOH, this peak disappears because the mobility of the OH^- ions is decreased, according to Dirkse and de Vries (138).

Tarasevich and co-workers (474) applied a triangular potential sweep pulse (0.04–300 V/sec) to a rotating (900–10,000 rpm) Ag disk in $1N$ NaOH solutions. The traces obtained showed regions for adsorbed oxygen, Ag_2O and AgO. At sweep rates greater than 1 V/sec, they observed a peak which they identified with Ag_2O_3. With slow sweep speeds this peak was not observed because any Ag_2O_3 formed had time to decompose. The reduction curve was the reverse of the oxidation curve. These studies confirm the results obtained from charging curves.

f. Fundamental Battery Studies. A detailed discussion of the construction and properties of Ag–Zn and Ag–Cd batteries will not be entertained in this section. Certain investigations concerning the properties of the oxides of Ag on sintered Ag battery plates, however, will be considered.

For light-weight and high-rate applications, the Ag–Zn battery is an excellent choice because it delivers current at a steady voltage even under conditions of high-rate discharge. A disadvantage is the requirement that the battery be charged at a slow (12- to 24-hr) rate in 30–45% KOH. In a series of investigations, Wales (508–511) has studied the charge acceptance of commercial sintered Ag battery plates as a function of a number of variables.

The coulombic capacity of the Ag plate decreased with increasing rate of charge. In 50% KOH, the loss in capacity was greater than in 35% KOH (508), and a suggested explanation may rest in the decreased mobility of OH^- ions in concentrated KOH solutions, as pointed out by Dirkse and de Vries (138). Apparently, at high rates a barrier oxide film builds up rapidly (508) and O_2 evolution occurs early, whereas at low rates, more time is allowed for the diffusion of electrolyte into the film and for the ionic diffusion processes to take place. At low rates of charging, virtually no increase in the capacity of the plate is realized during O_2 evolution, but with high charging rates, the capacity may be increased (509) up to 50%.

The production of high-rate battery electrodes has been discussed by Shepard and Langdon (452a). Metal plates with controlled porosity and high-surface area may be obtained by the electrolytic reduction of metal compounds subjected to controlled pressure during electrolysis. Coulomb capacities and efficiencies increase with increasing porosity. High rate electrodes have been successfully made out of Zn, Cd, Pb, Ag_2O, and AgO.

Dirkse (135) noted that extended periods of gassing contributed little to the capacity of a slow-charged, sintered Ag plate, although appreciable amounts of O_2 could be added to the active material. This excess oxygen is firmly held, and such charged plates did not lose weight over a period of a year or two when stored dry in the dark. When placed in solution, however, some plates still exhibited the Ag_2O/AgO potential, showing that not all the AgO had been converted to Ag_2O by Eq. 7.5. During discharge the excess oxygen was liberated as bubbles of O_2 and consequently was not part of the active material as such.

It was found (131,134,135,508) that the current efficiency for charge and discharge is virtually 100%, but with pasted Ag electrodes, in all cases, all of the Ag was never completely converted to the oxide. Kinoshita (285) noted that the O_2 fixed in the active material is consumed completely in the discharge. On discharge, most of the current is delivered at the lower potential plateau (135,508,509). Even though much of the AgO is converted to Ag_2O during dry storage in the dark, there is no loss of weight or coulombic capacity. It has been reported (135) that after long periods of storage some active oxygen is not available because interior oxygen is trapped by a deterioration of the oxide crystals, producing changes in the diffusion paths.

Another disadvantage of the Ag–Zn cell is the short cycle life. In 20% KOH (509), the loss of capacity with cycling increases rapidly with the number of cycles (20% after 10 cycles, 60% after 50 cycles) and is caused by a loss of active material (after 58 cycles, a loss of weight of 1 g). The best performance was found in 35% KOH. Using radioactive Ag, Dirkse and Van der Lugt (139) found that more Ag goes in when charging at the low-potential level (Ag/Ag_2O) than at the high level (Ag_2O/AgO). This finding agrees with the report by Rüetschi and Amlie (435) that AgO is less soluble than Ag_2O. The Ag can build up on the separators (139), and eventually shorting occurs. If the Ag migrates to the Zn electrodes, it may be deposited on the Zn (137); and corrosion of the Zn plate (7) may take place by local cell action. The use of ion-exchange membranes (447) as battery separators has minimized greatly the migration of material from one electrode chamber to the other without greatly increasing the internal resistance. Loss of capacity is not due solely to loss of active material, according to Langer and Patton (313a). Sealing of capillary passages in the sintered electrodes during cycling also contributes to the loss.

Wales (510,511) found that the capacity of sintered Ag plates could be increased by superimposing pulses of current on a steady constant-current bias, and the potential at which the electrode was charged was lowered. With a commercial battery, this technique did not affect the capacity much, but it did lower the potential at which the battery accepted charge.

2. Anodic Evolution of Oxygen on Silver

The anodic evolution of O_2 has been studied (7,24,39,103,210,251,259, 329,354,427,499) in a variety of investigations.

The rest potential of a Ag electrode in O_2-saturated alkaline solutions was found by Vielstich (499) to lie between 1 and 1.15 V vs. Pt/H_2. In strong alkaline solutions (pH 14), the potential of the Pt/H_2 electrode is 0.828 V, and the rest potential, therefore, is about 0.32 V, which is close to the Ag/Ag_2O potential. Bianchi and co-workers (39) report a value of 0.28 V in O_2 saturated 0.1M NaOH solution. Most likely, the rest potential on Ag in alkaline solutions is a true oxide potential and is determined by Eq. 7.3 since these potentials are independent of the partial pressure of O_2 (138).

When H_2O_2 is added to an alkaline solution in which a Ag electrode is immersed under open-circuit conditions, the rest potential falls to less noble values initially (38,39,251,499). In the range of H_2O_2 concentrations from 10^{-5} to 10^{-1}, the rest potential is virtually independent of the H_2O_2 concentration (38,39,251). It seems likely (251) that a potential-determining mechanism similar to the mixed potential one considered in the Pt/O_2, H_2O_2 system (239) may be applied to account for the behavior of the rest potential in the solutions of high H_2O_2 content.

Before O_2 is liberated at a Ag anode, a layer of AgO is formed, as shown by the data obtained from charging curves discussed above. Steady-state polarization curves (E vs. log i) contain a Tafel region with a slope of about 0.06 (24,251,499). In general, the η for the evolution of O_2 on a Ag anode in alkaline solution is about the same order of magnitude as on a Pt anode. Rozental' and Veselovskii (427) give values between 10^{-6} and 10^{-4} A/cm^2 for i_0.

Anbar and Taube (7) determined the ratio of $^{18}O_2$ to $^{16}O_2$ in the oxygen evolved at anodes made of Ag/Ag^{16}O and Ag/Ag^{18}O in normal solutions of KOH. In each case, within the precision of the measurements, the isotopic composition of the O_2 evolved was the same, from which fact they concluded that the oxygen evolved did not come from oxygen bound to the silver as AgO. They also found that the oxide oxygen was exchanged with oxygen from the electrolyte, particularly during anodic polarization. Using KOH containing ^{18}O, Rozental' and Veselovskii (427) obtained oxygen enriched in ^{18}O at Ag anodes and favored a mechanism in which O_2 is evolved from the decomposition of an unstable oxide.

According to the oxide mechanism, the anodic current is used to form the unstable oxide, such as

$$AgO + 2OH^- \rightarrow Ag_2O_3 + H_2O + 2e \qquad (7.6)$$

and O_2 is evolved by the chemical decomposition of the unstable oxide similar to Eq. 7.7,

$$Ag_2O_3 \rightarrow 2AgO + \tfrac{1}{2}O_2 \qquad (7.7)$$

However, the formation of Ag_2O_3 by Eq. 7.6 followed by a further discharge of electrons, Eq. 7.8,

$$Ag_2O_3 + 2OH^- \rightarrow AgO + O_2 + H_2O + 2e \qquad (7.8)$$

is preferred by Rozental' and Veselovskii (427). Using a path suggested by Krasil'shchikov (295), OH^- ions are discharged to form adsorbed O^{2-} ions which, in turn, react with the surface oxide to produce the unstable oxide,

$$2AgO + O^{2-} \rightarrow Ag_2O_3 + 2e \qquad (7.9)$$

Then, O_2 is evolved by the decomposition of Ag_2O_3, according to Eq. 7.7.

In view of the results obtained by Anbar and Taube (7), the oxide does not enter into the mechanism of O_2 evolution, and a mechanism involving the discharge of OH^- ions should be considered. In this case, the adsorbed OH radicals are produced from the discharge of OH^- ions,

$$OH^- \rightarrow (OH)_{ads} + e \qquad (7.10)$$

From the interaction of the $(OH)_{ads}$ radicals, adsorbed oxygen atoms are produced,

$$2(OH)_{ads} \rightarrow (O)_{ads} + H_2O \qquad (7.11)$$

from the combination of which O_2 is evolved. Apparently, the oxygen of the $(O)_{ads}$ atoms does not exchange with the AgO layers, and consequently, O_2 is evolved on the electronically conducting layer of AgO. Gorodetskii (210) has estimated the thickness of this AgO layer as ranging from 10 to 16 monolayers.

Probably, additional Ag is oxidized to AgO during the evolution of O_2. Such parallel processes could account for the low value (0.06) obtained for the Tafel slope. At very low temperatures ($-40°C$), Casey and Moroz (103) obtained polarization curves with a Tafel slope of about 0.16 below 60 $\mu A/cm^2$ and 0.1 above 120 $\mu A/cm^2$. They concluded that O_2 is evolved on the surface of a layer of Ag_2O_3 in the high current density range. Recently, Gorodetskii (210) found evidence for an unstable oxide, Ag_2O_3, and believed that O_2 is liberated on a Ag anode at least partially through the participation of the Ag_2O_3.

Before a mechanism and the identity of the rate-determining step can be established uniquely for the evolution of O_2 at a Ag anode in alkaline solution, one must await the accumulation of additional evidence from future investigations.

In acid solutions such as H_2SO_4, the evolution of oxygen takes place on a layer of AgO and $Ag_7O_8SO_4$ (68), but in $HClO_4$ solutions the anodic reaction is entirely the dissolution of silver (39).

3. Cathodic Reduction of Oxygen on Silver

Many investigators (1,3,31,38,39,127,251,294,357,372,376,393,395, 443,444,455,499) have carried out polarization measurements on the reduction of O_2 at Ag cathodes in acid and alkaline solutions. A plot of E vs. log i for O_2 reduction on Ag in alkaline solutions gives a curve with a Tafel region having a slope of 0.06 (3,39,294), and the η decreases with increasing temperature (3,357). Compared to Pt cathodes, the η for the reduction of O_2 is much lower on Ag cathodes in alkaline solutions (499). Over much of the potential range studied, the value for the number of electrons transferred in the overall electrode reaction on Ag cathodes is 4 according to Dehahay (127). The overvoltage varies with the partial pressure of O_2 (294) with a coefficient, $d\eta/d \log p_{O_2}$, equal to 0.04. As in the case of noble metal cathodes, H_2O_2 is detected as a Ag cathode during reduction of O_2.

Reduction of oxygen experiments were carried out on a rotating Ag disk with a Pt ring by Shumilova and co-workers (455) in KOH solutions. During the reduction of O_2, H_2O_2 is formed as a stable intermediate, but any oxygen which was adsorbed on the Ag surface (such as oxides or atomic oxygen) is reduced directly to OH^- since H_2O_2 could not be detected in this case. Some evidence that oxygen can penetrate the silver lattice was obtained.

Evidently the reduction of O_2 on Ag cathodes proceeds by two two-electron steps (393), the reduction of O_2 to peroxide,

$$O_2 + H_2O + 2e \rightarrow OH^- + HO_2^- \tag{7.12}$$

and the catalytic decomposition of HO_2^- ions,

$$2HO_2^- \rightarrow 2OH^- + O_2 \tag{7.13}$$

The O_2 liberated by Eq. 7.13 may then react according to Eq. 7.12 so that effectively 4 electrons are transferred, in agreement with Delahay's findings (127). When the Ag electrode is cathodized initially, the surface may be covered with a layer of oxide, but this layer is soon stripped away since HO_2^- ion produced in the reduction process can react with the oxide (39) by a reaction similar to that of Eq. 7.14,

$$HO_2^- + Ag_2O \rightarrow Ag + OH^- + O_2 \tag{7.14}$$

Once the oxide layer is removed, the reduction of O_2 takes place on oxide-free Ag sites. In this case, the peroxide is decomposed by a mixed potential mechanism similar to that described by Eq. 4.15 and 4.16 in Chapter IV.   For alkaline solutions, these equations would be

$$HO_2^- + H_2O + 2e \rightarrow 3OH^- \tag{7.15}$$

and

$$HO_2^- + OH^- \rightarrow O_2 + H_2O + 2e \tag{7.16}$$

Since the η is a function of temperature and P_{O_2} and since a Tafel region exists in the polarization curve, a reaction-controlled mechanism with an electron transfer rate-limiting step most likely prevails.   The two-electron step, Eq. 7.12, occurs by the one-electron steps

$$(O_2)_{ads} + e \rightarrow (O_2^-)_{ads} \tag{7.17}$$

$$(O_2^-)_{ads} + H_2O \rightarrow OH^- + (HO_2)_{ads} \tag{7.18}$$

and

$$(HO_2)_{ads} + e \rightarrow (HO_2^-)_{ads} \tag{7.19}$$

The discharge of the first electron, Eq. 7.17, is favored (294,357,393) as the rate-determining step which agrees with the observation (294) that the η is a function of P_{O_2}.   Krasil'shchikov (294,372) concluded from his studies that atomic oxygen is involved in the discharge step at high-current densities but molecular oxygen is involved at low-current densities.   It is difficult to see why atomic oxygen should be discharged at high-current densities and not at low current densities also.   Mazitov et al. (357) came to the conclusion that as the temperature is increased the rate-determining step shifts from the slow electron-transfer step (Eq. 7.17) to slow diffusion of oxygen.

Silver, apparently, is a good catalyst for the catalytic decomposition of H_2O_2 because the steady-state concentration of H_2O_2 which is reached during the cathodic reduction of O_2 at Ag is very low, 10^{-7} mole/ liter in $1M$ KOH, according to Sandler and co-workers (251).

Steady-state rest potentials on Ag in acid solutions are difficult to assess because the complication of metal dissolution is present.   At relatively high current densities, Ag no longer corrodes (39), and steady-state overvoltage curves may be obtained for the reduction of O_2 on Ag cathodes in acid solutions (39,294).   In this case, the Tafel slope is 0.112 (39,294), and the dependence on P_{O_2}, $dE/d \log P_{O_2}$, is 0.112 (294).   It is assumed that the mechanism for the reduction of O_2 at Ag cathodes in acid solutions is the same as that favored for Pt cathodes in acid solutions (see Chapter IV).

Polarization curves have been determined on Ag anodes and cathodes in peroxide-stabilized acid and alkaline solutions (38,39,251,499).   In the acid solutions of H_2O_2, the anodic process is the dissolution of silver, as reported by Bianchi (39). In alkaline solutions, the η is greatly lowered. It is interesting that the oxide plateaus observed in the charging curves on Ag anodes in alkaline solution in the absence of H_2O_2 also appear in the polarization curves obtained on Ag anodes in alkaline solutions to which H_2O_2 had been added (39). Bianchi and co-workers conclude that the formation of the oxides does not interfere with the oxidation of H_2O_2. On the cathodic side, the η is very low, and the reduction of H_2O_2 occurs by a diffusion-controlled process according to Sandler and co-workers (251). These low polarization characteristics have been attributed (39,251) to the very labile nature of H_2O_2 in alkaline solution.

Nikulin (376) has studied the reduction of O_2 on different crystal faces of a single crystal of Ag in $0.1N$ H_2SO_4 and $0.1N$ KOH solutions. Differences in the rate of reduction of O_2 on the different crystal faces have been reported to be as high as 10%. Such variations in the electrochemical properties of the crystal faces of a single crystal have been explained in terms of the atom packing of the given crystal face.

B. LEAD

In recent years an understanding of the nature of the oxides of lead formed on the positive plates of the lead–acid secondary battery has been greatly increased by intensive investigations of the interaction of Pb with oxygen. The history of the lead–acid battery goes back to the work of Planté in the middle of the nineteenth century and to his paper (405) in which he disclosed a new battery concept based on Pb.

The lead–acid battery is composed of a lead–dioxide positive plate and a pure lead negative plate immersed in sulfuric acid of a specific gravity range between about 1.26–1.30. According to Gladstone and Tribe (198) in a series of studies of the properties of the lead–acid system, the cell discharge reaction takes place by the conversion of PbO_2 to $PbSO_4$ at the positive plate and Pb to $PbSO_4$ at the negative plate. The cell reaction for this so-called double sulfation theory may be written as

$$PbO_2 + Pb + 2H_2SO_4 \rightleftarrows 2PbSO_4 + 2H_2O \qquad (7.20)$$

with the discharge reaction going from left to right and charge from right to left. Since H_2SO_4 is converted to H_2O during the discharge reaction, a measure of the specific gravity of the electrolyte is directly related to the level of charge of the battery (e.g., 80a), a most convenient circumstance.

The active material is formed on lead plates electrochemically by cycles of anodic and cathodic polarization. Such a formed plate is called a Planté plate, but the coulombic capacity of such a plate is limited. In 1881, Faure (166) demonstrated that greater capacity could be obtained if a paste of red lead which was held in a lead grid was reduced or oxidized electrochemically to form the two battery plates. These plates are referred to as Faure or pasted plates and account for nearly all the battery plates found in modern commercial lead–acid batteries. Usually antimony is alloyed with the lead of the grid to give rigidity to the structural material.

Lead–acid batteries have the disadvantage of weight, deterioration when stored, and poor acceptance of charge at low temperatures. Although the lead–acid battery may be dry-charged to remove the problem of self-discharge during distribution, the deterioration of these batteries still exists on stand in the activated state. In service, most batteries fail because of corrosion of the grids (307,432), leading to loss of active material, mechanical destruction of the grids, and electrical shorts due to growth. Attempts to reduce the effects of grid corrosion by the use of Pb alloys containing Ca, Ag, Co, and Sn among others have been carried out. To reduce weight, the use of plastic grids has been considered, but poor adhesion with the resulting shedding of active material has been a drawback to commercial production.

The great advantage of the lead–acid battery exists as a dependable and economic source of electrical power in relatively rugged applications requiring shallow cycling operation.

Cadmium plates of electroformed Cd sponge have been considered (411) as the negative plate of a battery using a PbO_2 positive plate in 39.5% H_2SO_4 (specific gravity, 1.30). Although a slightly higher discharge voltage and capacity have been claimed, the high self-discharge rate is a severe disadvantage.

The accepted authority on storage batteries is Vinal (500). Reviews of the lead–acid battery have been presented by White et al. (521) and Morehouse et al. (360).

1. The Oxides of Lead

A knowledge of the oxides of lead is important in the field of ceramics as well as the battery field. It is interesting to note that a Pb anode must be used as an insoluble anode in chromium plating (173). The lead anode must be preanodized to form a layer of PbO_2 because it appears that a principal role played by the anode is the oxidation of Cr^{3+}–Cr^{6+} on the PbO_2 surface.

a. Pb₂O. It was believed (60,396) that a suboxide, Pb_2O, was obtained from the thermal decomposition of lead oxalate according to

$$2Pb(OOC)_2 \rightarrow Pb_2O + CO + 3CO_2 \qquad (7.21)$$

and Bircumshaw and Harris (42) have presented a summary of the early history of conflicting evidence for and against its existence. From x-ray diffraction studies, Van Arkel (491) concluded that the thermal decomposition of lead oxalate gave only PbO and metallic Pb,

$$3Pb(OOC)_2 \rightarrow 2PbO + Pb + 2CO + 4CO_2 \qquad (7.22)$$

The x-ray diffraction evidence against the existence of Pb_2O has been confirmed (42,121). In addition, Bircumshaw and Harris measured the ratio of CO to CO_2 in the gases liberated by the decomposition of lead oxalate and obtained a value of 67% CO_2 in agreement with Eq. 7.22 rather than 75% required by Eq. 7.21. The low resistance of the velvet black ash (42) betrays the presence of metallic lead. As concluded by Sidgwick (ref. 456, p. 624), Pb_2O does not exist.

b. PbO. The monoxide of lead, PbO, exists in two modifications (14, 87,121,217,319,523), the red form (litharge) which is stable at low temperatures and the yellow form (massicot) which is stable at high temperatures. Because the conversion from one form to the other is a slow process, metastable conditions may be maintained, and the uncertainty which exists in the knowledge of the transition temperature is expressed by the range of temperatures (486–586°C) quoted (279,319,397,418) for its value.

PbO may be made by reacting $Pb(OH)_2$ with an alkali such as KOH, where the $Pb(OH)_2$ is obtained by precipitation from a solution of a soluble lead salt with carbonate-free alkali. What modification is obtained depends on the concentration of alkali used. If $Pb(OH)_2$ is dissolved in hot $15M$ KOH, red platelets separate out when the mixture is cooled, but in hot $10M$ KOH, yellow needles separate out (14a). With NaOH, the same dependency is observed (189) at lower concentrations ($10M$ and $5M$ NaOH, respectively). Red PbO may be converted to yellow PbO by heating, and the reverse transition may be made to take place by the long continued irradiation of the sample with light at room temperature (14a).

From optical observations and because he found that both forms of PbO were equally soluble, Glasstone (200) concluded that the only difference between red and yellow PbO is the state of subdivision. This, of course, has not proved to be true. Yellow PbO has been found to be nearly twice as soluble in water and alkaline solutions as the red form

(14a,189,436); 2.8 $\times$ 10^{-5} (red) and 5 $\times$ 10^{-5} (yellow) mole/liter in water (436); 0.0140 (red) and 0.237 (yellow) mole/liter in $1M$ NaOH (14a). Cells were constructed (14a,189) with Pb/PbO and Hg/HgO electrodes in $1M$ NaOH solution, and the cell potentials were recorded for both forms of PbO (0.5594) V (yellow) and 0.5668 V (red) at 25°C. This gives a value of 0.0074 V for ΔE, which agrees well (14a) with the value of 0.0066 V calculated from the difference in concentrations using the Nernst relationship, $(RT/nF) \log (C_{yel}/C_{red})$.

Conclusive evidence for the existence of two modifications of PbO rests in the results of x-ray diffraction studies (84,121,132,223,280,319, 330,359,410,472,523). It is generally agreed that red PbO belongs to a tetragonal crystal system, and yellow PbO to an orthorhombic one. The standard x-ray powder diffraction data for both red and yellow PbO are given by Swanson and Fuyat (472). However, there is some difficulty in determining the structure and the arrangement of the atoms in the crystal. As pointed out by Kay (283), the oxygen atoms cannot be "seen" in x-ray powder diffraction patterns, and one must resort to neutron diffracton studies. On the basis of an analysis of neutron diffraction patterns obtained on yellow PbO, Kay (238) and Leciejewicz (326) was able to correct the crystal structure proposed by Byström (84) for orthorhombic PbO. Leciejewicz (327) also presents the neutron diffraction data and crystal structure for tetragonal PbO.

When either red or yellow PbO is ground (15,108), a distorted tetragonal lattice is produced. The distortion may be removed with annealing. In the case of PbO, the high temperature form has the higher density (522), which is not the normal situation encountered in other oxide systems. When pressure is applied to red PbO ($p > 6000$ atm), it is converted to yellow PbO slowly. The transition may be accelerated with grinding or seeding.

Nuclear magnetic resonance studies were carried out on yellow and red PbO (399). The chemical shift for the orthorhombic form is large, whereas it is normal for the tetragonal form. Just what difference in the energy level diagram takes place when there is a symmetry change, such as a change from yellow to red PbO, according to Piette and Weaver (399), is not understood.

It is to be noted that color is not sufficient evidence for distinguishing between the two forms of PbO, because tetragonal PbO may have any color between red and yellow (111,417). Consequently, x-ray analysis is the most dependable means for detecting the two forms of PbO.

A black form of PbO may be obtained by heating a solution of Pb(OH)$_2$ in alkali. Garrett and co-workers (189) dissolved Pb(OH)$_2$ in

$6M$ NaOH and heated to 80°C.   Black crystals precipitated out when
cooled.   With $3N$ KOH, Appleby and Reid (14a) also obtained the
black crystals which they described as similar to the yellow crystals of
PbO.   X-ray diffraction studies (109) show only the lattice structure of
yellow PbO.   It is suggested (14,109,523) that the black color is a surface phenomenon caused by a thin coating of metallic lead produced by
the action of heat or light on the crystals.   A necessary condition for the
production of black PbO (14) is contact with alkali during the preparation of the crystals.

Anderson and Tare (10) have studied the oxidation of evaporated
lead films between 90 and 548°C and 10^{-2} to 10^{-5} atm of O_2.   The only
product detected with electron diffraction techniques was orthorhombic
PbO.   From thermal emf measurements, they concluded that yellow
PbO is an n-type semiconductor.

c. Pb_3O_4.   When litharge is heated in air or when lead dioxide is
thermally decomposed under controlled conditions, red lead, Pb_3O_4, is
formed.   Gross (216) describes the Pb_3O_4 crystals as transparent red
needles about 4 mm long.   By adding excess KOH to $Pb(NO_3)_2$ until
the hydroxide redissolved, Darbyshire (121) obtained Pb_3O_4 crystals by
precipitating them with sodium hypochlorite.   Glasstone (201) considered Pb_3O_4 a complex oxide containing Pb^{2+}, and Gross (216) regarded
the structure of red lead to be made up of tetragonal PbO and PbO_2
units.   Thermodynamic studies of Pb_3O_4 were carried out by Cariter
and Brenet (101).

X-ray diffraction investigations (88,121,186,216,280,523) show that
Pb_3O_4 is a distinct phase (121,280) and has a tetragonal crystal structure.   Standard x-ray powder diffraction patterns have been published
by Swanson and co-workers (473).   The crystal structure for tetragonal
Pb_3O_4 has been described by Byström and Westgren (88) and has been
confirmed by the analysis of neutron diffraction patterns obtained from
Pb_3O_4 by Fayek and Leciejewicz (167).

d. PbO_2.   It has been stated that lead dioxide, PbO_2, may be obtained by the thermal oxidation of PbO (ref. 456, p. 600) or Pb_3O_4(121),
but with the aid of x-ray analyses, White and Roy (523) found that an
oxide with an oxygen content higher than $Pb_{12}O_{19}$ ($PbO_{1.582}$) could not
be obtained.   The most reliable preparation is the anodic oxidation of a
solution of a plumbous salt, but even in this case, there is always a
deficiency of oxygen in the compound (83,86,109,151,278,281,392,431).
The oxygen content of the black anodic product has been reported as
$PbO_{1.99}$ by Katz and LeFaivre (281); $PbO_{1.988}$ by Byström (86); and
$PbO_{1.98}$ by Butler and Copp (83).   Eberius and LeBlanc (151) stated

that in vacuum, oxygen could be removed from PbO_2 down to $PbO_{1.66}$ without a change in phase, but this finding does not agree with other investigations. A new phase begins at an O/Pb atom ratio of 1.9 after Katz and LeFaivre (281); 1.92 after Byström (86); 1.935 after Butler and Copp (83); and 1.95 after Clark and Rowan (109).

Thomas (477) has recorded the resistance of PbO_2 in a pellet form as 2×10^{-4} ohm-cm and in battery plate active material as 74×10^{-4} ohm-cm. These values agree with the earlier measurement of 0.95×10^{-4} ohm-cm reported by Palmaer (392). Such an unusually high conductivity is associated with the excess Pb present in the nonstoichiometric lead dioxide (314,431,477). Hall effect measurements were carried out on PbO_2 samples (314,477), and a Hall coefficient of between -1.7 and -3.4×10^{-2} cm²/coulomb were found which indicates that the charge carriers are electrons. Carrier concentrations of from 10^{20} to 10^{21} electrons/cm³ are recorded.

Nuclear magnetic resonance (NMR) studies of PbO_2 have been reported (183,399,423) using the lead isotope ^{207}Pb. The value of $+0.63$ to $+0.65\%$ for the Knight shift with respect to metallic lead is consistent with the view that PbO_2 behaves as a metal. As pointed out by Piette and Weaver (399), this chemical shift for the magnetic resonances in PbO_2 is due to the conduction of electrons, because the lattice relaxation time for the resonances is short. Since the Knight shift in PbO_2 resonance is dependent on the density of electrons at the top of the Fermi distribution (183), it is a qualitative measure of the conductivity of the sample. Frey and Weaver (183) conclude that it is a deficiency of oxygen rather than impurity content of the sample that is responsible for the conductivity of PbO_2. As oxygen is removed from PbO_2, the Knight shift increases, showing a decrease in conductivity. Rüetschi and Cahan (431) point out that Kittel (287) observed the conductivity to decrease as oxygen was removed from PbO_2 and that the Hall coefficient increased showing a decrease in the number of charge carriers. This, of course, seems strange if the conductivity is caused by the deficiency of oxygen, since the opposite effect should have been observed. However, as mentioned above, the stability range of PbO_2 with respect to oxygen content is very narrow before a change of phase sets in. The appearance of a poorly conducting phase in the partially reduced PbO_2 could explain the loss of conductivity as oxygen is removed. Rüetschi and Cahan have reached this conclusion.

Although these investigations have shown that PbO_2 is an excellent electronic conductor, it has not been determined whether the conductivity of PbO_2 is the result of a highly doped semiconductor or a metallic conductor with overlapping bands. Lappe (314) made optical absorption measurements of thin films of PbO_2 supported on quartz as a function of the temperature and concluded from the results that PbO_2 is a highly doped semiconductor with excess Pb and a band width of about 1.5 eV.

From x-ray powder diffraction patterns, it was determined (121,249) that PbO_2 has a tetragonal crystal lattice. It is important to recognize that the PbO_2 examined in these studies was prepared in an acid medium, since this modification of PbO_2 is known as β-PbO_2. Powder diffraction data are given by Darbyshire (121) for β-PbO_2.

As early as 1946, Kameyama and Fukumoto (279) detected from x-ray diffraction studies a new crystal form of PbO_2 other than the ordinary tetragonal type. They suggested that the crystal structure was similar to fluorspar. Later, Zaslavskii and co-workers (534,535) reported that the structure of PbO_2 obtained from the oxidation of $PbCO_3$ in alkaline solution is orthorhombic and is a second modification of PbO_2. They suggested that the orthorhombic form be called α-PbO_2 and the tetragonal form β-PbO_2 in analogy with the modifications of MnO_2. Since that time, numerous studies of the structure and properties of α-PbO_2 have been undertaken (12,16,21,46,51,78,100,176,183,314,316,348,431, 434,519,522,523), and its presence in passive films on Pb (348), in the active material of battery plates (46,78,79,144,431), and in the thin supported films ($\sim$100 Å thick) produced by sputtering Pb in an O_2 atmosphere (314) has been confirmed. In passing, Lappe (314) sputtered thin films of Pb on quartz surfaces in various O_2–Ar mixtures. When the O_2 content was below 3%, the films were composed of pure Pb; when between 3 and 25%, low conducting film of Pb_3O_4 was formed; and when above 25%, a highly conducting film of PbO_2 was obtained containing both the tetragonal and orthorhombic modifications.

The x-ray powder diffraction data for α-PbO_2 may be found in the reports of White et al. (522) and Zaslavskii et al. (534). It has been demonstrated (21,51,78,431,432) that β-PbO_2 may be deposited anodically from a solution of $Pb(NO_3)_2$ in HNO_3 or $Pb(ClO_4)_2$ in $HClO_4$ and α-PbO_2 from a solution of lead acetate ($PbAc_2$) in KOH containing some NaAc. Using the electron microscope, Astakhov and co-workers (16) found that α-PbO_2 plated out as a low surface area deposit of densely packed large crystals ($\sim$1 μ in diameter), whereas the β-PbO_2 formed as a high surface area deposit of a porous mass of needles.

It appears that in many cases the oxygen content of α-PbO_2 is less than that of β-PbO_2 (12). If oxygen deficiency is the cause of the conductivity of PbO_2, the α form should be a slightly better electronic conductor than the β form. This conclusion is supported by the NMR measurements of Frey and Weaver (183). The Knight shift for the α form is 0.48%, whereas the shift for the β form is 0.63%, indicating that the conductivity of α-PbO$_2$ is slightly better than that of β-PbO$_2$.

Under ordinary laboratory conditions, β-PbO$_2$ is the more stable phase, but under pressure, β-PbO$_2$ may be transformed to α-PbO$_2$ (432, 522). Rüetschi and Cahan (432) record a pressure of 125,000 psi which is required to produce the transformation. When the pressure was released, the β form had not reappeared even after a year at room temperature, according to White and co-workers (522). Temperature had an effect because at 100°C some β-PbO$_2$ was detected after 2 weeks; however, at 290°C (11,253,523), PbO$_2$ begins to lose oxygen. White (522) records the heat of transition of α to β as 11 cal/mole at 1 atm pressure and 32°C. According to Burbank (78), α-PbO$_2$ is converted to β-PbO$_2$ just before the β form is thermally decomposed and the conversion temperature lies between 296 and 301°C.

e. Pb$_{12}$O$_{19}$. As noted above, PbO$_2$ loses oxygen when heated, and when the O/Pb atomic ratio falls below about 1.92, a second less conducting phase appears. In fact, between PbO$_2$ and Pb$_3$O$_4$ a nearly continuous series of oxides with decreasing oxygen content has been reported. A large number of investigators (11,12,83,86,101,109,243,244,253,280, 282,319,417,431,523) have studied these intermediate oxides, and it is generally concluded that at least one distinct phase exists between PbO$_2$ and Pb$_3$O$_4$. This intermediate phase has been variously identified as: Pb$_2$O$_3$ (O/Pb = 1.50) (417); Pb$_7$O$_{11}$ (Pb/O = 1.572) (244); PbO$_x$ (Pb/O $\sim$ 1.58) (83,86,101); Pb$_{12}$O$_{19}$ (Pb/O = 1.582)(11,523); and Pb$_5$O$_8$ (Pb/O = 1.60) (109,253). As mentioned by White and Roy (523), the differences in the O/Pb ratio are so small and the atomic weight of Pb is so large that precise analysis is very difficult. In recent times the very careful work of Anderson and Sterns (11) and Byström (86) has helped clarify the situation.

Byström's investigations indicate that two phases exist between PbO$_2$ and Pb$_3$O$_4$ to which he assigned the notation α-PbO$_x$ and β-PbO$_x$. These phases were believed to have an orthorhombic crystal lattice. The range of oxygen content over which α-PbO$_x$ exists is $1.67 > x > 1.15$ and over which β-PbO$_x$ exists is $1.51 > x > 1.47$. From differential thermal analysis, Butler and Copp (83) found three phase transitions in the thermal decomposition of PbO$_2$ to Pb$_3$O$_4$: PbO$_2$ to α-PbO$_x$ to β-PbO$_x$ to Pb$_3$O$_4$. They identified α-PbO$_x$ with Holterman's (244) Pb$_7$O$_{11}$ and Clark and Rowan's (109) Pb$_5$O$_8$.

Anderson and Sterns (11) concluded that the best descriptions of these phases, according to x-ray and electron-diffraction studies, are $Pb_{12}O_{19}$ corresponding to α-PbO_x and $Pb_{12}O_{17}$ corresponding to β-PbO_x. This second phase, $Pb_{12}O_{17}$, has a very narrow stability range in air (83, 86,523) and is difficult to prepare (523). A series of decomposition temperatures for the thermal decomposition of PbO_2 has been given by Butler and Copp (83) and by Anderson and Sterns (11) and has been confirmed by White and Roy (523) as follows:

$$PbO_2 \xrightarrow{293°} Pb_{12}O_{19} \xrightarrow{351°} Pb_{12}O_{17} \xrightarrow{374°} Pb_3O_4 \xrightarrow{605°} PbO$$

Katz (280) and Burbank (78) report that α-PbO_2 is converted to β-PbO_2 before decomposing to the lower oxides after which the series of decomposition products formed with increasing temperature is the same as that obtained with β-PbO_2.

In the literature, as mentioned by Butler and Copp, the crystal structure for $Pb_{12}O_{19}$ has been identified as cubic, tetragonal, orthorhombic, and monoclinic. Since some lines were neglected in reaching these conclusions, Butler and Copp feel that all these treatments of the data are oversimplifications. The latest x-ray powder diffraction data for $Pb_{12}O_{19}$ is recorded by White and Roy (523), who conclude that this phase is monoclinic. Data are not given for $Pb_{12}O_{17}$ which has such very narrow stability limits. Angstadt et al. (12) and Reuter and Töpler (419) maintain that a conclusive distinction between orthorhombic and monoclinic structures cannot be made from powder diffraction data alone. Byström has suggested an orthorhombic structure, and Butler and Copp a cubic one for $Pb_{12}O_{17}$.

A word about nomenclature may be considered. The modifiers α and β have been used to describe the red and yellow forms of PbO as α- and β-PbO (280) and the $Pb_{12}O_{19}$ and $Pb_{12}O_{17}$ phases as α- and β-PbO_x (83, 86). As pointed out by White and Roy (523), Rüetschi and co-workers (12) have confused the nomenclature even further by referring to $Pb_{12}O_{19}$ obtained from α-PbO_2 as α-PbO_x and from β-PbO_2 as β-PbO_x. However, they (12) have shown from x-rays that their α- and β-PbO_x are identical. Any use of α and β for descriptions of oxide modifications other than α- and β-PbO_2 is discouraged. White and Roy have eliminated the α and β designations altogether and refer to α-PbO_2 as PbO_2II and to β-PbO_2 as PbO_2I.

f. Pb_2O_3. Clark and co-workers (110) heated various amounts of PbO_2, NaOH, and H_2O in a bomb, and depending on the experimental conditions, various compounds were detected from chemical analysis and x-ray diffraction studies. When 15 g of PbO_2, 5 g of NaOH, and

20 cc of H_2O were heated in a bomb for 3 days at 260–275°C, large black crystals of triclinic Pb_2O_3 were obtained; when heated for 3 days at 295–310°C, small black crystals of tetragonal Pb_5O_8 were obtained; and when heated for 3 days at 355–375°C, large crystals of monoclinic Pb_3O_4 were obtained. Gross (216) studied these crystals and showed that Pb_3O_4 is tetragonal and not monoclinic, and Pb_2O_3 is monoclinic and not triclinic. Confirmation of the monoclinic structure for Pb_2O_3 was obtained by Byström (85). Only under high pressures does Pb_2O_3 (522) exist as a stable phase. Darbyshire (121) was not able to detect Pb_2O_3 under ordinary laboratory conditions, but White and Roy (523) concluded that Pb_2O_3 is stable only above 1000 atm. According to White et al. (522), pressures between 10,000 and 14,000 atm are required to stabilize the Pb_2O_3 phase between 200 and 500°C. At 500°C and 13,000 atm, Pb_3O_4 begins to appear. X-ray powder diffraction data are given for monoclinic Pb_2O_3 by Clark and Roy (523). Clark's tetragonal Pb_5O_8 may be identified with monoclinic $Pb_{12}O_{19}$ (523).

Thermodynamic equilibrium P_{O_2}–T data for the equilibria in the Pb–O_2 system and powder diffraction data for the oxides have been reported by White and Roy (523).

g. Basic Lead Sulfates. Compounds of PbO and $PbSO_4$ are known, and Jaeger and Germs (257) and Schenck (448) made melts of various ratios of PbO to $PbSO_4$ to determine the chemical formulation of these compounds. They reported the existence of $PbSO_4 \cdot PbO$, $PbSO_4 \cdot 2PbO$, and $PbSO_4 \cdot 3PbO$. Using wet methods of forming the basic lead sulfates, Clark and co-workers (107) reported the detection of the compounds $PbSO_4 \cdot 2PbO$, $PbSO_4 \cdot 3PbO$, and $PbSO_4 \cdot 4PbO$ from x-ray diffraction studies.

Because of the existence of certain conflicts in the reported data, Lander (306a) made a systematic study of these compounds and used x-ray powder diffraction methods for identification of them. From the results of this work, he was able to construct a temperature–composition phase diagram for the PbO/$PbSO_4$ system. The mono-, di,- and tetrabasic sulfates may be obtained from melts of PbO and $PbSO_4$ in a ratio corresponding to the composition of the given basic sulfate. Below 450°C, $PbSO_4 \cdot 2PbO$ is unstable and decomposes to a mixture of the mono- and tetrabasic lead sulfate. By boiling a mixture of PbO and $PbSO_4$ in stoichiometric amounts in water for 7 to 8 hr, the mono- and tetrabasic sulfate may be made. In addition, the monohydrate of the tribasic lead sulfate, $PbSO_4 \cdot 3PbO \cdot H_2O$, may be formed, but on heating at 210°C, it also decomposes to a mixture of the mono- and tetrabasic sulfates. Powder diffraction patterns for these compounds are given.

A number of investigations of the basic lead sulfates (41,47,161a,271, 351,384,400,437,450) have been reported and confirm the conclusion that only the mono- and tetrabasic lead sulfates are stable in the anhydrous state at room temperature. From an analysis of x-ray data, it was suggested that $PbSO_4 \cdot PbO$ is tetragonal (437) and $PbSO_4 \cdot 4PbO$ is monoclinic (41,47,437).

For all preparations of the hydrates of PbO, Clark and Tyler (111) showed from x-ray data that the product is always $5PbO \cdot 2H_2O$. Electron diffraction data for $Pb(OH)_2$ are given by Fordham and Tyson (182) and for $3PbO \cdot 2Pb(OH)_2$ by Clark and Tyler (111).

2. The Electrochemistry of the Oxides of Lead

In the previous section, the physical and chemical properties of the lead oxides were considered. A discussion of the electrochemical conditions under which various oxides of lead may exist at lead anodes follows.

a. Standard Electrode Potentials. In acid solution lead dissolves as Pb^{2+} ions according to the reaction

$$Pb^{2+} + 2e \rightleftarrows Pb \qquad E_0 = -0.126 \text{ V} \qquad (7.23)$$

Carmody (97) measured the potential of a lead amalgam electrode in a saturated solution of $PbCl_2$ against a Ag/AgCl electrode. From a knowledge of the potential of a lead amalgam electrode against a Pb electrode in lead acetate solution which was recorded by Gerke (194) as 0.0057 V, a potential of -0.1263 V was obtained for Eq. 7.23. There was some disagreement (95,98) of about 0.006 V with Randall and Cann (415), but the results published by Fromherz (185) are in agreement with those of Carmody. The accepted E_0 value (126; ref. 315, p. 151) is -0.126 V.

In strong alkaline solutions, lead goes into solution as the plumbite ion, $HPbO_2^-$,

$$HPbO_2^- + H_2O + 2e \rightleftarrows Pb + 3OH^- \qquad E_0 = -0.54 \text{ V} \qquad (7.24)$$

for which the accepted E_0 value (126; ref. 315, p. 151) is -0.54 V.

The $PbSO_4$ electrode is important in battery research, and the solubility of $PbSO_4$ in H_2SO_4 solutions was found (145) to go through a minimum at a concentration of one mole of H_2SO_4 per mole of H_2O. In solutions from 0.5 to 10.5N H_2SO_4 (specific gravity from 1.17 to 1.31), the solubility is about 2 mg/liter (277); and in water, it is 47.1 mg/liter (277). As acid is added to water, the solubility of $PbSO_4$ decreases.

A cell composed of a lead amalgam electrode in contact with lead sulfate and a Pt/H_2 reference electrode in H_2SO_4 solutions was studied by Shrawder and Cowperthwaite (454).   The E_0 value of the $Pb(Hg)/PbSO_4$ electrode was determined as -0.3505 V.   Using Gerke's (194) value for the $Pb/Pb(Hg)$ cell, one arrives at a potential of -0.3562 V for the $Pb/PbSO_4$ electrode,

$$PbSO_4 + 2H^+ + 2e \rightleftharpoons Pb + H_2SO_4 \qquad E_0 = -0.356 \text{ V} \qquad (7.25)$$

This system was studied more thoroughly by Harned and Hamer (230). Latimer (ref. 315, p. 154) and Delahay et al. (129) quote a value of -0.356 V, but de Bethune and Loud (126) prefer a value of -0.3588 V for Eq. 7.25.

In the literature (126,129,311,315) a number of couples involving the oxides of lead are calculated from the thermodynamic data.   The Pb/PbO electrode,

$$PbO + H_2O + 2e \rightleftharpoons Pb + 2OH^- \qquad E_0 = -0.580 \text{ V} \qquad (7.26)$$

using red PbO has an E_0 in alkaline solutions of -0.580 V (126) and in acid solutions 0.248 V (129).   Ekler (155) gives a value of 0.253 V for Eq. 7.26 using yellow PbO in acid solutions.   For the PbO/Pb_3O_4 couple (126,315) in alkaline solution,

$$Pb_3O_4 + H_2O + 2e \rightleftharpoons 3PbO + 2OH^- \qquad E_0 = 0.248 \text{ V} \qquad (7.27)$$

when the value of 1.076 V quoted (129) in acid solutions is corrected for the hydrogen electrode (-0.828 V) in strong base.   Similarly, for the Pb_3O_4/PbO_2 couple,

$$3PbO_2 + 2H_2O + 2e \rightleftharpoons Pb_3O_4 + 4OH^- \qquad E_0 = 0.294 \text{ V} \qquad (7.28)$$

from the E_0 value in acid, 1.22 V (129).   The potential of the Pb/PbO_2 electrode,

$$PbO_2 + 4H^+ + 4e \rightleftharpoons Pb + 2H_2O \qquad E_0 = 0.666 \text{ V} \qquad (7.29)$$

is set at 0.666 V by Lander (311) and at 0.665 V by Rüetschi and Cahan (432).   An E_0 value of 0.242 V is given (129,311) for the $Pb/Pb(OH)_2$ electrode,

$$Pb(OH)_2 + 2H^+ + 2e \rightleftharpoons Pb + 2H_2O \qquad E_0 = 0.242 \text{ V} \qquad (7.30)$$

Other couples along with the pH dependency may be found in the report of Delahay and co-workers (129).

In the Pb electrode system, probably the $PbO_2/PbSO_4$ electrode,

$$PbO_2 + SO_4^{2-} + 4H^+ + 2e \rightleftharpoons PbSO_4 + 2H_2O \qquad E_0 = 1.68 \text{ V} \qquad (7.31)$$

commands the most interest. Vosburgh and Craig (505) describe the construction of such an electrode. A paste is made in H_2SO_4 solution with about equal quantities of $PbSO_4$ and PbO_2 obtained from the electrolysis of a nitric acid solution of $Pb(NO_3)_2$. Electrical contact is made with a Pt wire, and the E_0 potential of Eq. 7.31 at 25°C is recorded by Vosburgh and Craig as 1.681 V.

Hamer (224) maintained that earlier determinations of the potential of the $PbO_2/PbSO_4$ electrode were unreliable because reference electrodes of uncertain behavior were used. Using a Pt/H_2 reference electrode, he determined the potential of the $PbO_2/PbSO_4$ electrode as a function of temperature (0–60°C) and H_2SO_4 concentration (0.0005–7M). The results are recorded as $E_0 = 1.67699 + 2.85 \times 10^{-4}T + 1.2467 \times 10^{-6}T^2$. At 25°C, $E_0 = 1.68597$ V.

If the activities of H_2SO_4 and H_2O are determined (229) from the emf data reported by Hamer (224) for a series of H_2SO_4 solutions and compared to the corresponding values calculated from vapor pressure measurements, a discrepancy of about 2 mV is found (451,466). Beck and co-workers (29) suggest that Hamer's emf data are in error, and they investigated the potential behavior of the $PbO/PbSO_4$ electrode over a range of H_2SO_4 concentrations from 0.1 to 8 molality and over a range of temperature from 5 to 55°C. Their results obey the Nernst relationship, and the temperature coefficient agrees with calorimetric data (119). Using the activities obtained from Stokes data (466), a constant value of 1.687 V was determined (29) for E_0. The electrode system is reversible over the experimental range studied, and in fact, the $PbO_2/PbSO_4$ electrode is a good reference electrode (ref. 256, p. 399). On the basis of of experimental technique, Beck and co-workers could not offer an explanation for discrepancy in Hamer's data.

At first, Bode and Voss (46) reported that the potential of $PbO_2/PbSO_4$ was different for α-PbO_2 than for β-PbO_2. This difference in potential amounted to about 30 mV with the β form having the more noble potential. Later, Rüetschi and co-workers (430,431) found a potential of 1.7085 V for the α-$PbO_2/PbSO_4$ electrode and 1.7015 V for the β-$PbO_2/PbSO_4$ electrode with respect to a Pt/H_2 reference electrode in 4.4M H_2SO_4. They report (434) E_0 values of 1.698 V for α-PbO_2 and 1.690 V for β-PbO_2. From considerations of the physical and chemical properties of α- and β-PbO_2, they (430,431,434) conclude that the results obtained by Bode and Voss are in error. In confirmation of this conclusion, Bone and co-workers (51) found the potential of α-PbO_2 electrodes to be about 10 mV more noble than that of β-PbO_2 electrodes.

Latimer (ref. 315, p. 155) accepts a value of 1.685 V for the E_0 of the $PbO_2/PbSO_4$ electrode, but de Bethune (126) quotes 1.682 V.   Probably, the value of 1.687 V found by Beck et al. (29) should be favored.

Ohse (384) measured the potential of the $PbO \cdot PbSO_4/PbSO_4$ electrode,

$$PbO \cdot PbSO_4 + 2H^+ + 2e \rightleftharpoons Pb + PbSO_4 + H_2O \qquad E_0 = 0.034 \text{ V} \quad (7.32)$$

The potential–pH diagrams for lead in the presence and absence of sulfate are given by Delahay and co-workers (129).   Rüetschi and Angstadt (429) have attempted to modify these diagrams by adding stability regions for the basic lead sulfates.

b. Charging Curves.   Constant current charging curves on Pb electrodes have been investigated (12,28,76,78,128,155,168,170,176,252, 254, 270,275,335,346,347,352,353,369,401,429,431,432,526).   A typical charging curve for a Pb anode in H_2SO_4 solutions is given in Fig. 7.7.   As soon as the electrical circuit is closed, the potential rises rapidly to the first plateau AB, during which the lead surface becomes covered with a layer of $PbSO_4$.   From the length of the plateau, Rüetschi and Cahan (432) calculated that a layer about 640 Å thick is built up by the end of the $PbSO_4$ plateau.   Intermittent charging curves obtained by Nagel and co-workers (369) show that during the first plateau the rest potential corresponds to the $Pb/PbSO_4$ potential, as seen in Fig. 7.8.

Since the resistance of $PbSO_4$ is very large (about 10^8 ohms-cm) (177, 432), the lead surface becomes passivated as the film grows, and the potential rises to more noble values.   Double layer capacity measurements made during the charging curve by Kabanov and co-workers (276) indicate that the Pb surface becomes passive when more than 90% of the surface is covered with an insulating layer of $PbSO_4$.   At this point

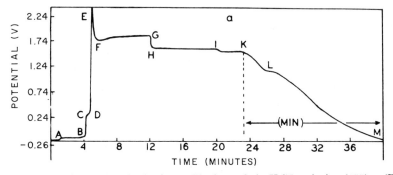

Fig. 7.7. Charging curves obtained on a Pb electrode in H_2SO_4 solution (155).   (By permission of The National Research Council of Canada.)

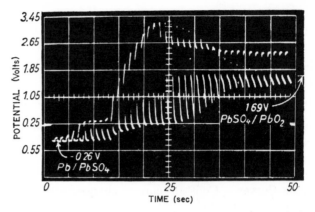

Fig. 7.8. Intermittent charging curve (369) obtained on a Pb anode in $2N$ H_2SO_4 solution at 0.71 mA/cm². (By permission of the Bunsengesellschaft.)

the double layer capacity falls from 50 to 5 $\mu F/cm^2$. Borisova and co-workers (56) have also made double layer capacity measurements on Pb. As this insulating layer builds up, Fig. 7.8 shows how quickly the potential builds up. Supporting evidence for the buildup of an insulating film of $PbSO_4$ was obtained by Maeda (346) using radioactive tracer techniques. Lead anodes were studied in solutions of Na_2SO_4 labeled with [35]S.

According to the date in Fig. 7.7, a very small inflection is observed at a potential of about 0.3 V at CD. The break in the curve disappeared at low H_2SO_4 concentrations but became more prominent as the concentration increased from 0.09 to $10.5N$ (155). In Fig. 7.8, the rest potential rises to the vicinity of 0.3 V just before a maximum in the polarization is reached. This potential corresponds to the potential of the Pb/PbO couple, Eq. 7.26, or the $Pb/Pb(OH)_2$ couple, Eq. 7.30. Lander (308) and Burbank (78) have detected the presence of PbO in partially charged lead–acid battery plates. Possibly the break in the curve of Fig. 7.7 at CD reflects the presence of PbO. A poorly defined inflection noted by Rüetschi and Cahan (432) was attributed to the presence of PbO_2 since they estimated that the break occurred at about 0.8 V. According to Ekler (155), the length and position of the inflection depends on current density, and it is possible that Rüetschi and Cahan charged at a rate with which it is difficult to locate the break.

After the break in the curve at CD, a sharp peak is observed at F in Fig. 7.7. At this point, the polarization is at a maximum value (Fig. 7.8). Since the surface is passivated by the $PbSO_4$ layer, the potential must rise until another process can take place.   This process is the formation of PbO_2, and the sharp potential peak in the charging curve is related (176,177) to the difficulty with which PbO_2 is nucleated in the $PbSO_4$ layer.   Fleischmann and co-workers (176,177) discuss the theory and mechanism of the nucleation and growth of films.   Ikari and Yoshizawa (254) favor the suggestion that the formation of poorly conducting oxides lower than PbO_2 causes the peak to appear.   Conversion of the $PbSO_4$ to PbO_2 by reaction with OH radicals produced by the discharge of H_2O is offered as a mechanism by Feitknecht and Gaumann (107).   In this way, the peak may be associated with the overvoltage required to discharge H_2O on the $Pb/PbSO_4$ electrode.   Once the highly conducting PbO_2 layer begins to form, the potential falls (Fig. 7.7), and the polarization decreases (Fig. 7.8).

After the potential peak is passed, the potential goes through a slight minimum before slowly decreasing toward less noble values with time. During this time, the open-circuit potential of the system as noted in Fig. 7.8 corresponds to the $PbO_2/PbSO_4$ potential, and the polarization slowly decreases with time.   In this region, O_2 is evolved on the PbO_2 surface, and PbO_2 is continuously formed by anodic attack of the metallic Pb (76,78,155,177,429,432).

In alkaline solutions, the minimum following the peak in the charging curves on Pb is more pronounced than in acid solutions, as seen in Fig. 7.9.   Of course, the initial plateau is the formation of a layer of PbO (51,270).   According to Burbank, the PbO obtained by the potentiostatic anodic polarization of Pb in alkaline solutions has a distorted tetragonal lattice and is more reactive than litharge or massicot.   (Note that these common names for tetragonal and orthorhombic PbO are reversed from the usual notation in Table I of ref. 78.)   She refers to the distorted PbO as PbO_t.   At higher potentials, Pb_3O_4 and $Pb_{12}O_{19}$ are formed so that a layer composed of a mixture of these three oxides is obtained before the potential peak is reached.   After the peak in Fig. 7.7 is reached, PbO_2 nuclei are formed, and the PbO is converted to α-PbO_2 in the high potential range.

When lead is anodized in KOH solution, a loosely held film of PbO_2 is found which may be flaked off to expose an orange layer of the lower oxides underneath.   Such an observation indicates (78) that Pb corrodes in alkaline solutions by a divalent mechanism (203,217,307,308),

$$Pb + 2OH^- \rightarrow PbO + H_2O + 2e \qquad (7.33)$$

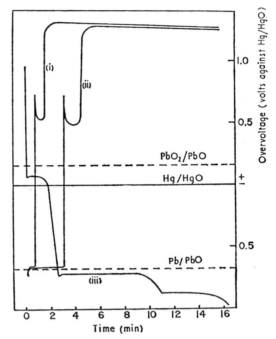

Fig. 7.9. Charging curves on Pb electrodes (270) in $1N$ KOH: (i) at 45 mA/cm^2 on Pb anode; (ii) at 23.9 mA/cm^2 on Pb anode; and (iii) at 5 mA/cm^2 on Pb/PbO$_2$ cathode. (By permission of The Faraday Society.)

Part of the PbO is then oxidized to α-PbO$_2$ by oxygen discharged from water (151,280),

$$H_2O \rightarrow (O)_{ads} + 2H^+ + 2e \qquad (7.34)$$

$$PbO + (O)_{ads} \rightarrow PbO_2 \qquad (7.35)$$

Because the oxide films formed on Pb in KOH are not protective (78), in contrast to the highly protective film of PbSO$_4$ solutions, Pb corrodes at a high rate in alkaline solutions.

Above a potential which roughly corresponds to the O$_3$/H$_2$O couple,

$$O_3 + 6H^+ + 6e \rightleftharpoons 3H_2O \qquad E_0 = 1.51\text{--}0.059 \text{ pH} \qquad (7.36)$$

lead corrodes directly to PbO$_2$ by a tetravalent mechanism. In acid solutions, Lander (307,308) concluded that below 1.51 V nearly all the current goes into a divalent mechanism, and only above this potential does the tetravalent corrosion path become important. Because PbSO$_4$ forms a protective layer permeable to H$^+$ ions and H$_2$O molecules but not to $SO_4{}^{2-}$ ions (58,307,526), an alkaline condition may be maintained next to the Pb interface which explains the detection (78,307,308) of PbO on lead anodes in acid solutions. At first, Rüetschi and coworkers were not able to detect PbO on Pb anodes (432), but later (429) agreed that alkaline conditions can be set up behind the $PbSO_4$ film so that PbO or basic lead sulfates may be formed at the Pb interface.

Once the tetravalent formation of α-PbO_2 takes place, O_2 evolution is observed. Both α- and β-PbO_2 are observed on Pb anodes in acid solutions (12,78,168,431,432). It is agreed that the β-PbO_2 is formed by the oxidation of $PbSO_4$, whereas the α-PbO_2 is formed next to the lead interface in a region of high pH either by the oxidation of PbO through a divalent mechanism or by the direct oxidation of Pb through a tetravalent mechanism.

To explain the appearance of PbO in the oxide films of Pb anodes in H_2SO_4 solution, Lander (307) suggested that it was formed by a solid state mechanism,

$$PbO_2 + Pb \rightarrow 2PbO \qquad (7.37)$$

This reaction is possible, as shown by Lander (307,308). He sandwiched PbO_2 between two weighed Pb plates. After passing a current of 15 mA/cm^2 for 400 hr, the resistance in the circuit had increased with time, and the lead plates had lost weight. As added evidence, granulated Pb was mixed with PbO_2 and glass beads. This mixture was rotated for several days. X-ray examination of the mixture initially showed the presence of only PbO_2 and Pb. At the end of the experiment, the PbO_2 lines were very faint, but the lines for tetragonal PbO were very strong. Although Eq. 7.37 is possible, it seems improbable, as noted by Feitknecht (168), that Eq. 7.37 occurs at the Pb anode. A more likely process is the oxidation of lead according to an equation similar to Eq. 7.33.

The cathodic discharge curve obtained by Ekler (155) on a Pb/PbO_2 electrode is shown in Fig. 7.7. from G to M. As soon as the cathodic circuit is closed, the potential drops quickly from the oxygen overvoltage value to a plateau HK, corresponding to the $PbO_2/PbSO_4$ potential. In the middle of this plateau, there is a step of about 30 mV (432) at I. This step has also been observed by Rüetschi et al. (431,432) and Ikari (252). The upper part corresponding to HI has been ascribed to the reduction of α-PbO and the lower part IK to the reduction of β-PbO, by Rüetschi and Cahan (431,432). A small minimum or overshoot has been observed (252,270,352,353,431) just before the PbO_2 plateau.

Mark (352) studied the discharge behavior of pure α- and β-PbO$_2$ and laminated layers of the two oxides plated on Pt electrodes. The discharge characteristics of α- and β-PbO$_2$ are different, as seen in Fig. 7.10. Similar curves were obtained by Angstadt et al. (12). On a pure α-PbO$_2$ electrode, the minimum preceding the PbO$_2$/PbSO$_4$ plateau was missing, and the length of the PbO$_2$ plateau was much too short compared to the calculated amount of PbO$_2$ plated on the electrode. Decay of the potential after the PbO$_2$ step was slow. On the other hand, a minimum preceded the PbO$_2$ plateau on the discharge of a pure β-PbO$_2$ electrode, and the number of coulombs corresponding to the long flat plateau accounted for up to 90% of the calculated PbO$_2$ plated on the electrode.

If the discharge of α-PbO$_2$ was halted at point e in Fig. 7.10, a white film of PbSO$_4$ was found to cover the electrode surface. An analysis of the film with KI showed the presence of appreciable amounts of unreacted PbO$_2$. When this film was removed by washing in a NH$_4$Ac solution, the electrode was reactivated and another discharge at the PbO$_2$ potential was obtained. By repeating this procedure, the calculated coulomb capacity could be realized experimentally. After the in-

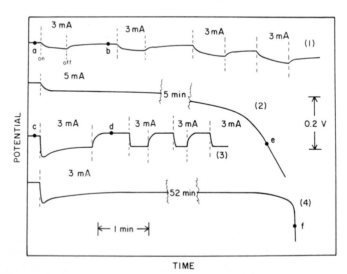

Fig. 7.10. Constant current discharge curves (352) of α- and β-PbO$_2$ cathodes in 0.1M H$_2$SO$_4$ solution: (1) α-PbO$_2$ under successive short discharges of 3 mA; (2) α-PbO$_2$ under continuous discharge of 5 mA; (3) β-PbO$_2$ under successive short discharges of 3 mA; and (4) β-PbO$_2$ under continuous discharge of 5 mA. (By permission of The Electrochemical Society.)

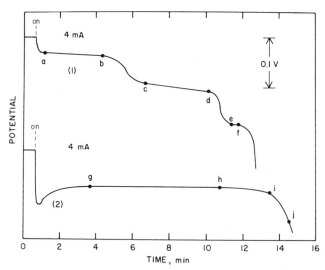

Fig. 7.11. Constant current discharge curves (352) obtained on laminar α- and β-PbO$_2$ films plated on Pt in 0.1M H$_2$SO$_4$ solution at 4 mA: (1) a layer of α-PbO$_2$ plated over a layer of β-PbO$_2$; (2) a layer of β-PbO$_2$ plated over a layer of α-PbO$_2$. (By permission of The Electrochemical Society.)

terruption of the discharge of β-PbO$_2$ at point f in Fig. 7.10, in the electrode could not be rejuvenated with NH$_4$Ac, and reaction with KI showed very little unreacted PbO$_2$ present. White crystals of PbSO$_4$ were scattered over the electrode surface, but a white film as found on the surface of α-PbO$_2$ was not observed. Mark (352) concludes that the poor discharge capacity of α-PbO$_2$ electrodes may be traced to a complete, tightly adherent film of PbSO$_4$.

Electrodes on which a layer of β-PbO$_2$ was plated over a layer of α-PbO$_2$ were discharged under constant current conditions (252,352). Also investigated was the converse arrangement with β-PbO$_2$ underneath the α-PbO$_2$. The results are given in Fig. 7.11. As found by Ikari and co-workers (252), the outside layer discharges first followed by the second (352). It is possible that the two steps in the PbO$_2$ plateau of Fig. 7.7 (155,431,432) may be associated with the discharge of a mixture of α- and β-PbO$_2$ produced by the anodization of a Pb electrode in H$_2$SO$_4$ solution. However, in the light of the data in Fig. 7.11, it is likely that the upper plateau corresponds to the reduction of β-PbO$_2$ since it is at the solution interface even though α-PbO$_2$ has the higher rest potential (431,432). Mark (352) suggests that the mechanisms of the discharge of the two oxides are different in 0.1M H$_2$SO$_4$ solution.

Astakov and co-workers (16) observed that α-PbO$_2$ preparations contained large crystals, whereas β-PbO$_2$ samples were composed of very fine crystallites. As concluded by Angstadt and co-workers (12), the poor discharge capacity of α-PbO$_2$ electrodes (less than one-half of a β-PbO$_2$ electrode) may be traced to the low surface area of α-PbO$_2$ deposits. Only a thin layer about 10^{-4} cm deep of α-PbO$_2$ reacted (12, 252).

Lorenz and Lauber (335) reported at least 14 arrests in the cathodic discharge curve of a Pb/PbO$_2$ electrode, but Wynne-Jones and co-workers (27,28) could not detect all of these arrests. They concluded that the discharge process was the reverse of the charge process which corresponds to the reaction associated with the reversible PbO$_2$/PbSO$_4$ electrodes since plateaus corresponding to the reduction of PbO$_2$ and PbSO$_4$ were the only ones observed. Although Burbank (76) observed that PbO$_2$ was reduced to PbSO$_4$, several arrests in the potential range between -0.1 and -0.2 V which were associated with PbO, PbSO$_4$·PbO, or Pb(OH)$_2$ were detected. By private communication, she explained that these intermediate species arise from corrosion reactions of the underlying lead which take place before the PbO$_2$ is totally discharged to PbSO$_4$. The potential arrests are mixed potentials which may be associated with the different corrosion products.

An arrest in the potential range indicated by L of Fig. 7.7 has not been recorded anywhere else, and Ekler (155) offers no explanation for this behavior.

Cahan and Rüetschi (91) studied the polarization of a Pb anode by superimposing an ac square wave current on the steady direct current. In this way, the impedance of the electrode was determined along with the potential as a function of time. The results for the discharge curve of a Pb/PbO$_2$ electrode are given in Fig. 7.12. When the cathodic circuit was closed, the potential dropped from the oxygen η value to the PbO$_2$/PbSO$_4$ potential, and the double layer capacity was very high (500–1000 μF/cm^2). After the PbO$_2$ step, a short arrest appeared between -0.1 and -0.2 V, and the capacity fell to a very low value. Such behavior agrees with the contention that this step in the discharge curve may be associated with the presence of the poorly conducting PbO, Pb(OH)$_2$, or basic lead sulfates. Finally, a long PbSO$_4$ plateau is reached with a capacity value of about 75 μF/cm^2 followed by the slow descent to H$_2$ evolution.

In alkaline solutions, the discharge curve, as shown by curve iii of Fig. 7.9, exhibits steps corresponding to the PbO/PbSO$_4$ and the Pb/PbO potentials before H$_2$ evolution is reached. A small overshoot is

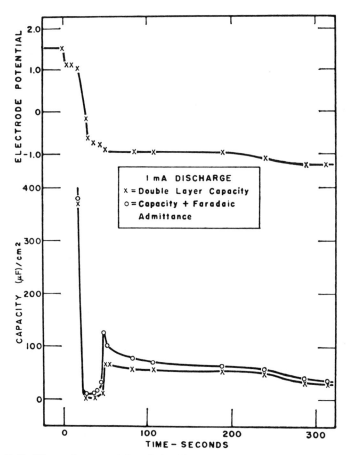

Fig. 7.12. Electrode potential and double layer capacity (91) as a function of time during the discharge of the anodic oxide film on a Pb electrode at 1 mA/cm². (By permission of The Electrochemical Society.)

detected just ahead of the PbO plateau.   Jones and co-workers (270) suggest that a plateau following the PbSO₄ step is associated with a lead hydride.

3. Oxygen Overvoltage Studies on Lead

Many investigators (5,102,156,176,177,203,235,268,270,295,353,430, 433,434) have studied the mechanism of the evolution of O₂ on Pb anodes.   It is generally agreed that a layer of PbO₂ is formed on the Pb surface before O₂ evolution takes place, and this anodic process occurs on the PbO₂ surface.

a. Rest Potentials.   When a dry sample of Pb is plunged into H_2SO_4 solutions (about 30%), the open-circuit potential is nearly zero volt (76). From electron diffraction data it is determined that a film of $Pb(OH)_2$ exists on the Pb surface.   With time, the rest potential drifts to a steady value within a few millivolts of the $Pb/PbSO_4$ potential.   If the Pb sample is etched with $HAc-H_2O_2$ or HNO_3 etching solution, a film of $3PbO \cdot 2Pb(OH)_2$ cover the lead surface as determined from electron diffraction measurements, but this film may be removed by washing in a solution of NH_4Ac.   After a Pb sample whose surface is free of absorbed films is placed in H_2SO_4 solution, the potential is about -0.5 V initially but then drifts to the $Pb/PbSO_4$ potential with time.

This $PbSO_4$ potential at about -0.35 V may be maintained in N_2 stirred acid solutions (452).   If, however, the acid solution is aerated or stirred with O_2, the $PbSO_4$ electrode passivates, and the potential rises to a steady, poorly reproducible value between 0.5 and 0.65 V more noble than the $Pb/PbSO_4$ potential (5,102,226,367).   The time required for a given electrode to passivate seems to vary over a relatively wide range, as shown by the dashed lines in Fig. 7.13.   The rate of passivation is a diffusion-controlled process (226,367) and increases with increasing P_{O_2}.

After the rest potential had remained constant at the $PbSO_4$ potential for a number of hours, it rose quickly to a potential in the neighborhood

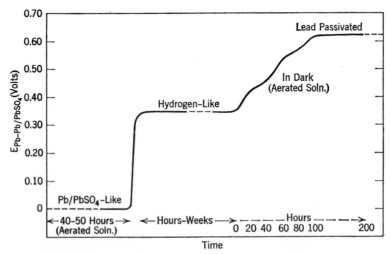

Fig. 7.13. Rest potential of Pb in oxygen-saturated H_2SO_4 solution (102) as a function of time.   (By permission of The Electrochemical Society.)

of 0 V (Fig. 7.13). Following a stay at this potential for some time, the rest potential slowly drifts to the passivated potential of about 0.3 V. Casey and Campney (102) have shown that irradiation of a Pb electrode of 0 V with ultraviolet light decreases the time for passivation in proportion to the intensity of the radiation. According to photomicrographs (102), the only apparent physical difference between a passivated and unpassivated Pb electrode is the number and depth of the small crystals of $PbSO_4$ on the metal surface. If H_2O_2 was added to the solution (102), the passivation time was reduced nearly to zero.

Potential-determining mechanisms have been offered for the open-circuit corrosion of Pb in H_2SO_4 solutions. On the $Pb/PbSO_4$ plateau, the rest potential is determined by the $Pb/PbSO_4$ reaction, Eq. 7.25. However, the passivated potential is not so easily explained. Müller (367) proposed that the ennobled potential was caused by an IR drop in the pores of the $PbSO_4$ layer produced by local cell corrosion currents. The local corrosion cell is made up of anodic sites at which Pb dissolves to form Pb^{2+} ions which react with the $SO_4{}^{2-}$ ions producing $PbSO_4$, and cathodic sites at which H_2 is evolved. In aerated solutions, the H_2 could react with adsorbed O_2 on the electrode surface, and the mixed potential would lie somewhere between the H_2 and the O_2 potentials, thus giving an overall mixed potential above 0.6 V more noble than the $PbSO_4$ potential (226). From the studies of the effect of H_2SO_4 and H_2O_2 concentrations on the rest potential, Casey and Campney (102) concluded that O_2 can be dissolved in the $PbSO_4$ layer and the passivated potential is determined by an unspecified equilibria between $SO_4{}^{2-}$, $S_2O_8{}^{2-}$, $SO_5{}^{2-}$ ions and H_2O_2 molecules.

Since the passivated potential ($\sim$0.3 V) does not correspond to any known lead–lead oxide or oxygen potential, it is most likely a mixed potential. The oxygen depolarized hydrogen electrode scheme of Haring and Thomas seems plausible. Possibly, the intermediate potential at 0 V could be associated with the $PbO \cdot PbSO_4$ potential measured by Ohse (384). The presence of basic lead sulfate in the $PbSO_4$ layer may correspond to a form of the dissolved oxygen suggested by Casey and Campney.

The rate at which a lead electrode reached the passive potential, $\sim$0.3 V, was greatly increased (about 6 times) by alloying the Pb with 9% Sb. The presence of 0.08% Ca delayed the time required to reach the passive potential as compared to pure Pb (226). Antimony alloys hasten self-discharge, whereas Ca alloys hinder it, as will be discussed later.

b. Steady-State Polarization Studies. As early as 1922, Glasstone (203) observed that the evolution of oxygen takes place only after a layer of PbO_2 has been formed. The oxygen overvoltage on Pb is very high (434); e.g., η has values between 0.6 and 0.7 V at a current density of only 10^{-7} A/cm². Some workers (235,268,270) found that the η varied with time and was difficult to reproduce unless the polarization values were recorded after a fixed arbitrary time.

Oxygen may be occluded by PbO_2 (156,168,274,430), and its presence increases the potential to more noble values. As O_2 is removed (168) the potential falls. To get reproducible η curves, Rüetschi and co-workers (5,429,431,433) had to preanodize the electrode at potentials above oxygen evolution to get a stable and reproducible surface. Once this was done, reproducible polarization data were obtained. Overvoltage curves obtained on both α- and β-PbO_2 anodes for the evolution of oxygen are given in Fig. 7.14.

The η on β-PbO_2 is higher than on α-PbO_2, and the slopes are different, which indicates that the mechanism of O_2 evolution is different on the two forms. On β-PbO_2, the normal slope of 0.12 found in acid

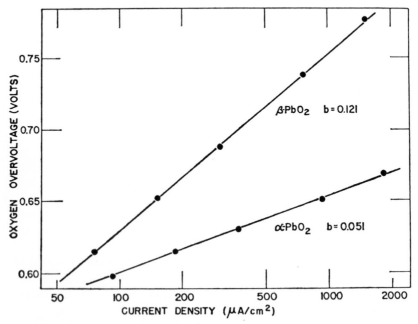

Fig. 7.14. Oxygen overvoltage (431) on α- and β-PbO_2. (By permission of the Electrochemical Society.)

solutions on the noble metals is obtained, but on α-PbO_2, a 0.05 slope typical for many metals in alkaline solutions is observed. Since i_0 for β is 6.2×10^{-10}, whereas i_0 for α is 1.7×10^{-16} A/cm^2, either the β form offers a better catalytic surface for the oxygen evolution process than the α form or there is a large change in roughness factor between an α- and a β-PbO_2 surface. As mentioned above, Astakhov and co-workers (16) noticed that the deposit of α-PbO_2 consisted of much coarser crystals than the β-PbO_2 deposits.

Recently, Makrides (349) reported that η measurements made on low surface area α- and β-PbO_2 anodes were identical in the two cases. Normally prepared PbO_2 electrodes are porous. As pointed out by Frumkin (187), polarization measurements obtained on porous and smooth electrodes may lead to different results because of a nonuniform current distribution over the electrode surface. No details were given (349) as to how the compact PbO_2 electrodes were obtained, how the measurements were made, or what the shape of the curves were. It was merely reported that no difference in the kinetic parameters for oxygen evolution were found. Until more complete reports are received, further comment must be withheld.

With decreasing current density, the potential of the system does not go to the reversible O_2 potential (1.23 V) but becomes steady on open circuit at the $PbO_2/PbSO_4$ potential (433,434). Such behavior is the result of a film of $PbSO_4$ which forms on the PbO_2 surface in the vicinity of the $PbO_2/PbSO_4$ potential because of the self-discharge of PbO_2 in H_2SO_4.

c. Mechanisms. The fact that the steady-state polarization curve obtained on Pb anodes for the evolution of oxygen has a Tafel region with a slope between 0.1 and 0.14 (5,176,219,268,430,431,469) indicates that the process is activation controlled. A mechanism with an electron-transfer step as rate determining is not incompatible with these findings. As the temperature is increased, the overvoltage decreases (268), which is to be expected if the process is activation controlled.

According to Krasil'shchikov (295), the evolution of O_2 proceeds through the discharge of OH^- ions or H_2O molecules to form adsorbed OH radicals which, in turn, are discharged to form absorbed O^{2-} ions. These O^{2-} ions convert the PbO_2 to a higher oxide which decomposes (202) to yield O_2 and PbO_2. However, PbO_2 cannot be oxidized to a higher state of oxidation (433), and this mechanism is not favored for the evolution of O_2 on Pb anodes.

Burbank (78) observed that O_2 was evolved on Pb anodes only at potentials more noble than a value corresponding roughly to the O_3/H_2O

potential (1.5 V in strong acid solutions). Below this potential limit, the Pb corrodes by a divalent mechanism, H_2O is discharged, and PbO_2 is formed by the reaction of PbO and adsorbed oxygen, Eq. 7.33–7.35, called the water reaction. As a result, oxides are formed in preference to O_2 evolution. If, however, a Pb anode is corroded at a fixed potential below the O_3/H_2O limit and the surface is scratched so that direct contact is made between PbO_2 and Pb, O_2 is evolved immediately. It is concluded (78) that the Pb surface must be in contact with the electronically conducting PbO_2 before O_2 evolution will occur. If the poorly conducting layer of PbO separates the Pb from the PbO_2 layer, the water mechanism of oxide formation takes place instead of O_2 evolution. Above the O_3/H_2O potential, the tetravalent corrosion of Pb takes place, and O_2 is evolved.

Since PbO_2 is such an excellent conductor, and a higher oxide does not exist, the surface of a Pb/PbO_2 electrode acts as an inert electrode, and evolution of oxygen most likely proceeds by a mechanism similar to that on the noble metals (see Chap. III). Water molecules or OH^- ions are discharged to form adsorbed OH radicals which either react or are discharged to form adsorbed oxygen atoms. Oxygen is evolved by the combination of O atoms. Some of this oxygen could diffuse into the PbO_2 layer as observed experimentally (156,168,430). It it not unlikely that the discharge of the first electron, either

$$H_2O \rightarrow (OH)_{ads} + H^+ + e \tag{7.38}$$

or

$$OH^- \rightarrow (OH)_{ads} + e \tag{7.39}$$

is rate determining, as suggested by Jones and co-workers (268).

In alkaline solutions, the oxygen overvoltage curves obtained (270) on Pb anodes were similar to those in acid solutions. A Tafel region with a slope of about 0.12 was found, but negative deviations from the Tafel behavior were observed at low current densities because self-discharge processes become important in this region.

Burbank (78) points out that O_2 could be evolved by the discharge of H_2O to H_2O_2, which has an E_0 of 1.77 V and could take place at these high overvoltage values. Oxygen would then be evolved by the catalytic decomposition of the H_2O_2.

4. Fundamental Battery Studies

Because commercial battery plates for lead–acid batteries are pasted plates and because most battery failures occur as the result of grid corrosion, basic studies of pasted electrodes must be carried out to study possible approaches to improved battery performance and technology.

a. Cell Voltage. It is generally agreed (30,112,230) that the cell reaction for the lead–acid battery is the so-called double sulfate theory of Gladstone and Tribe (198), Eq. 7.20.   The potential of this cell has been determined as a function of the H_2SO_4 concentration, the temperature, and the pressure (30,112,118,230,284,476,505).   As a rule, the emf of the cell increases with increasing H_2SO_4 concentration, increasing temperature, and decreasing pressure.   In a range of solutions from 0.2 to 1.6M, the temperature coefficient, dE/dT increases with increasing concentration.   Between 1.6M and 4M, dE/dT is relatively constant at about 0.26 V/°C, and then decreases again at higher H_2SO_4 concentrations (476).   In about 21% H_2SO_4 (1.15 specific gravity), Harned and Hamer (230) give the voltage of the lead–acid battery as $E = 2.00795 + 1.237 \times 10^{-3}T + 2.536T^2$ V.   Beck and Wynne-Jones (30) give a summary of these measurements.

The relationship between specific gravity, per cent by weight, and molarity of H_2SO_4 solutions may be found in Table XIII of ref. 500, p. 107.   The usual battery acid has a specific gravity of 1.26 which is about 35% H_2SO_4 by weight or between 3.5M and 4M in H_2SO_4.

Cohen and Overdijkink (112) and Beck and Wynne-Jones (30) have determined the pressure coefficient dE/dp of the lead–acid cell.   At 1 atm and 0°C, dE/dp is $-3.07 + 10^{-6}$ V/atm.

b. Battery Paste. The active material of a battery plate is formed from a paste of lead oxides in sulfuric acid which is applied to a suitable grid structure made of an alloy of lead (usually between 5 and 12% Sb for the sake of rigidity).   Ordinarily, the oxide mixture is obtained by burning lead in a furnace and is a mixture of Pb_3O_4 and tetragonal PbO (212) from which a paste is made with a solution of H_2SO_4.   Basic lead sulfates are formed (421,450) in the paste and have been identified (363) with x-ray diffraction measurements.   After the paste has been applied to the grid, it is dried in an oven under controlled conditions of atmosphere and temperature.   Finally, the plate is formed electrochemically in acid solution, anodized to PbO_2 for the positive plate, or cathodized to sponge Pb for the negative plate.

The temperature at which the plate is formed affects the performance of the final battery plate.   How much the oxygen content of the formed active material deviates from the stoichiometric ratio (O/Pb = 2) for pure PbO_2, is called the amount of apparent PbO.   The higher the forming temperature, the lower is the amount of apparent PbO (143,212, 231) which is desirable from two standpoints.   As the O/Pb ratio approaches the ratio for PbO_2, the coulombic capacity of the plate is greater, and as the temperature of formation increases, the rate of self-discharge of the formed plates is lowered (212,231). Hatfield and Brown (231) observed that the shedding of active material on a plate in service decreased with an increase in the forming temperature.

It has been reported by Ishikawa and Tagawa (255) that positive battery paste made with H_3PO_4 rather than H_2SO_4 increased the life of the positive plate by 40% with respect to mechanical failure. This behavior was attributed to the increased density of the paste.

When a plate was formed electrochemically, Dittmann and Sams (142) observed that increased capacity of the positive plate was obtained if the formed plate was permitted to stand for a given time after which it was recharged. They attributed this effect to a decrease in the density of the active material. Dobson reported that the amount of α-PbO_2 in the formed plate increased with increasing density of paste, increasing temperature of the electrolyte, and decreasing specific gravity of H_2SO_4 solution, and it depended on the rate of formation (current density). The presence of PbO in the paste causes the formation of α-PbO_2. For a dense plate formed rapidly in low specific gravity H_2SO_4, a film of $PbSO_4$ creates a high pH region next to the Pb where α-PbO_2 is formed from unreacted PbO. With a less dense plate, at lower current densities the SO_4^{2-} ions can penetrate the paste and convert most of the PbO to $PbSO_4$ which then oxidizes to β-PbO_2.

With anodization, α-PbO_2 is formed by the direct oxidation of the metallic Pb (432) or from the PbO of the paste (143), and β-PbO_2 is formed by the oxidation of $PbSO_4$ (432). Self-discharge occurs through the reverse of Eq. 7.20 by converting both α- and β-PbO_2 to $PbSO_4$ (168, 198). When the electrode is recharged, the $PbSO_4$ resulting from the self-discharge of α-PbO_2 is converted to β-PbO_2 and consequently, the amount of β-PbO_2 increases at the expense of the α-PbO_2 with recharging the electrode. Also, since the oxygen content of α-PbO_2 is less than β-PbO_2 (30) and the discharge capacity of β-PbO_2 plated on Pt is greater than α-PbO_2 plated on Pt (12, 352), it is expected that a plate with a high content of α-PbO_2 would be more dense and have less capacity, as found by Dittmann and Sams (142).

To test these conclusions, Dobson (144) made pure α-PbO_2 pasted plates by soaking the cured plate in $1N$ NaOH solution for 2 hr, followed by anodic formation in the NaOH solution, and pure β-PbO_2 pasted plates by soaking the cured plate in 1.4 specific gravity H_2SO_4 for 18 hr before anodic formation in 1.135 specific gravity H_2SO_4. He found that the coulomb capacity of the β-PbO_2 plates was about 1.7 times that of the α-PbO_2 plates. In addition, the amount of β-PbO_2 in the active material of a given positive lead plate increased with increased cycling of the plate. These findings support the conclusions presented above.

Simon and co-workers (81,458,459) have studied the microstructure of the active material of lead–acid battery plates during formation and during anodic corrosion with photomicrographic techniques. Other studies of the microstructure of the lead plates have been carried out by Feitknecht (168,170), Burbank (79), Shibasaki (453), and Astakhov and co-workers (16). Apparently, those battery plates exhibiting a prismatic character maintain a higher capacity. Interlocking of the crystallities reduces shedding and loss of active material with service.

c. Grid Corrosion. One of the most serious problems concerning the lead–acid battery in service is failure of the battery because of mechanical disintegration of the battery plate grids due to the action of corrosion processes. A number of basic studies have been carried out (78–80, 122,144,212,226,307,308,310,311,429,431,432,458a,526) on the corrosion of Pb and alloys of lead.

Because the nature of the corrosion product on Pb depends on the potential, the lead corrosion studies were carried out at constant potential as early as 1941 by Wolf and Bonilla (526). Weighed samples of pure lead were corroded at constant potential by Lander (307,311) in H_2SO_4 concentrations over a range from 1 to 40% in a temperature range from 20 to 50°C for various periods of time as a function of the potential. The amount of corrosion was determined from the loss in weight of the corroded sample after the corrosion film had been stripped with NH_4Ac and the sample dried and weighed. Results of this work are shown in Fig. 7.15. With increasing potential, the rate of corrosion increases until a potential of about 1.57 V is reached after which the corrosion rate falls to very low values in the vicinity of the $PbO_2/PbSO_4$ potential. Further increases in potential cause the rate to increase again.

At potentials between 1 and 0 V, a loosely held film of $PbSO_4$ covers a brown or black layer of tetragonal PbO which was observed as low as 0.1 V. Such a compound can exist in 40% H_2SO_4 because a protective film of $PbSO_4$ is formed. In alkaline solutions, no protective film is formed over the entire range of anodic potentials (78), and high rates of corrosion occur at all potential values. Unless the electrolyte contains an ion which will form a protective layer, corrosion cannot be controlled. Wolf and Bonilla (526) have shown that many ions, such as PO_4^{3-}, I^-, and SO_4^{2-}, form protective films. Lander found that the corrosion rate increased with increasing temperature and decreasing H_2SO_4 concentration, but the position of the peak corrosion rate was independent of

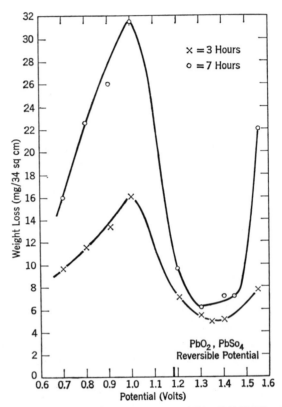

Fig. 7.15. Corrosion of Pb (307) at constant potential in 40% H_2SO_4 at 30°C (potentials vs. $HgSO_4$). (By permission of The Electrochemical Society.)

concentration of H_2SO_4. Similar results were obtained by Rüetschi and Angstadt (429).

It appears, then, that deep cycling or prolonged overcharging of a lead–acid battery is deleterious to the life of the battery because of the enhanced corrosion of the grids in these potential regions. Life of the battery may be shorter in tropical than in temperate regions. Finally, a fully charged battery (strong H_2SO_4 electrolyte) should suffer less corrosion.

Antimony, in concentrations between 5 and 12%, is added to lead from which the grids are made to give the lead rigidity and to improve the casting properties of lead (ref. 500, p. 15). Burbank (80) studied the effect of Sb on the electrochemical properties of the positive plate using a shallow cycling procedure coupled with electron microscopic and x-ray diffraction techniques.   The active material of positive plates with Pb–Sb alloy grids contained significant amounts of α-PbO$_2$ and was made up of many complex prismatic crystals giving a firm texture and good capacity to the plates.   On the other hand, the active material on Sb-free grids had very little α-PbO$_2$ and few prismatic crystals.   With cycling, the active material became softer and lost capacity in a shorter time.   These beneficial effects of the presence of Sb in the positive plate grids are caused by the Sb since the paste and the plate formation were the same in experiments using either pure Pb or Pb–Sb grids.

A disadvantage of the presence of Sb in the positive plate is an increase in the self-discharge rate.   Rüetschi and Cahan (431) found that the corrosion of the grid increased with increasing Sb content at a constant current charging rate.   For a given discharge rate, the time for discharge decreased with higher Sb content.   Such behavior is traced to an increase in the rate of self-discharge with increasing Sb content of the grid alloy.

To alleviate the problems of grid corrosion and self-discharge, metals other than Sb have been considered as alloying agents.   Metal ssuch as Ca, Sn, Ag, and As have been studied (77,79,80,122,123,226,307,308, 310,429,431,432,478).   Haring and Thomas (226) first reported that the substitution of about 0.1% Ca for the Sb greatly reduced the rate of self-discharge and should be more resistant to corrosion of the grid in batteries on float service, such as employed in communication applications.   In general these findings were confirmed (75,242,461,504), but variable results prompted Thomas and co-workers (478) to study the system further.   For float service, the presence of Ca in the grid is a definite advantage, but the amount of Ca added must be carefully controlled.   Below a concentration of 0.1%, little or no grid corrosion was observed, and the amount of growth of the grid was much less than for Pb–Sb grids.   For deeper cycling applications, such as automotive service, Pb–Ca grids form a high resistance film at the metal interface which severely impairs the usefulness of such a battery for this application.

At a positive grid, the presence of an oxide film on the metal interface causes a tensile stress on the metal surface (308).   If the metal is soft, elongation of the metal grid members may take place, which limits the life of the plate.   When about 5% elongation is reached (478), the active material becomes loosened, loss of electrical contact with the active material may take place (308), and the grid frame may be broken (478).   With Ca content greater than 0.1%, growth of the grid is increased with increasing Ca content.

Lander (308,310) studied the effect of the presence of Sn in the grid. The Sn alloys (4.5% Sn) showed greater resistance to corrosion than either Pb or Pb–Sb alloys (6% Sb). There is a minimum tensile stress below which growth does not occur, but once the oxide film becomes thick enough and the stress threshold is passed, a linear relationship exists between elongation and the depth of corrosion. Since the Sn alloys have about one-half the tensile strength of Sb alloys, the rate of growth of Sn-containing grids is greater. Although Pb–Sn alloy grids are subject to less corrosion, the higher rate of growth precludes their use in place of Pb–Sb.

The grain size is an important factor in the corrosion of Pb alloys since the finer the grain size the lower is the rate of corrosion as Dasoyan (122) suggested. He concluded that the presence of Ag or Ca in the grid metal reduced the size of the metal crystallites, and as a result, lowered the corrosion rate. The good corrosion characteristic of Ag–Pb alloys was first reported by Fink and Dornblatt (171).

It is known (13,17,21,288,291,296,312,501) that addition of $CoSO_4$ to the electrolyte lowers the overvoltage on overcharge and inhibits grid corrosion. The presence of Co^{2+} ion is effective (291,312) only at potentials exceeding the Co^{2+}/Co^{3+} potential, 1.842 V, according to Latimer (ref. 315, p. 213) and 1.808 V according to de Bethune and Loud (126). Lander (312) explains the behavior of the system containing Co^{2+} ion in terms of an alternate path for the evolution of oxygen. At these high anodic potentials, Co^{2+} ion is oxidized to Co^{3+} ion,

$$Co^{2+} \rightarrow Co^{3+} + e \qquad (7.40)$$

which in turn oxidizes H_2O to H_2O_2,

$$2H_2O + 2Co^{3+} \rightarrow H_2O_2 + 2Co^{2+} + 2H^+ \qquad (7.41)$$

Finally, the H_2O_2 may be catalytically decomposed in the acid solution to form O_2 and H_2O. Koch (291) has reached similar conclusions.

Silver ion may also be effective (312) but in a higher range of potentials because the Ag^+/Ag^{2+} potential is 1.98 V (126,315), and its behavior is explained by equations similar to Eqs. 7.40 and 7.41. Grid corrosion studies made by Lander (312) showed that only 6% of grid corrosion occurred under overcharge conditions, whereas 89% occurred during standing conditions. Consequently, additions of Co^{2+} ion or Ag^+ ion to the electrolyte have little or no effect in increasing automotive battery performance, since such additions are effective only during overcharge.

It is interesting that in the presence of Co^{2+} the layer of PbO_2 formed is thinner than in its absence and this oxide layer is composed only of α-PbO_2. Kiseleva and Kabanov (286) have concluded that the presence of absorbed SO_4^{2-} on the electrode surface causes the formation of β-PbO_2 and Co^{2+} ion displaces the SO_4^{2-} which brings about the formation of α-PbO_2. The presence of adsorbed SO_4^{2-} ions influences the oxygen η and the anodic corrosion according to Rüetschi and co-workers (431,433). From double layer studies carried out on PbO_2 plated on Au wires, the point of zero charge on a PbO_2 surface was determined by Kabanov et al. (274) at 1.8 V. However, Baker (21) pointed out that control of pH is the important factor in preparation of α- and β-PbO_2 as mentioned above in the section on PbO_2. He concluded that a more likely process is the replacement of H^+ ions with Co^{2+} ions. Under such conditions of high pH, only α-PbO_2 would be formed.

By studying single crystals, Rüetschi and Cahan (432) determined that self-discharge rates were very different for the various crystal faces but that little difference in the cathodic discharge behavior could be detected between the faces. The 110 plane was least resistant to corrosion. Some basic investigations of the self-discharge process may be mentioned (77,168,309,428,429,434,533). Photochemical studies of lead oxides have been reported (40,64), and the effects of ultrasound on the passivity of Pb in H_2SO_4 and the evolution of O_2 on PbO_2 were given by Kukoz et al. (301).

C. NICKEL

Interest in the electrochemistry of nickel and its oxides stems from the highly active research in alkaline batteries throughout the world since 1900. Such storage batteries as the Edison or Ni–Fe battery discovered by Edison (152) and the Ni–Cd battery discovered by Jungner (273) at about the same time are characterized by their ability to withstand severe mechanical and electrical abuse (9,36,158,360,479). These batteries do not lose active material with vibration, are mechanically rugged, can remain in a discharged state indefinitely without harming their performance, and can withstand overcharge and short-circuiting.

The Edison battery has been used only for heavy-duty industrial use and for railroad and marine applications. Much of its mechanical ruggedness comes from the construction of the battery plates. The active material of the negative plate, a mixture of finely divided iron oxide and metallic iron, is held in perforated pockets made of nickel-plated steel. For the positive plate, layers of pure $Ni(OH)_2$ separated by layers of flake nickel are pressed into perforated nickel-plated steel tubes arranged in the form of a battery plate. Since the Ni–Fe battery gives poor performance at low temperatures and has a low voltage at high discharge rates, other battery systems are preferred for most applications other than for heavy-duty industrial service.

On the other hand, Ni–Cd batteries can be discharged at high rates at a steady voltage and have good low temperature characteristics (36, 158). Under open-circuit conditions, this battery has a very low rate of self-discharge and no gas evolution. These batteries may be sealed (120,175). Major disadvantages to the use of the Ni–Cd batteries in competition with lead–acid batteries are their high cost and the lack of availability of Cd.

As pointed out by Fleischer (174), the hydroxides or oxides of Ni and Cd are unsuitable for making pasted plates. The active materials may be held in perforated pocket plates (36,232) similar to the Edison battery plates or in the pores of sintered nickel plaque (158,174). Porous Ni plaque is soaked in the nitrate of Ni or Cd, formed with cathodic polarization in alkaline solution, and finally washed and dried. Apparently, a number of cyles of anodic and cathodic polarization of the battery are required (174) to get the desired coulombic capacity.

During the charging process, the active material of the positive plate is oxidized from the Ni(II) to the Ni(III) state, and the oxides of the negative plate material are reduced to the free metal (45,206) according to Eq. 7.42 for the Edison and Eq. 7.43 for the Ni–Cd batteries.

$$2NiOOH + Fe + 2H_2O \rightarrow 2Ni(OH)_2 + Fe(OH)_2 \qquad (7.42)$$

$$2NiOOH + Cd + 2H_2O \rightarrow 2Ni(OH)_2 + Cd(OH)_2 \qquad (7.43)$$

Under charge, the cell reaction is the reverse of Eqs. 7.42 or 7.43. The electrolyte of these cells is about 21% KOH with a specific gravity of about 1.19. Unlike the lead–acid cell, the specific gravity of the electrolyte does not change significantly with the level of charge. Consequently, the danger of battery case fracture because of freezing at low temperatures is not a factor with Ni–Cd as with lead–acid batteries. In fact, specific gravities of KOH up to 1.30 (36) may be used in cold climates so that these batteries will not freeze. In the cell reaction, oxygen or hydroxide ions are transferred from one plate to the other with the electrolyte acting only as a transport medium. Additions of LiOH to the electrolyte improve the performance of the battery (36), about which more will be said later.

The independence of the specific gravity on the level of charge makes it difficult to determine the level of charge of a nickel battery quickly and simply as in the case of the lead–acid battery. Research work continues in the search for a rapid and uncomplicated method of assessing the level of charge of Ni–Cd batteries (339,371,527).

Studies of the electrochemical behavior of Cd have been reported in the literature (163,247,305,383,529). The final product of the discharge of the nagative plate is $Cd(OH)_2$, although intermediates such as CdO (305) and $(HCdO_2)^-$ (529) may be involved.

1. The Oxides of Nickel

When Ni metal is heated in oxygen, green NiO is formed but not in the pure state (ref. 456, p. 1430), since some metallic Ni is always present. If, however, Ni(OH), is heated in the absence of air, pure NiO may be obtained (58). The hydroxide is made by precipitation of $Ni(NO_3)_2$ with KOH and has a hexagonal crystal structure (92,370). Nickel monoxide may also be made by the thermal decomposition of $NiCO_3$ (323). The decomposition of $Ni(OH)_2$ is not a reversible process since $Ni(OH)_2$ cannot be made by the interaction of NiO with H_2O (248), and, in fact Cairns and Ott (92) could not detect any change in the x-ray diffraction pattern of a mixture of NiO and H_2O after having been kept in a bomb at 150°C for 5 days.

From x-ray diffraction studies, it has been determined (49,65,124,299) that NiO has a face-centered cubic lattice with a NaCl structure. In the literature (34,233,304,409), reports of the existence of two modifications of NiO may be found, but Cairns and Ott (92) have shown that this is not true on the basis of x-ray diffraction analysis. Rooksby (426) examined many samples of NiO from various sources and obtained diffraction patterns in each case identical with face-centered cubic NiO. When green NiO is in contact with O_2, it can take up relatively large amounts of oxygen (66,150,306,321) without a phase change, and the color of the oxide changes to black. With increasing oxygen content the conductivity of the nonconducting green NiO increases, but only the lines of NiO can be detected with x-rays. NiO is considered (66,150, 426a) to be p-type semiconductor. Lander (306) observed that the material becomes more amorphous with increasing oxygen content.

A number of investigators (25,48–50,105,106,199,338,492,525) have reported that higher oxides than NiO may be obtained by carefully heating $Ni(NO_3)_2$ or $Ni(OH)_2$ under controlled conditions. With such procedures, Ni_2O_3 and NiO_2 have been claimed to have been made. Bogatskii (49) found that all products were face-centered cubic, and

Lunde (338) observed that x-rays showed only the lines of NiO. According to LeBlanc and Sachse (324), the oxide Ni_2O_3 reported by Lunde is probably NiO combined with absorbed oxygen and H_2O, in agreement with the conclusions of Hendricks and co-workers (234). If any Ni is present in a higher valence state than 2, it is less than 10% of the Ni present (234,322–324). From later investigations (66,93,234,248,321–323), it is generally agreed that anhydrous oxides in a higher oxidation state than NiO do not exist. Lebat (317) has presented an excellent review of the oxides of nickel.

Electron diffraction studies of evaporated Ni films on a hot rock salt substrate in vacuum gave diffraction patterns which could not be attributed to Ni or NiO (2) alone. Because of a similarity to Co_2O_3, Ni_2O_3 was suggested. Germer and MacRae (197) studied the adsorption of oxygen on the faces of clean Ni single crystals with low energy electron diffraction techniques and observed the presence of Ni-O structures before a monolayer of adsorbed oxygen was formed. Other workers (4,345) observed extra spots corresponding to forbidden reflections in the low energy diffraction patterns obtained from the oxidation of thin Ni films. Such reflections were attributed to metastable Ni-O structures. The adsorption of O_2 on Ni powders (363a), NiO and lithium-doped NiO (524a) have been studied. According to Farnsworth and co-workers (164,165,394), a definite structure designated as Ni_3O is formed as an intermediate before a layer of NiO appears when clean Ni is exposed to oxygen. Oxygen absorbs as an amorphous layer of O_2 molecules which diffuse to defect sites where they dissociate (165). The oxygen atoms form a lattice structure at the defect site, and the oxide layer thickens by replacing Ni atoms with O atoms. The oxidation of Ni in air or oxygen has been investigated thoroughly (e.g., 26,188, 221,358,507), and the oxidation product is NiO. For an understanding of the mechanisms of film growth on metals, one may consult the literature (e.g., 89,178,362,398,532).

Although only one oxide of Ni, NiO, is known in the anhydrous state, hydrated oxides of Ni(III) and possibly Ni(IV) may be obtained. By using oxidizing agents such as $K_2S_2O_8$, Br_2, Cl_2, HClO, and HBrO at various temperatures, Belluci and Clavari (32) made a series of oxides from the oxidation of nickel salts having O/Ni atom ratios from 1.3 to 1.9. Since they found no evidence for the existence of an intermediate oxide, Ni_2O_3, they considered the series of nonstoichiometric oxides to be mixtures of NiO and NiO_2. This higher oxide, NiO_2, was found to be very unstable and lost O_2 rapidly in aqueous suspensions even at room temperature.

Howell (246) oxidized $Ni(OH)_2$ with $Ca(ClO)_2$ to a higher oxide equivalent to $NiO_{1.6}$ and concluded that two oxides are present: NiO_2 which loses O_2 rapidly, and Ni_2O_3 which loses oxygen slowly. The formation and decomposition of NiO_2 and Ni_2O_3 are parallel processes so that Ni_2O_3 is not an intermediate in the oxidation of $Ni(OH)_2$ to NiO_2 or in the reverse reduction process. Although LeBlanc and co-workers (320,321) found that all oxides higher than NiO were amorphous to x-rays, Cairns and Ott (94) found evidence from x-ray studies that Ni is in a higher valence state than Ni(II) and concluded that this state is Ni(III).

Hüttig and Peter (248) and Goralevich (209) presented evidence for the existence of oxides with waters of hydration such as $Ni_2O_3 \cdot nH_2O$, where $1 < n < 4$. From x-ray studies, Cairns and Ott (93) obtained evidence for $Ni_2O_3 \cdot H_2O$, $Ni_2O_3 \cdot 2H_2O$, and $Ni_3O_4 \cdot 2H_2O$, but these compounds do not have the properties of true hydrates. In these cases, the water is bound as hydroxyl groups. In a series of investigations, Besson (37) describes methods of preparation and the properties of Ni_2O_3 and Ni_3O_4 oxides but could find no direct evidence for existence of NiO_2.

According to Glemser and Einerhand (204), various higher oxides of Ni involving Ni(III) may be made from a choice of oxidizing agents and nickel salts. An analysis of the x-ray patterns shows that the Ni(III) oxide can be best described as NiOOH. There are at least three modifications of NiOOH which have a layered structure and which have been designated by Glemser and Einerhand as α-, β-, and γ-NiOOH. X-ray diffraction data and electron micrographs were obtained for these oxides (205) and reviewed by Labat (317). These oxides together with Ni_3O_2-$(OH)_4$, the hydrated form of Ni_3O_4, appear to have a hexagonal crystal structure (205). Oxidation of $Ni(OH)_2$, $Ni(NO_3)_2$, or $NiSO_4$ with Br_2, NaClO, or $NaBrO_2$ gives β-NiOOH, which is observed in the active material of Ni positive plates of alkaline storage batteries. The interaction of Na_2O_2 on Ni gives γ-NiOOH, and the oxidation of $K_2Ni(CN)_4$ with $K_2S_2O_8$ gives α-NiOOH. The results of Glemser and Einerhand (204,205) have been confirmed by Feitknecht and co-workers (169) and Bode (45).

Because oxygen-rich, nonstoichiometric oxides with an oxygen content greater than that required for Ni_2O_3 (O/Ni = 1.5) may be made (32,37,73a,246,321), it may be assumed that some Ni is oxidized to the Ni(IV) state. In this state, the sytsem is very unstable, and as pointed out by Howell (246), pure $NiO_2 \cdot nH_2O$ cannot be obtained. Since the presence of NiO_2 dissolved in Ni_2O_3 (β-NiOOH) cannot be detected with x-rays, direct evidence for its existence is difficult to find. Recently, however, Labat and Pacault (318,391) have measured the magnetic susceptibility of $Ni(OH)_2$ and recorded these measurements while $Ni(OH)_2$ was oxidized and reduced electrochemically. In this way, the presence of Ni(IV) has been demonstrated.

2. The Electrochemistry of the Oxides of Nickel

In addition to the chemical formation of Ni oxides, the hydrated higher oxides of Ni are also formed at a Ni anode and are present in the active material of Ni positive battery plates of alkaline cells. A consideration of the electrochemical behavior of nickel oxides will be discussed now.

a. **Standard Electrode Potentials.** The potential value of -0.250 V for the E_0 of the Ni/Ni^{2+} couple,

$$Ni^{2+} + 2e \rightleftharpoons Ni \qquad E_0 = -0.250 \text{ V} \qquad (7.44)$$

accepted by Latimer (ref. 315, p. 200), de Bethune (126), and Pourbaix, (408) is not supported very well by direct experimental determination. Because the presence of O_2 interferes with these measurements (191,408), the experimental difficulty experienced in obtaining the required data is reflected in the range over which the reported values of E_0 are spread. Murata (368) reported -0.232 V; Haring and Vanden Bosche (227), -0.231 V; Carr and Bonilla (99), -0.248 V; and Lopez-Lopez (334), -0.246 V.

The standard potential of the $Ni/Ni(OH)_2$ couple,

$$Ni(OH)_2 + 2e \rightleftharpoons Ni + 2OH^- \qquad E_0 = -0.72 \qquad (7.45)$$

is recorded as -0.72 V in basic solutions (126,315,408). A number of couples may be calculated or estimated from thermal and electrochemical data, and some of these which are listed by Pourbaix and co-workers (408) are given as follows: Ni/NiO, $E_0 = 0.116-0.059$ pH; $Ni(OH)_2/Ni_3O_2(OH)_4$, $E_0 = 0.897-0.059$ pH; $Ni(OH)_2/\beta$-NiOOH, $E_0 = 1.032-0.059$ pH; $Ni_3O_2(OH)_4/\beta$-NiOOH, $E_0 = 1.305-0.059$ pH; β-NiOOH/NiO_2, $E_0 = 1.434-0.059$ pH. As pointed out by Conway and Gileadi (117), the E_0 quoted by Latimer (ref. 315, p. 202) and de Bethune (126) for the Ni^{2+}/NiO_2 couple (1.68 V) is not valid since NiO_2 has not been identified.

Hickling and Spice (236) made a slurry of powdered Ni and the oxide obtained from the oxidation of $Ni(NO_3)_2$ with KBrO and measured the potential of a Pt wire in contact with the slurry against a hydrogen reference electrode using a number of electrolytes. In $1M$ NaOH they obtained a value of 0.56 V which they associated with the $Ni(OH)_2$/ β-NiOOH couple.   By adding 0.83 V to correct for the hydrogen electrode in basic solutions, on obtains 1.39 V, which is in closer agreement with the β-NiOOH/NiO_2 couple, as pointed out by Pourbaix et al. (408). The steady open-circuit potential of a charged Ni battery positive plate lies between 0.47 and 0.49 V, according to the investigations of Zedner (536) and Foerster (179,180).   This potential is associated with NiO_2. Conway and co-workers (59,117) determined this reversible potential of charged, sintered Ni plates by extrapolating the linear portions of the anodic and cathodic polarization curves to the point of crossing.   They recorded a value of 0.424 V (vs. HgO) for the Ni^{2+}/Ni^{3+} couple.   Earlier determinations are less valid because mixed potentials were studied instead of the truly reversible potential.   Since NiO_2 is so unstable, any steady-state measurement must be associated with Ni(III) and not with Ni(IV). Accordingly, one obtains in alkaline solutions

$$\beta\text{-NiOOH} + H_2O + e \rightleftharpoons Ni(OH)_2 + OH^- \qquad E_0 = 0.522 \text{ V} \qquad (7.46)$$

for an anodized Ni electrode.

b. Charging Curves.   In alkaline solutions, the constant current charging curves obtained on Ni (69–72,125,159,161,163,169,192,236, 502,503,524) are similar to the one shown in Fig. 7.16.   When Ni is first placed in an alkaline solution, such as $1M$ KOH, a passivating film of $Ni(OH)_2$ forms on the Ni surface (125,159,169,193,236,517).   Weininger and Breiter (517) found that one or two layers of $Ni(OH)_2$ are formed on the Ni surface immediately after immersion in the KOH, but the film can grow to a thickness of up to about 15 atoms after 16 hr on open circuit.

If an oxygen-free Ni surface is polarized anodically at a constant current (potential changed from H_2 to O_2 evolution), a plateau at about -0.6 V (125,159) corresponding to the oxidation of Ni to $Ni(OH)_2$ precedes the plateau shown in Fig. 7.16.   This is because Wynne-Jones and co-workers (69) polarized a $Ni(OH)_2$-impregnated, porous, Ni electrode so that the charging curve of Fig. 7.16 started at the $Ni(OH)_2$ potential. It requires the deposition of about two layers of $Ni(OH)_2$ (125,159,169, 502) before the potential rises to the slowly rising step between 0.4 and 0.5 V shown in Fig. 7.16, which corresponds to the oxidation of Ni^{2+} to Ni^{3+}.

Nickel hydroxide is a poor electronic conductor (125,168,218,300,438, 487) but appears to be a good proton conductor.   Wynne-Jones and co-workers (69,71,267) concluded that proton transfer through a film of $Ni(OH)_2$ was rapid, and Sagoyan (438) found from tracer studies with

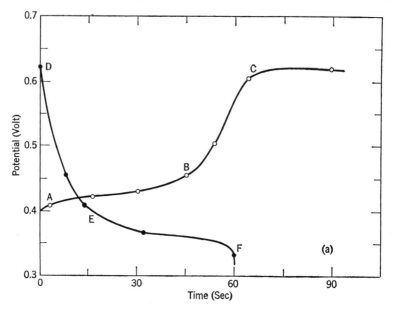

Fig. 7.16. Constant current charging curves (69) on Ni electrodes in $1N$ KOH solution at 2 mA/cm²: (○) anodic; (●) cathodic. (By permission of The Faraday Society.)

D_2O and $H_2{}^{18}O$ that the rate of diffusion of oxygen through the lattice was insignificant but that the exchange of hydrogen in the lattice was complete. The poorly conducting film of $Ni(OH)_2$ passivates the Ni surface, and under galvanostatic conditions, the potential rises until a point is reached ($\sim$0.4 V) where a new electrochemical reaction takes place. This electrode process is the oxidation of Ni^{2+} to Ni^{3+} by a proton transfer process (69,71,125,169,267,336,438).

$$Ni(OH)_2 \rightarrow \beta\text{-}NiOOH + H^+ + e \qquad (7.47)$$

Grube and Vogt (218) observed a brown film of higher oxide which formed on a Ni anode in NaOH solution while the potential slowly increased with time. When current was passed through a grain of $Ni(OH)_2$ held between two Pt wires, Kuchinskii and Ershler (300) found that the grain darkened at the point of contact with the anode first, and then, gradually, the coloration spread over the entire grain. The oxidized grain became conducting which was attributed to active oxygen present in the oxidized grain. Complete reduction of the grain could not be carried out because of a film of nonconducting $Ni(OH)_2$ insulated the cathode electrically.

After a potential of 0.4 V is reached, the $Ni(OH)_2$ is oxidized to β-NiOOH, and it is likely, as proposed by Wynne-Jones and co-workers (69,71,72,267) that a series of solid solutions of $Ni(OH)_2$ and β-NiOOH is formed while the amount of β-NiOOH in the oxide layer increases at the expense of $Ni(OH)_2$. If there were definite oxide phases corresponding to $Ni_3O_2(OH)_4$ and β-NiOOH, one should expect two steps (125), but only a slowly rising plateau is obtained (Fig. 7.16). The species $Ni_3O_2(OH)_4$ corresponds to the situation where the ratio of $Ni(OH)_2$ to β-NiOOH in the film is unity. Besson (37), Glemser and Einerhand (206), and El Wakkad and Emara (159) reported that a very short step preceeding the long plateau between 0.4 and 0.5 V could be observed if the charging current was very low (5 $\mu A/cm^2$). However, Davies and Baker (125) could not find any evidence for the $Ni_3O_2(OH)_4$ step at current densities as low as 2.5 $\mu A/cm^2$. The nucleation and growth of the β-NiOOH film is discussed by Briggs and Fleischmann (68a).

X-ray diffraction studies (71–73,163,297,439,487) have been carried out on Ni electrodes while they were being charged anodically and discharged cathodically. Falk (163) placed the electrodes in a polyethylene bag filled with electrolyte. The test electrode was moved to the side of the bag, and x-ray measurements were made on the test electrode during polarization. Although polyethylene peaks were observed, the patterns for $Ni(OH)_2$ and β-NiOOH were distinguished. Similar studies were also made by Salkind and Bruins (439). By comparing a series of patterns taken at a number of points along the anodic charging curve, there is a disappearance of lines (439) from the $Ni(OH)_2$ pattern rather than the appearance of new lines during the change from the $Ni(OH)_2$ to the β-NiOOH pattern. Only those patterns associated with $Ni(OH)_2$ and β-NiOOH are observed. There seems to be considerable justification in assuming that the composition of the anodic film on Ni electrodes in the potential range of the plateau in the charging curve of Fig. 7.16 is a solid solution of $Ni(OH)_2$ and β-NiOOH.

Although β-NiOOH is the highest oxidation state of Ni observed by x-rays even after prolonged anodization, it does not mean that higher states such as NiO_2 are not present since these species may be amorphous to x-rays (163). NiO_2 has been reported to have been made anodically at Ni, Pt, and PbO_2 electrodes in buffered solutions of Ni salts by Haissinsky and Quesney (222). Because compounds with ratios of O/Ni in the range 1.7–1.9 (32,71,72,94,125,206) have been detected, oxidation of Ni beyond the point corresponding to the complete conversion of $Ni(OH)_2$ to β-NiOOH (O/Ni = 1.5) is possible. A probable explanation proposed by Jones and Wynne-Jones (267) is the possibility that further oxidation produces a conversion of Ni(III) to Ni(IV) in the

β-NiOOH lattice leading to the formation of nonstoichiometric oxides, which are very unstable. According to this viewpoint, it is incorrect to consider such oxide films as solid solutions of β-NiOOH and NiO_2, as suggested by Glemser and Einerhand (204–206). Pure NiO_2 cannot be obtained (267) since an oxide layer with an O/Ni ratio between 1.7 and 1.8 is the final product of anodic polarization. This layer is composed of a single lattice containing Ni^{3+}, Ni^{4+}, OH^-, and O^{2-} ions. Because Davies and Barker (125) observed that the ratio of Ni^{3+} to Ni^{4+} was dependent on potential and not on time at a given potential, they preferred to think of the higher oxide as a solid solution of β-NiOOH and NiO_2. However, the single lattice point of view suggested by Jones and Wynne-Jones would seem to be in better accord with the results of x-ray studies.

At higher current densities (potential rises above 0.6 V, as shown in Fig. 7.16), steady evolution of O_2 is observed.

When a previously anodized electrode is discharged at constant current, one plateau (Fig. 7.16) is obtained at about 0.45 V (69,236), corresponding to the reduction of β-NiOOH to $Ni(OH)_2$. Only under special conditions was a second small step near 0 V observed corresponding to the step reported by Besson (37), Rollet (425), El Wakkad and Emara (159), and Glemser and Einerhand (206) as reduction to Ni_3O_2 $(OH)_4$. The final reduction product is $Ni(OH)_2$. At potentials below -1.1 V (69), H_2 is evolved. The potential of the anodic plateau is more noble than that of the cathodic plateau, indicating that activation energies are involved in the formation and reduction of β-NiOOH. The high value of the plateau potentials (~ 0.8 V) obtained by Davis (125) is unexplainable since the reference potential is not noted on their diagrams.

Broad lines corresponding to β-NiOOH are obtained (71) on a fully charged nickel oxide electrode, and with reduction, no change in the pattern occurs until the plateau at 0.45 V is reached where new lines appear. The new lines grow in intensity as cathodic reduction proceeds. Finally, at the end of the plateau only the pattern for $Ni(OH)_2$ remains. Along the plateau, there is a mixture of lines corresponding to a mechanical mixture of $Ni(OH)_2$ and β-NiOOH.

The oxide layers on an anodized Ni electrode are very unstable. On open circuit, the potential drops quickly to the β-NiOOH potential and then approaches the $Ni(OH)_2$ potential more slowly (159). As determined from x-ray studies, anodized Ni electrodes which had been stored for 27 hr did not show a change in the β-NiOOH pattern (439). After 7 days, $Ni(OH)_2$ lines were evident, and the complete $Ni(OH)_2$ pattern was obtained eventually after a longer period of time.

Wynne-Jones and co-workers (70,72) observed an aging effect on the electrochemical properties of $Ni(OH)_2$. The plateaus of the charging curves taken on fresh $Ni(OH)_2$ electrodes are less sloping than those on aged (12–60 days) electrodes, and also, the oxygen overvoltage is greater on aged $Ni(OH)_2$ electrodes. From x-ray patterns taken on these electrodes, both structures appear to be laminar, but the fresh structure is open and disordered, producing a more reactive species than the normal ordered structure of the aged $Ni(OH)_2$. As expected, the freshly prepared $Ni(OH)_2$ electrode is more easily oxidized than the aged one. Okada et al. (385) observed that freshly prepared $Ni(OH)_2$ was more easily oxidized chemically than aged $Ni(OH)_2$.

As pointed out by Briggs and Wynne-Jones (72), the layer of oxide on an anodized Ni electrode is rich in the number of hydroxyl groups. Such a compound should show ion exchange properties as well as water absorption producing variations in potential with KOH and H_2O activities. Indeed, such effects have been reported by Bourgault and Conway (59), Ershler and co-workers (161), Kornfeil (293), and Bradshaw (61).

The first to use optical methods employing polarized light in the study of active and passive Ni surface was Tronsted (482). However, manual control of the potential was not precise enough to follow the changes in the surface films during the transition region between the passive and active state. Reddy and co-workers (416), using a modern electronically controlled potentiostat, were able to carry out more precisely controlled measurements of the changes in the elliptically polarized light beam reflected from a bright Ni surface as the Ni was polarized through the transition from the active to the passive state. They (416) reported that a film about 60 Å thick formed on the nickel surface at potentials less noble than the passive potential. Since the optical adsorption coefficient may be related directly to the electrical conductivity of the film, it was concluded that the onset of passivity coincides with the formation of a good conducting film because the adsorption coefficient rose rapidly after going through a minimum just at the potential where the Ni becomes passive. A poorly conducting film formed by precipitation of a Ni salt from the supersaturated solution layer next to the corroding electrode surface appears first, which is then converted to the conducting passive layer. That a good conducting film is the passivating mechanism has been suggested by Weil (516) and Vetter (495) and is preferred to the proposal made by Günther (220) that passivity is due to the presence of a monolayer of absorbed oxygen. For a more comprehensive discussion of the mechanism of passivity, one may consult the literature (e.g., 160,162,196,238,292,365,490,496).

In acid solutions, Ni is active initially and goes into solution as Ni^{2+} (267,298). The Ni surface becomes passive in sulfuric acid solutions beginning at a potential of about 0.15 V (292), but some corrosion of the Ni continues up to a potential of about 1.35 V. According to Sato and Okamoto (441), the actual transition from the active to the passive state occurs in the potential range between 0.3 and 0.7 V. Müller and co-workers (366) reported the detection of a layer of basic nickel sulfate below 1.5 V. In $5N$ H_2SO_4, the layer is $NiSO_4 \cdot 7H_2O$ and becomes more stable with increasing acid concentration. This layer is converted to an oxide layer (Ni_2O_3) above 2 V, and it is the stability of this oxide layer to which passivity is attributed (364,366,387,449,484).

According to the viewpoint offered by Turner (488), a basic salt of Ni produced by the corrosion of the metal accumulates near the anode surface and is eventually deposited on the Ni surface. This nonconducting layer shields the surface so that the current density, and hence the potential, is increased at the uncovered sites. Under such conditions, a point is reached where water may be discharged, producing an oxide layer. Other workers (191,192,313,482) have concluded that a fixed value of pH must be reached before passivation can occur.

Although passivation may begin with the adsorption or precipitation of compounds on the surface (236,488), Piontelli and Serravalle (402) in agreement with Kolotyrkin (292) maintain that such salt layers do not provide complete passivity. Because the layer of higher oxides (Ni_2O_3) is a good conductor, such a film does render a high degree of passivity to the electrode surface. Results of the ellipsometric studies of the passivity of Ni (416) support these conclusions.

When the Ni becomes passivated with the conducting oxide layer, O_2 is evolved on the surface. Some information can be obtained from the open-circuit decay curves. If the circuit is opened at 2.2 V where steady evolution of O_2 occurs, the potential drops quickly to a step at about 1.6 V (441,488), followed by a longer step at about 2.6 V before reaching a steady rest potential about 0.02 V. The initial rapid drop is the decay of the overvoltage; the step at 1.6 V corresponds to the decomposition of a higher oxide of Ni, possibly a mixture of Ni_2O_3 and NiO_2 (488) or Ni_3O_4 and NiO_2 (441); the step at 0.26 V is attributed (441) to a film of NiO which is not considered to be the passivating oxide; and finally, a steady-state rest potential is reached (488) when the precipitated Ni salt film is formed as fast as it is dissolved.

c. **Potential-Sweep Traces.** Potential-sweep studies have been carried out on bright Ni electrodes in KOH solutions by Weininger and

Breiter (517,158).   Single crystals of Ni were used.   Between the
first and second cycles of polarization, considerable changes in the
shapes of the curves take place as seen in Fig. 7.17.   Oxygen is adsorbed
at different rates, and the anodic sweeps (from −0.2 to 1.6 V at 0.11
V/sec) are different for the first cycle on the three faces.   The rate of
oxide formation increases with a decrease in the packing of the atoms in
a given plane; (110) > (100) > (111).   Since the closest packed face
has the strongest bonds in the planes, it follows that this face would have
the weakest bonds for a constituent external to the plane.   For potential
sweeps following the first, the traces are identical for each face and do
not change sensibly with further cycling.   This is explained by a dis-
ordering of the Ni surface produced by the anodization followed by the
cathodization of the electrode during the first cycle of polarization.
However, cycling at potentials below 0.51 V did not disorder the surface.

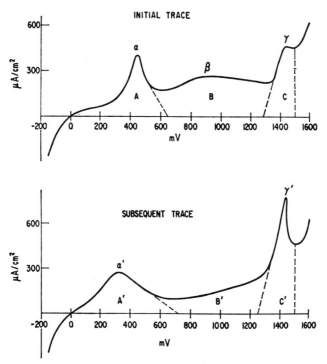

Fig. 7.17. Schematic current–potential sweep curves (517) obtained on Ni single
crystals with ordered surfaces (initial trace) and disordered surfaces (subsequent
trace).   (By permission of The Electrochemical Society.)

In the trace obtained from the first cycle, Weininger and Breiter distinguish three peaks. The first peak between 0.2 and 0.6 V corresponds to the formation of $Ni(OH)_2$, and the peak above 1.4 V to the formation of β-NiOOH. At first, Ni is oxidized to $Ni(OH)_2$,

$$Ni + 2OH^- \rightarrow Ni(OH)_2 + 2e \qquad (7.48)$$

followed by the oxidation of $Ni(OH)_2$ to β-NiOOH at higher potentials,

$$Ni(OH)_2 + OH^- \rightarrow \beta\text{-NiOOH} + H_2O + e \qquad (7.49)$$

At sufficiently high potentials β-NiOOH may be obtained directly from the metal,

$$Ni + 3OH^- \rightarrow \beta\text{-NiOOH} + H_2O + 3e \qquad (7.50)$$

Above 1.5 V, O_2 is evolved and the presence of Ni(IV) cannot be ruled out. Lukovtsev and Slaidin (336) estimate a current efficiency of 1 to 2% for Ni(IV) formation. For traces obtained from subsequent cycles, the middle peak disappears. In the middle region, both $Ni(OH)_2$ and β-NiOOH may be formed by Eqs. 7.48 and 7.49, producing a solid solution of the two hydroxides. Oxidation of Ni(III) to Ni(IV) cannot be detected because it is masked by O_2 evolution.

The cathodic sweep (518) is the reverse of the anodic, showing reduction peaks for both β-NiOOH and $Ni(OH)_2$.

Double layer capacity curves as a function of potential are also recorded (157,518) and are given in Fig. 7.18. The first peak at about 0.2 V is attributed (518) to the formation of about two layers of $Ni(OH)_2$ followed by a minimum where the electrode begins to passivate. This minimum in the double layer capacity corresponds to an electrode resistance maximum, indicating that $Ni(OH)_2$ is poorly conducting (157). As soon as β-NiOOH begins to form (in the vicinity of 0.7 V), the capacity rises to very high values because of the formation of the conducting layer of β-NiOOH and eventual evolution of O_2.

3. Anodic Evolution of Oxygen

Oxygen overvoltage curves have been determined (6,115,157,172,218, 295,337,442,462,480,486,489,497,502,503) for the evolution of O_2 on Ni anodes in alkaline and acid solutions. The steady-state anodic polarization curves for several values of pH are given in Fig. 7.19 for acid solutions and in Fig. 7.20 for alkaline solutions.

As determined by Ammar and Darwish (6), the anodic current in acid solutions is produced by corrosion processes up to a potential of about 1.72 V, after which the corrosion current falls off with increasing potential while the rate of O_2 evolution increases until all the anodic current

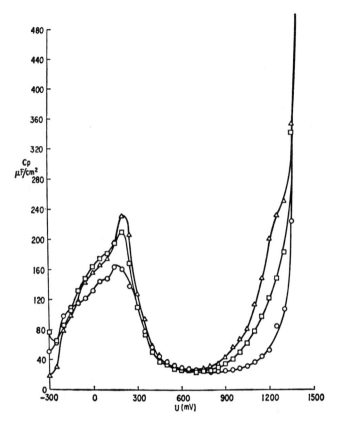

Fig. 7.18. A plot of capacity C_p (518) as a function of potential for a polycrystalline Ni electrode in $4N$ KOH saturated with Ar and measured at 100 cps: (○) first sweep; (□) second sweep; (△) third sweep.   (By permission of The Electrochemical Society.)

may be accounted for by O_2 evolution processes.   This behavior produces the inflection in the polarization curves of Fig. 7.19 in acid solutions.   The partial polarization curves are shown in Fig. 7.21.   As the pH increases, the point at which O_2 evolution becomes the determining process of the electrode kinetics shifts to lower current densities and less noble potentials (Fig. 7.19), and is independent of the rate of stirring (442).

Evolution of O_2 takes place on the electronically conducting oxide surface, and as seen in Fig. 7.19, the η is independent of pH.   Tafel slope values range from 0.083, reported by Vetter and Arnold (497) to

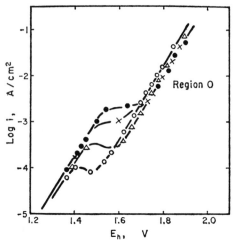

Fig. 7.19. Anodic polarization curve (442) for the evolution of O_2 on a Ni electrode in sulfuric acid solutions of four different pH values: (●) 1.55; (×) 1.80; (△)2.80; (○) 4.10. (By permission of Pergamon.)

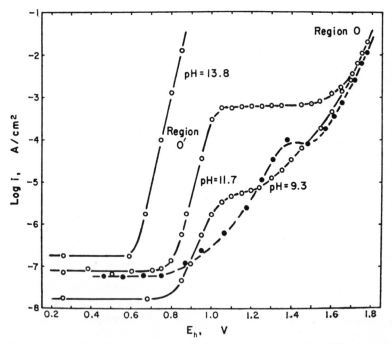

Fig. 7.20. Anodic polarization curves (442) for the evolution of O_2 of Ni electrodes in NaOH solutions of three different pH values: 13.8, 11.8, and 9.30. The dotted line is the curve in an acid solution of pH 3.10 shown for comparison. (By permission of Pergamon.)

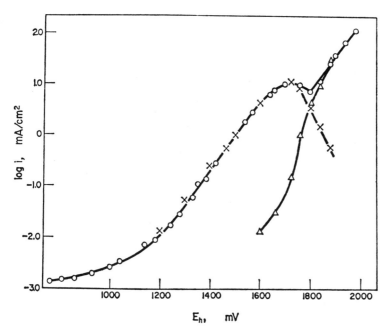

Fig. 7.21. Steady-state polarization curve (6) obtained on a Ni anode showing the total polarization (○), the partial corrosion (×), and O_2 evolution (△) curves.   (By permission of Pergamon.)

0.14, reported by Ammar and Darwish (6).   A mechanism in which the discharge of H_2O to produce absorbed OH radicals as the rate-determining step is consistent with the observed Tafel slopes.

In alkaline solutions, two Tafel regions are found (157,172,337,442, 486,489,503) as shown in Fig. 7.20.   The Tafel region at low current densities has a slope of about 0.05, and at high current densities, a slope of about 0.1.   The transition region separating the two Tafel regions has the properties of a limiting current since its position was highly dependent on the rate of stirring (442).   Although the polarization curves at high current densities are independent of pH as in the case of acid solutions, those at low current density depend strongly on pH (Fig. 7.3).   In the high current density region, the discharge of H_2O is favored as the rate-determining step for the mechanism of oxygen overvoltage, which is identical with the process preferred in acid solutions. From the data in Fig. 7.3, it is seen that the η in acid and alkaline solutions is the same in the high current density region.

At low-current densities, OH^- ions are discharged to form adsorbed OH radicals (157,172,295,442,486,489). Sato and Okamoto (442) suggest that the atom–ion (so-called electrochemical) step,

$$(OH)_{ads} + OH^- \rightarrow (O)_{ads} + H_2O + e \qquad (7.51)$$

is the rate-determining step in order to account for the low slope. As the current density is increased, a limiting current of OH^- ions is reached, and the system goes to the discharge of H_2O where the electron transfer step is rate limiting, in agreement with a slope of 0.12. The transition from the two-Tafel-region behavior of alkaline solutions to the one-Tafel-region behavior in acid solutions occurs at about pH 11, caused by the OH^- ion limiting current. Such a viewpoint can explain the data of Fig. 3.

Another approach may be taken in terms of the participation of the oxides in the O_2 evolution process. Grube and Vogt (218) suggested that O_2 was evolved by the decomposition of NiO_2 to form Ni_2O_3 and O_2. According to Krasil'schchikov (295), OH radicals obtained from the discharge of OH^- ions may interact with β-NiOOH to form NiO_2 from which O_2 is evolved by the decomposition of the oxide. The following series of reactions represents a possible path.

$$OH^- \rightarrow (OH)_{ads} + e \qquad (7.52)$$

$$(OH)_{ads} + OH^- \rightarrow (O^-)_{ads} + H_2O \qquad (7.53)$$

$$2\beta\text{-NiOOH} + (O^-)_{ads} \rightarrow 2NiO_2 + H_2O + e \qquad (7.54)$$

$$NiO_2 + (OH)_{ads} \rightarrow \beta\text{-NiOOH} + (O)_{ads} \qquad (7.55)$$

$$2(O)_{ads} \rightarrow O_2 \qquad (7.56)$$

Any one of these steps may be rate determining. Elina and co-workers (157) suggest that at low current densities the concentration of NiO_2 on the surface would be small so that the rate of spontaneous decomposition of NiO_2 (Eq. 7.55) would be slow and, hence, rate determining. At high-current densities, relatively large amounts of NiO_2 are present, causing the rate of decomposition of NiO_2 to be high. Under these circumstances, the discharge of OH^- ions, Eq. 7.52, is the slow step, in agreement with a Tafel slope of 0.12. Instead of the steps involving $(O^-)_{ads}$, Eqs. 7.53 and 7.54, Tsinman (486) substituted a step involving oxidation by OH radicals,

$$\beta\text{-NiOOH} + OH \rightarrow Ni_2 + H_2O \qquad (7.57)$$

Whether Eqs. 7.53 and 7.54 or Eq. 7.57 were used, he (486) found that the two mechanisms give the same kinetics.

Possibly a combination of the mechanisms suggested by Sato and Elina is close to the true situation. According to Jones and Wynne-Jones (267) and Kuchinskii and Ershler (300), the oxide layer on an anodized Ni electrode is composed of a nonstoichiometric oxide composed of Ni^{3+} and Ni^{4+} ions which are good electronic conductors. At an atomic ratio, O/Ni, equal to 1.75 where equal numbers of Ni^{3+} and Ni^{4+} ions are present, the oxide layer should have the greatest conductivity (267). Such a layer offers the most favorable surface on which O_2 may be evolved. Since $NiO_2(O/Ni = 2)$ is never reached because the Ni(IV) state is very unstable, it seems reasonable that the evolution of O_2 could proceed through an oxide mechanism described by Eqs. 7.52–7.56. Current goes into the formation of the higher oxide, and O_2 is evolved by its decomposition. At low current densities, a reaction such as an ion–atom reaction, Eq. 7.51, or a chemical reaction, Eq. 7.55 or Eq. 7.57, may be rate limiting, as suggested by Elina et al. (157) in accordance with the observed low Tafel slope. When higher currents are reached or when the pH is low, a limiting OH^- ion current sets in, and the kinetics are governed by a water discharge mechanism. If the electron transfer step, Eq. 7.58,

$$H_2O \rightarrow OH + H^+ + e \qquad (7.58)$$

is rate determining, then the kinetics are in agreement with the experimentally observed fact that the polarization curves are the same in acid and alkaline solutions with a Tafel slope of about 0.12.

It is known (9,36,158,360) that additions of LiOH to the electrolyte of alkaline cells are beneficial to the performance of the cells. The effect of the alkali cation on oxygen overvoltage has been studied (337, 462,489) for NaOH, KOH, and LiOH. For a given concentration of alkali, the η increases as one proceeds from KOH to NaOH to LiOH. Slaidin and Lukovtsev (462) concluded that the alkali ion replaces Ni^{2+} or Ni^{3+} ions in the oxide lattice, and consequently, influences the rate at which protons can diffuse through the oxide layer. The rate of diffusion of protons is greatest for Li^+ and least for K^+. Since the presence of Li^+ increases the η for O_2 evolution on Ni in alkaline solutions, it is seen that a beneficial effect on the cycle life of an alkaline battery may be expected. More will be said about this matter in a later section.

Conway and Bourgault (115) report that the electrode reactions, both for steady-state polarization and open-circuit decay studies, were faster in solutions made from D_2O and KOD than in ones made from H_2O and KOH. This behavior is considered as evidence that the O_2 reaction on Ni is rate controlled rather than diffusion limited.

4. Cathodic Reduction of Oxygen

Little work has been carried out (295a,336,356,444) on the reduction of O_2 on Ni cathodes. Mazitov et al. (356) report that the overvoltage is much higher on Ni than on Pt or Pd and that it is reduced by increasing the temperature. Erratic and irreproducible results were obtained by Sawyer and Interrante (444) for the reduction of O_2 on bright Ni, but when an oxide film was formed on the surface by preanodization reproducible data could be obtained. Because a nonconducting film of $Ni(OH)_2$ exists on a Ni surface exposed to an alkaline solution, one would expect the poor results obtained by Sawyer and Interrante on Ni electrodes which had not been preanodized. Preanodization produces a good electronically conducting film of a nonstoichiometric higher oxide of Ni. In this case, reproducible results may be obtained; however, as noted by Sawyer and Interrante, reduction of the oxide film accompanies O_2 reduction. If the oxide reduction current could be separated from the O_2 reduction current, more could be said about the mechanism.

A mechanism for the reduction of β-NiOOH is offered by Lukovtsev and Slaidin (336). The process occurs through a proton transfer mechanism. On the surface of a grain of β-NiOOH, reduction to $Ni(OH)_2$ takes place by a transfer of a proton,

$$\beta\text{-NiOOH (surf)} + H_2O + e \rightarrow Ni(OH)_2 \text{ (surf)} + OH^- \qquad (7.59)$$

Then a proton diffuses into the bulk of the grain producing another β-NiOOH site,

$$Ni(OH)_2 \text{ (surf)} + \beta\text{-NiOOH (bulk)} \rightarrow \beta\text{-NiOOH (surf)} + Ni(OH)_2 \text{ (bulk)} \qquad (7.60)$$

This process is repeated until the grain is converted to $Ni(OH)_2$. It is considered that Eq. 7.60 is the rate-limiting step.

An interesting experiment was carried out (336) on a Ni diaphragm. One side of the Ni foil was anodized, and the diffusion of protons through the $Ni(OH)_2$ film was detected by observing the increase in anodic current required to maintain the film after switching on a cathodic current on the other side of the diaphragm.

Huq et al. (250) studied the reduction of oxygen on Ni and a number of Ni alloys, such as NiSb and NiAs, in a search for more active catalysts for the reduction of O_2 than the noble metals. In all cases, these materials were inferior to the noble metals.

5. Fundamental Battery Studies

Since 1900, a considerable number of investigations of the nature and properties of pocket (or tubular) and sintered Ni positive plates of alkaline batteries have been made (33,45,59,61,69,71,72,114,117,163,179, 180,206,245,267,293,386,403,439,470,487,524,536).

In the discharged state the active material of the Ni positive plate is $Ni(OH)_2$ (72,163,206,293) which is oxidized to β-NiOOH (71,163,206) during the charging process. When fully charged, the active material consists of a nonstoichiometric oxide composed of Ni^{3+}, Ni^{4+}, OH^-, and O^{2-} in a single lattice (267). The Ni matrix of the sintered Ni battery plates acts only as a convenient economical structural material, and only the oxides and hydroxides participate in the electrochemical process occurring in battery operation. As pointed out by Wynne-Jones and co-workers (69), the Ni could be replaced by any suitable material, such as Pt, without interfering with the performance of the electrode.

A fully charged positive Ni plate evolves oxygen when placed under open-circuit conditions (179,180,524) which results from the spontaneous decomposition of the Ni(IV) content of the nonstoichiometric oxide to β-NiOOH,

$$2NiO_2 + H_2O \rightarrow 2\beta\text{-NiOOH} + \tfrac{1}{2}O_2 \qquad (7.61)$$

From studies of the change in weight of positive Ni plates with time during charge and discharge at constant current, Winkler (524) found that the O_2 formed by Eq. 7.61 remained occluded in the electrode-active material. Occluded oxygen in the active material of the positive plate was also detected by Zedner (536) and Pitman and Work (403). The self-discharge of the β-NiOOH is slow (360), and the shape of the self-discharge curve is similar to the cathodic discharge curve (114,116).

Salkind and Bruins (439) measured the potential of Ni–Cd cells as a function of temperature (-17.8 to $51.7°C$) and concentration of KOH (18.1–34.7%). At 25°C, the potential of the cell was 1.27 V, and the temperature coefficient, dE/dt, over the temperature range studied was -0.0004 V/°C for sintered Ni plates. This value agrees well with -0.0003 V/°C reported by Falk (163) for Ni pocket electrodes and with -0.0004–0.0008 V/°C, obtained by Hosono and Watanabe (245) for tubular Ni electrodes. The electrochemical reaction for the Ni–Cd cell (439) is Eq. 7.34.

The effect on the performance of Ni–Cd batteries by the addition of LiOH to the electrolyte has been studied (33,386,470,481,487). Although LiOH added to the KOH of sealed Ni–Cd batteries did not improve the coulombic capacity for a given cycle (33), it did help maintain the voltage level in a continuous cycling program.   Sugita (470) found that batteries containing LiOH retained 88–97% of their capacity after 5000 cycles, whereas those without LiOH retained only 57% of capacity after 3000 cycles.   Electronmicrographs showed that the decrease in capacity resulted from the crystallization of the active material with a consequent loss in the active surface.   Sugita considers the role of Li to be the prevention of this crystallization.   When LiOH was added to the electrolyte of the battery, Okada et al. (386) found that the potential for charging the Ni plate increased and the discharge potential was higher.   They suggest that the LiOH raises the oxygen η, and consequently, increases the coulombic capacity of the electrode.

A lithium-doped NiO is a p-type semiconductor (481).   During rapid discharge of a Ni positive plate, Kuchinskii and Ershler (300) detected a film of NiO at the metal surface which seriously limited the discharge capacity of the plate because the poorly conducting film isolated the active material electrically from the positive plate.   Tichenor (481) suggests that the Li^+ ion enters this film, thus imparting a high conductivity to it.

Tuomi (487) concludes that LiOH additions not only affect the oxygen overvoltage, but also cause changes in phases present in the active material.   In concentrated KOH or NaOH, the β-NiOOH obtained from the oxidation of $Ni(OH)_2$ may be further oxidized to a laminar species containing Ni^{3+} and Ni^{4+} ions.   He designates this oxide as the α-phase higher oxide and β-NiOOH as the β-phase higher oxide.   In pure KOH, the α phase (in addition to the β phase) is formed, which contributes to the electrochemical capacity only if the discharge current is low.   If the discharge current is large, the α phase does not cycle and effectively lowers the capacity of the plate.   Only the β phase is formed in pure LiOH, and although α phase is formed in a mixture of LiOH and KOH, it disappears with cycling.   Even extensive overcharge of the cell when Li^+ is present does not produce the α phase in a fully charged plate.   According to Tuomi, the increased cycle life produced by the addition of LiOH to the KOH of the alkaline battery may be traced to an inhibition of the α phase formation.   The observed data were interpreted in terms of a model based on a solid-state mass and charge-transport processes.

The effect of ultrasound on the properties of a Ni-oxide electrode has been reported by Kukoz and Skalozubov (302).

When a nonstoichiometric Ni oxide electrode was irradiated with neutrons from a reactor, the amount of adsorbed oxygen increased (355). It is thought that irradiation produces positive holes in the oxide film, and the electrons from weakly adsorbed oxygen atoms are trapped in the acceptor levels producing strongly bound oxygen.

Quinn and Roberts (412) have measured the work function of a Ni surface as a function of the partial pressure of O_2 between -195 and 25°C.

References

1. Afanas'ev, A. S., A. N. Burmistrova, V. I. Sotnikova, and E. N. Chankova, *Ukr. Khim. Zh.*, **28**, 492 (1962).
2. Aggarwal, P. S., and A. Goswanie, *J. Phys. Chem.*, **65**, 2105 (1961).
3. Akopyan, A. U., *Zh. Fiz. Khim.*, **33**, 82 (1959).
4. Alessandrini, E. I., and J. F. Freedman, *Acta Cryst.*, **16**, 54 (1963).
5. Amile, R. F., J. B. Ockerman, and P. Rüetschi, *J. Electrochem. Soc.*, **108**, 377 (1961).
6. Ammar, I. A., and S. Darwish, *Electrochim. Acta*, **11**, 1541 (1966).
7. Anbar, M., and H. Taube, *J. Am. Chem. Soc.*, **78**, 3252 (1956).
8. Andre, H. G., Fr. Pat. 786,276 (1935); Br. Pat. 491,782 (1938); U. S. Pat. 2,317,711 (1941); *Bull. Soc. Franc. Elec.*, **1**, 132 (1941); *Chem. Abstr.*, **37**, 1939 (1943).
9. Anderson, F. C., *J. Electrochem. Soc.*, **99**, 244C (1952).
10. Anderson, J. R., and V. B. Tare, *J. Phys. Chem.*, **68**, 1482 (1964).
11. Anderson, J. S., and M. Sterns, *J. Inorg. Nucl. Chem.*, **11**, 272 (1959).
12. Angstadt, R. T., C. J. Venuto, and P. Rüetschi, *J. Electrochem. Soc.*, **109**, 177 (1962).
13. Antropov, L. I., *Tr. Soveshch Elektrokhim. Akad. Nauk SSSR*, **1950**, 549.
14. Appleby, M. P., and H. M. Powell, *J. Chem. Soc.*, **133**, 2821 (1931).
14a. Appleby, M. P., and R. D. Reid, *J. Chem. Soc.*, **121**, 2129 (1922).
15. Appelt, K., M. Elbanowski, and A. Nowacki, *Electrochim. Acta*, **6**, 207 (1962).
16. Astakhov, I. I., I. G. Kiseleva, and B. N. Kabanov, *Dokl. Akad. Nauk SSSR*, **126**, 1041 (1959).
17. Astakhov, I. I., E. S. Vaisberg, and B. N. Kabonov, *Dokl. Akad. Nauk SSSR*, **154**, 1414 (1964).
18. Austin, P. C., *J. Chem. Soc.*, **99**, 262 (1911).
19. Baborovski, G., and B. Kusma, *Z. Elektrochem.*, **14**, 196 (1908); *Z. Physik. Chem. (Leipzig)*, **67**, 48 (1909).
20. Bagg, J., and L. Bruce, *J. Catalysis*, **2**, 93 (1963).
21. Baker, R. A., *J. Electrochem. Soc.*, **109**, 337 (1962).
22. Barbieri, G. A., *Ber.*, **B60**, 2424 (1927); *Atti. Acad. Lincei*, **13**, 882 (1931).
23. Barbieri, G. A., and A. Malaguti, *Atti. Acad. Nazl. Lincei*, **8**, 619 (1950); *Chem. Abstr.*, **45**, 55 (1951).
24. Barradas, R. G., and G. H. Fraser, *Can. J. Chem.*, **42**, 2488 (1964); **43**, 446 (1965).
25. Baubigny, H., *Compt. Rend.*, **87**, 1082 (1878); **141**, 1232 (1905).
26. Baumback, H. H., and C. Wagner, *Z. Physik. Chem. (Frankfurt)*, **B24**, 59 (1934).
27. Beck, W. H., P. Jones, and W. F. K. Wynne-Jones, *Trans. Faraday Soc.*, **50**, 1249 (1954).

28. Beck, W. H., R. Lind, and W. F. K. Wynne-Jones, *Trans. Faraday Soc.*, **50**, 147 (1954).
29. Beck, W. H., K. P. Singh, and W. F. K. Wynne-Jones, *Trans. Faraday Soc.*, **55**, 331 (1959).
30. Beck, W. H., and W. F. K. Wynne-Jones, *Trans. Faraday Soc.*, **50**, 136 (1954).
31. Beer, S. Z., and Y. L. Sandler, *J. Electrochem. Soc.*, **112**, 1133 (1965).
32. Belluci, I., and E. Clavari, *Gazz. Chim. Ital.*, **36**, 58 (1906); **37**, 409 (1907).
33. Belove, L., and I. M. Schulman, *Proc. Ann. Power Sources Conf.*, **13**, 82 (1959).
34. Bennett, O. G., R. W. Cairns, and E. Ott, *J. Am. Chem. Soc.*, **53**, 1179 (1931).
35. Benton, A. F., and L. C. Drake, *J. Am. Chem. Soc.*, **54**, 2186 (1932); **56**, 255 (1934).
36. Bergstrom, S., *J. Electrochem. Soc.*, **99**, 248C (1952).
37. Besson, J., *Compt. Rend.*, **219**, 130 (1944); **220**, 320 (1945); **222**, 390 (1946); **223**, 28, 288 (1946); *Ann. Chim.*, **2**, 527 (1947).
38. Bianchi, G., *Corrosion Anti-Corrosion*, **5**, 146 (1957).
39. Bianchi, G., G. Caprioglio, F. Mazza, and T. Muzzini, *Electrochim. Acta*, **4**, 232 (1961).
40. Bigelow, J. E., and K. E. Haq, *J. Appl. Phys.*, **33**, 2980 (1962).
41. Binnie, W. P., *Acta Cryst.*, **4**, 471 (1951).
42. Bircumshaw, L. L., and I. Harris, *J. Chem. Soc.*, **141**, 1637 (1939).
43. Bitton, H. T. S., *J. Chem. Soc.*, **127**, 2956 (1925).
44. Blocher, J. M., and A. B. Garrett, *J. Am. Chem. Soc.*, **69**, 1594 (1947).
45. Bode, H., *Angew. Chem.*, **73**, 553 (1961).
46. Bode, H., and E. Voss, *Z. Elektrochem.*, **60**, 1053 (1956).
47. Bode, H., and E. Voss, *Electrochim. Acta*, **1**, 318 (1959).
48. Bogatskii, D. P., *Zh. Obshch. Khim.*, **7**, 1397 (1937).
49. Bogatskii, D. P., *Zh. Obschch. Khim.*, **21**, 3 (1951); *Chem. Abstr.*, **46**, 7861 (1952).
50. Bogatskii, D. P., and I. A. Mineeva, *Zh. Obshch. Khim.*, **29**, 1328 (1959).
51. Bone, S. J., K. P. Singh, and W. F. K. Wynne-Jones, *Electrochim. Acta*, **4**, 288 (1961).
52. Bonk, J. F., and A. B. Garrett, *J. Electrochem. Soc.*, **106**, 612 (1959).
53. Bonnemay, M., G. Bonoël, and E. Levart, *Electrochim. Acta*, **9**, 727 (1964).
54. Boreskov, G. K., and A. V. Khasin, *Kinetika Kataliz.*, **5**, 956 (1964).
55. Boreskov, G. K., A. V. Khasin, and T. S. Starostina, *Dokl. Akad. Nauk SSSR*, **164**, 606 (1965).
56. Borisova, T. I., B. V. Ershler, and A. N. Frumkin, *Zh. Fiz. Khim.*, **22**, 925 (1948); **24**, 337 (1950).
57. Borisova, T. I., and V. I. Veselovskii, *Zh. Fiz. Khim.*, **27**, 1195 (1953).
58. Boswell, M. C., and R. K. Iler, *J. Am. Chem. Soc.*, **58**, 924 (1936).
59. Bourgault, P. L., and B. E. Conway, *Can. J. Chem.* **38**, 1557 (1960).
60. Boussingault, J., *Ann. Chim. Phys.*, **54**, 264 (1833).
61. Bradshaw, B. C., *Proc. Ann. Power Sources Conf.*, **12**, 18 (1958).
62. Braekken, H., *Kgl. Norske Videnskab. Selskal Forh.*, **7**, 143 (1935); *Chem. Abstr.*, **29**, 4647 (1935).
63. Brauner, B., *Bull. Acad. Belg.*, **18**, 81 (1889)
64. Brcec, B. C., M. Bulc, J. Siftar, and A. Urbane, *Monatsh.*, **95**, 248 (1963).
65. Brentano, J., *Proc. Phys. Soc. (London)*, **37**, 184 (1925).
66. Brewer, L., *Chem. Rev.*, **52**, 1 (1953).

67. Briggs, G. W. D., *Electrochim. Acta*, **1**, 297 (1959).
68. Briggs, G. W. D., I. Dugdale, and W. F. K. Wynne-Jones, *Electrochim. Acta*, **4**, 55, (1961).
68a. Briggs, G. W. D., and M. Fleischmann, *Trans. Faraday Soc.*, **62**, 3217 (1966).
69. Briggs, G. W. D., E. Jones, and W. F. K. Wynne-Jones, *Trans. Faraday Soc.*, **51**, 1433 (1955).
70. Briggs, G. W. D., G. W. Stott, and W. F. K. Wynne-Jones, *Electrochim. Acta*, **7**, 244 (1962).
71. Briggs, G. W. D., and W. F. K. Wynne-Jones, *Trans. Faraday Soc.*, **52**, 1272 (1956).
72. Briggs, G. W. D., and W. F. K. Wynne-Jones, *Electrochim. Acta*, **7**, 241 (1962).
73. Briggs, G. W. D., *J. Chem. Soc.*, **1957**, 1846.
73a. Bro, P., and D. Cogley, *J. Electrochem. Soc.*, **113**, 521 (1966).
74. Brow, M. J., *J. Phys. Chem.*, **20**, 680 (1916).
75. Bückle, H., and H. Hanemann, *Z. Metallk.*, **32**, 120 (1940).
76. Burbank, J., *J. Electrochem. Soc.*, **103**, 87 (1956).
77. Burbank, J., *J. Electrochem. Soc.*, **104**, 693 (1957).
78. Burbank, J., *J. Electrochem. Soc.*, **106**, 369 (1959).
79. Burbank, J., *J. Electrochem. Soc.*, **111**, 765 (1964).
80. Burbank, J., *J. Electrochem. Soc.*, **111**, 1112 (1964).
80a. Burbank, J., and E. Goldstein, *J. Electrochem. Soc.*, **113**, 1329 (1966).
81. Burbank, J., and A. C. Simon, *J. Electrochem. Soc.*, **100**, 11 (1953).
82. Burbank, J., and C. P. Wales, *J. Electrochem. Soc.*, **111**, 1002 (1964).
83. Butler, G., and J. L. Copp, *J. Chem. Soc.*, **1956**, 725.
84. Byström, A., *Arkiv. Kemi Mineral. Geol.*, **B17**, No. 8 (1943).
85. Byström, A., *Arkiv. Kemi Mineral. Geol.*, **A18**, No. 23 (1945); *Chem. Abstr.*, **41**, 5356 (1947).
86. Byström, A., *Arkiv. Kemi Mineral. Geol.*, **A20**, No. 11 (1945); *Chem. Abstr.*, **41**, 4053 (1947).
87. Byström, A., *Arkiv. Kemi Mineral. Geol.*, **A25**, No. 13 (1947); *Chem. Abstr.*, **43**, 451 (1949).
88. Byström, A., and A. Westgren, *Arkiv. Kemi Mineral. Geol.*, **B16**, No. 14 (1943); *Chem. Abstr.*, **38**, 5707 (1944).
89. Cabrera, N., and N. F. Mott, *Rept. Progr. Phys.*, **12**, 163 (1949).
90. Cahan, B. D., J. B. Ockerman, R. F. Amlie, P. Rüetschi, *J. Electrochem. Soc.*, **107**, 725 (1960).
91. Cahan, B. D., and P. Rüetschi, *J. Electrochem. Soc.*, **106**, 543 (1959).
92. Cairns, R. W., and E. Ott, *J. Am. Chem. Soc.*, **55**, 527 (1933).
93. Cairns, R. W., and E. Ott, *J. Am. Chem. Soc.*, **55**, 534 (1933).
94. Cairns, R. W., and E. Ott, *J. Am. Chem. Soc.*, **56**, 1094 (1934).
95. Cann, J. Y., and E. LaRue, *J. Am. Chem. Soc.*, **54**, 3456 (1932).
96. Carman, P. C., *Trans. Faraday Soc.*, **30**, 566 (1934).
97. Carmody, W. R., *J. Am. Chem. Soc.*, **51**, 2905 (1929).
98. Carmody, W. R., *J. Am. Chem. Soc.*, **54**, 210 (1932).
99. Carr, D. S., and C. F. Bonilla, *J. Electrochem. Soc.*, **999** 475 (1952).
100. Cartier, P., and J. Brenet, *Compt. Rend.*, **255**, 100 (1962).
101. Cartier, P., and J. Brenet, *Compt. Rend.* **256**, 1976 (1963); **258**, 4495 (1964).
102. Casey, E. J., and K. N. Campney, *J. Electrochem. Soc.*, **102**, 219 (1955).
103. Casey, E. J. and W. J. Moroz, *Can. J. Chem.*, **43**, 1199 (1965).

104. Chevillot, *Ann. Chim. Phys.*, **13**, 299 (1820).
105. Chufarov, G. I., M. G. Zhuravleva, and E. P. Tatievskaya, *Dokl. Akad. Nauk SSSR*, **73**, 1209 (1950).
106. Clark, G. L., W. C. Asbury, and R. M. Wick, *J. Am. Chem. Soc.*, **47**, 2661 (1925).
107. Clark, G. L., J. N. Megudick, and N. C. Schieltz, *Z. Anorg. Allgem. Chem.*, **229**, 401 (1936).
108. Clark, G. L., and R. Rowan, *J. Am. Chem. Soc.*, **63**, 1302 (1941).
109. Clark, G. L., and R. Rowan, *J. Am. Chem. Soc.*, **63**, 1305 (1941).
110. Clark, G. L., N. C. Schieltz, and T. T. Quirke, *J. Am. Chem. Soc.*, **59**, 2305 (1937).
111. Clark, G. L., and W. P. Tyler, *J. Am. Chem. Soc.*, **61**, 58 (1939).
112. Cohen, E., and G. W. R. Overdijkink, *Z. Physik. Chem. (Leipzig)*, **188**, 316 (1941).
113. Coles, R. E., *Brit. J. Appl. Phys.*, **14**, 342 (1963).
114. Conway, B. E., and P. L. Bourgault, *Can. J. Chem.*, **37**, 293 (1959).
115. Conway, B. E., and P. L. Bourgault, *Can. J. Chem.*, **40**, 1690 (1962).
116. Conway, B. E., and P. L. Bourgault, *Trans. Faraday Soc.*, **58**, 593 (1962).
117. Conway, B. E., and E. Gileadi, *Can. J. Chem.*, **40**, 1933 (1962).
118. Craig, D. N., and G. W. Vinal, *J. Res. Natl. Bur. Std.*, **14**, 449 (1935).
119. Craig, C. D., and G. W. Vinal, *J. Res. Natl. Bur. Std.*, **24**, 482 (1940).
120. Cupp, E. B., *Proc. Ann. Power Sources Conf.*, **18**, 44 (1964).
121. Darbyshire, J. A., *J. Chem. Soc.*, **134**, 211 (1932).
122. Dasoyan, M. A., *Dokl. Akad. Nauk SSSR*, **107**, 863 (1956).
123. Dasoyan, M. A., *Zh. Prikl. Khim.*, **29**, 1827 (1956).
124. Davey, W. P., and E. O. Hoffman, *Phys. Rev.*, **15**, 333 (1920); **17**, 402 (1921).
125. Davies, D. E., and W. Barker, *Corrosion*, **20**, 47t (1964).
126. de Bethune, A. J., and N. A. S. Loud, *Standard Aqueous Electrode Potentials and Temperature Coefficients*, Clifford A. Hampel, Skokie, Ill., 1964.
127. Dalahay, P., *J. Electrochem. Soc.*, **97**, 205 (1950).
128. Delahay, P., C. F. Pillon, and D. Perry, *J. Electrochem. Soc.*, **99**, 414 (1952).
129. Delahay, P., M. Pourbaix, and P. Van Rysselberghe, *J. Electrochem. Soc.*, **98**, 57 (1951).
130. Delahay, P., M. Pourbaix, and P. Van Rysselberghe, *J. Electrochem. Soc.*, **98**, 65 (1951).
131. Denison, I. A., *Trans. Electrochem. Soc.*, **90**, 387 (1946).
132. Dickinson, R. G., and J. B. Friauf, *J. Am. Chem. Soc.*, **46**, 2457 (1924).
133. Dirkse, T. P., *J. Electrochem. Soc.*, **106**, 453 (1959).
134. Dirkse, T. P., *J. Electrochem. Soc.*, **106**, 920 (1959).
135. Dirkse, T. P., *J. Electrochem. Soc.*, **107**, 859 (1960).
136. Dirkse, T. P., *J. Chem. Eng. Data*, **6**, 538 (1961).
137. Dirkse, T. P., and F. de Haan, *J. Electrochem. Soc.*, **105**, 311 (1958).
138. Dirkse, T. P., and D. B. de Vries, *J. Phys. Chem.*, **63**, 107 (1959).
139. Dirkse, T. P., and L. A. Van der Lugt, *J. Electrochem. Soc.*, **111**, 629, (1964).
140. Dirkse, T. P., and G. J. Werkema, *J. Electrochem. Soc.*, **106**, 88 (1959).
141. Dirkse, T. P., and B. J. Weirs, *J. Electrochem. Soc.*, **106**, 284 (1959).
142. Dittmann, J. F., and J. F. Sams, *J. Electrochem. Soc.*, **105**, 553 (1958).
143. Dodson, V. H., *J. Electrochem. Soc.*, **108**, 401 (1961).
144. Dodson, V. H., *J. Electrochem. Soc.*, **108**, 406 (1961).

145. Dolezalek, F., and K. Finckh, Z. Anorg. Allgem. Chem., **51**, 320 (1906).
146. Domanski, W., Arch. Hutnictwa, **3**, 81 (1958); Chem. Abstr., **52**, 15995 (1958).
147. Donnan, F. G., and T. W. A. Shaw, J. Soc. Chem. Ind., (London), **29**, 987 (1910).
148. Dugdale, I., M. Fleischmann, and W. F. K. Wynne-Jones, Electrochim. Acta, **5**, 229 (1961).
149. Dumas, J., Compt. Rend., **86**, 65 (1878).
150. Duquesnoy, A., and F. Marion, Compt. Rend., **256**, 2862 (1963).
151. Eberius, E., and M. LeBlanc, Z. Anal. Chem., **89**, 81 (1932).
152. Edison, T. A., U. S. Pats. 678,722 (1901); 1,083,356 (1914).
153. Eichenauer, W., and G. Meuller, Z. Metallk., **53**, 321 (1962).
154. Eidensohn, S., J. Electrochem. Soc., **99**, 252C (1952).
155. Ekler, L., Can. J. Chem., **42**, 1355 (1964).
156. Elbs, K., and J. Forsell, Z. Elektrochem., **8**, 760 (1902).
157. Elina, L. M., T. I. Borisova, Ts. I. Zalkind, Zh. Fiz. Khim., **28**, 785 (1954).
158. Ellis, G. B., H. Mandel, and D. Linden, J. Electrochem. Soc., **99**, 205C (1952).
159. El Wakkad, S. E. S., and S. E. Emara, J. Chem. Soc., **1953**, 3504.
160. Ershler, B. V., Dokl. Akad. Nauk, **37**, 258, 262 (1942).
161. Ershler, B. V., G. S. Tyurikov, and A. D. Smirnova, Zh. Fiz. Khim., **14**, 985 (1940).
161a. Esdaile, J. D., J. Electrochem. Soc., **113**, 71 (1966).
162. Evans, U. R., Z. Elektrochem., **62**, 619 (1958).
163. Falk, S. U., J. Electrochem. Soc., **107**, 661 (1960).
164. Farnsworth, H. E., Appl. Phys. Letters, **2**, 199 (1963).
165. Farnsworth, H. E., and H. H. Madden, J. Appl. Phys., **32**, 1933 (1961).
166. Faure, A., Electrician, **6**, 323 (1881); **7**, 122, 249 (1881); U. S. Pat. 252,002 (1882).
167. Fayek, M. K., and J. Leciejewicz, Z. Anorg. Allgem. Chem., **336**, 104 (1965).
168. Feitknecht, W., Z. Elektrochem., **62**, 795 (1958).
169. Feitknecht, W., H. R. Christen, and H. Studer, Z. Anorg. Allgem. Chem., **283** 88 (1956).
170. Feitknecht, W., and A. Gaumann, J. Chim. Phys., **49**, C135 (1952).
171. Fink, C. G., and A. J. Dornblatt, Trans. Electrochem. Soc., **79**, 269 (1931).
172. Fiseiskii, V. N., and Ya. I. Tur'yan, Zh. Fiz. Khim., **24**, 567 (1950).
173. Fishlock, D. J., Metal Finishing, **57**, 55 (1959).
174. Fleischer, A., Trans. Electrochem. Soc., **94**, 289 (1948).
175. Fleischer, A., Proc. Ann. Power Sources Conf., **13**, 78 (1959).
176. Fleischmann, M., and M. Liler, Trans. Faraday Soc., **54**, 1370 (1958).
177. Fleischmann, M., and H. R. Thirsk, Trans. Faraday Soc., **51**, 71 (1955); J. Electrochem. Soc., **110**, 688 (1963).
178. Fleischmann, M., and H. R. Thirsk, J. Electrochem. Soc., **100**, 688 (1963).
179. Foerster, F., Z. Elektrochem., **13**, 414 (1907); **14**, 17 (1908).
180. Foerster, F., and F. Krüger, Z. Elektrochem., **33**, 406 (1927).
181. Fogel, Ya. M., B. T. Nadykto, V. I. Shvachko, and V. F. Rybalko, Zh. Fiz. Khim., **38**, 2397 (1964).
182. Fordham, S., and J. T. Tyson, J. Chem. Soc., **139**, 483 (1937).
183. Frey, D. A., and H. E. Weaver, J. Electrochem. Soc., **107**, 930 (1960).
184. Fried, F., Z. Physik. Chem. (Leipzig), **123**, 406 (1926).
185. Fromherz, H., Z. Physik. Chem. (Leipzig), **153**, 382 (1931).
186. Frounfelker, R. E., and W. H. Hirthe, Trans. AIME, **244**, 196 (1962).

187. Frumkin, A. N., *Zh. Fiz. Khim.*, **23**, 1477 (1949).

188. Fueki, K., and J. B. Wagner, *J. Electrochem. Soc.*, **112**, 384 (1965).

189. Garrett, A. B., S. Vellenga, and C. M. Fontana, *J. Am. Chem. Soc.*, **61**, 367 (1939).

190. Gay-Lussac, J. L., *Ann. Chim. Phys.*, **45**, 211 (1830).

191. Georgi, K., *Z. Elektrochem.*, **38**, 681 (1932).

192. Georgi, K., *Z. Elektrochem.*, **38**, 714 (1932).

193. Georgi, K., *Z. Elektrochem.*, **39**, 736 (1933).

194. Gerke, R. H., *J. Am. Chem. Soc.*, **44**, 1684 (1922).

195. Gerkin, R. E., and K. S. Pitzer, *J. Am. Chem. Soc.*, **84**, 2662 (1962).

196. Gerischer, H., *Angew. Chem.*, **70**, 285 (1958).

197. Germer, L. H., and A. U. MacRae, *J. Appl. Phys.*, **33**, 2923 (1962).

198. Gladstone, J. H., and A. Tribe, *Nature*, **25**, 221, 461 (1882); **26**, 251, 602 (1882); **27**, 583 (1883).

199. Glaser, F., *Z. Anorg. Allgem. Chem.*, **36**, 1 (1903).

200. Glasstone, S., *J. Chem. Soc.*, **119**, 1689, 1914 (1921).

201. Glasstone, S., *J. Chem. Soc.*, **121**, 1457 (1922).

202. Glasstone, S., *J. Chem. Soc.*, **121**, 1469 (1922).

203. Glasstone, S., *J. Chem. Soc.*, **121**, 2091 (1922).

204. Glemser, O., and J. Einerhand, *Z. Anorg. Allgem. Chem.*, **261**, 26 (1950).

205. Glemser, O., and J. Einerhand, *Z. Anorg. Allgem. Chem.*, **261**, 43 (1950).

206. Glemser, O., and J. Einerhand, *Z. Elektrochem.*, **54**, 302 (1950).

206a. Glicksman, R., and C. K. Morehouse, *J. Electrochem. Soc.*, **104**, 589 (1957).

207. Göhr, H., and E. Lange, *Z. Physik. Chem. (Frankfurt)* **17**, 100 (1958).

208. Göhr, H., and H. Reinhard, *Z. Elektrochem.*, **64**, 414 (1960).

209. Goralevich, D. K., *Zh. Obshch. Khim.*, **1**, 973 (1931).

210. Gorodetskii, Yu. S., *Elektrokhim.*, **1**, 681 (1965).

211. Graff, W. S., and H. H. Stadelmair, *J. Electrochem. Soc.*, **105**, 446 (1958).

212. Greenburg, R. H., and B. P. Caldwell, *Trans. Electrochem. Soc.*, **80**, 71 (1941).

213. Gregor, L. V., *Chem. Abstr.*, **56**, 12378 (1962).

214. Gregor, L. V., and K. S. Pitzer, *J. Am. Chem. Soc.*, **84**, 2664 (1962).

215. Gregor, L. V., and K. S. Pitzer, *J. Am. Chem. Soc.*, **84**, 2671 (1962).

216. Gross, S. T., *J. Am. Chem. Soc.*, **63**, 1168 (1941); **65**, 1107 (1943).

217. Grube, G., *Z. Elektrochem.*, **28**, 273 (1922).

218. Grube, G., and A. Vogt, *Z. Elektrochem.*, **44**, 353 (1938).

219. Grünbaum, O. S., and J. V. Irbane, *Anales Asoc. Quim Argentina*, **39**, 62 (1951); *Chem. Abstr.*, **46**, 839 (1952).

220. Günther, P., *Z. Elektrochem.*, **62**, 619 (1958).

221. Gulbransen, E. A., and K. F. Andrews, *J. Electrochem. Soc.*, **101**, 128 (1954).

222. Haissinsky, M., and M. Quesney, *Compt. Rend.*, **223**, 792 (1946); **224**, 831 (1947); *J. Chim. Phys.*, **49**, 302 (1952).

223. Halla, F., and F. Pawlek, *Z. Physik. Chem. (Leipzig)*, **128**, 49 (1927).

224. Hamer, W. J., *J. Chem. Soc.*, **57**, 9 (1935).

225. Hamer, W. J., and D. N. Craig, *J. Electrochem. Soc.*, **104**, 206 (1957).

226. Haring, H. E., and U. B. Thomas, *Trans. Electrochem. Soc.*, **48**, 293 (1935).

227. Haring, M. M., and E. G. Vanden Bosche, *J. Phys. Chem.*, **33**, 161 (1929).

228. Harned, H. S., *J. Am. Chem. Soc.*, **51**, 416 (1929).

229. Harned, H. S., and W. J. Hamer, *J. Am. Chem. Soc.*, **57**, 27 (1935).

230. Harned, H. S., and W. J. Harmer, *J. Am. Chem. Soc.*, **57**, 33 (1935).

231. Hatfield, J. E., and O. W. Brown, *Trans. Electrochem. Soc.*, **72**, 361, (1937).
232. Hauel, A. P., *Trans. Electrochem. Soc.*, **76**, 435 (1939).
233. Hedvall, J. A., *Z. Anorg. Allgem. Chem.*, **92**, 392 (1915).
234. Hendricks, S. B., M. E. Jefferson, and J. F. Schultz, *Z. Krist.*, **73**, 376 (1930).
235. Hickling, A., and S. Hill, *Discussions Faraday Soc.*, **1**, 236 (1947).
236. Hickling, A., and J. E. Spice, *Trans. Faraday Soc.*, **43**, 762 (1947).
237. Hickling, A., and D. Taylor, *Discussions Faraday Soc.*, **1**, 277 (1947).
238. Hoar, T. P., *Modern Aspects of Electrochemistry*, Vol. 2, J. O'M. Bockris, Ed., Butterworths, London, 1959, p. 262.
239. Hoare, J. P., *J. Electrochem. Soc.*, **112**, 608 (1965).
240. Hoare, J. P., *Electrochim. Acta*, **9**, 1289 (1964).
241. Hoare, J. P., *Nature*, **204**, 71 (1964).
242. Hoehne, E., *Z. Metallk.*, **30**, 52, (1938).
243. Holtermann, C. B., *Ann. Chim.*, **14**, 21 (1940).
244. Holtermann, C. B., and P. Laffitte, *Compt. Rend.*, **204**, 1813 (1937).
245. Hosono, T., and K. Watanabe, *J. Electrochem. Soc. Japan*, **19**, 14 (1951).
246. Howell, O. R., *J. Chem. Soc.*, **123**, 669, 1772 (1923).
247. Huber, K., *J. Electrochem. Soc.*, **100**, 376 (1953); *Z. Elektrochem.*, **62**, 675 (1958).
248. Hüttig G. F., and A. Peter, *Z. Anorg. Allgem. Chem.*, **189**, 183, 190 (1930).
249. Huggins, M. L., *Phys. Rev.* **21**, 719 (1923).
250. Huq, A. K. M. S., A. J. Rosenberg, and A. C. Makrides, *J. Electrochem. Soc.*, **111**, 278 (1964).
251. Hurlen, T., Y. L. Sandler, and E. A. Pantier, *Electrochim. Acta*, **11**, 1463 (1966).
252. Ikari, S., S. Yoshizawa, and S. Okada, *J. Electrochem. Soc. Japan*, **27**, E-223, E-247 (1959).
253. Ikari, S., and S. Yoshizawa, *J. Electrochem. Soc. Japan*, **28**, E-138 (1960); *Denki Kagaku*, **28**, 432, 503, 576, 596 (1960).
254. Ikari, S., and S. Yoshizawa, *Denki Kagaku*, **28**, 675 (1960); *Chem. Abstr.*, **62**, 2504 (1965).
255. Ishikawa, T., and H. Tagawa, *Kogyo Kagaku Zasshi*, **94**, 264 (1961); *Chem. Abstr.*, **57**, 4451 (1962).
256. Ives, D. J. G., and G. J. Janz, *Reference Electrodes*, Academic Press, New York, 1961, p. 333.
257. Jaeger, F. M., and H. C. Germs, *Z. Anorg. Allgem. Chem.*, **119**, 145 (1921).
258. Jellinek, K., and H. Gordon, *Z. Physik. Chem. (Leipzig)*, **112**, 207 (1924).
259. Jirsa, F., *Z. Elektrochem.*, **25**, 146 (1919).
260. Jirsa, F., *Chem. Listy*, **19**, 3, 300 (1925); *Z. Anorg. Allgem. Chem.*, **148**, 130 (1925); **158**, 33 (1926).
261. Jirsa, F., *Collection Czech. Chem. Commun.*, **14**, 445 (1949).
262. Jirsa, F., *Collection Czech. Chem. Commun.*, **14**, 451 (1949).
263. Jirsa, F., and J. Jelinik, *Z. Anorg. Allgem. Chem.*, **158**, 61 (1926).
264. Jirsa, F., J. Jelinik, and J. Srbek, *Chem. Listy*, **19**, 114 (1925).
265. Jirsa, F., and K. Schneider, *Z. Elektrochem.*, **33**, 129 (1927).
266. Johnston, H. L., F. Cuta, and A. B. Garrett, *J. Am. Chem. Soc.*, **55**, 2311 (1933).
267. Jones, E., and W. F. K. Wynne-Jones, *Trans. Faraday Soc.*, **52**, 1260 (1956).
268. Jones, P., R. Lind, and W. F. K. Wynne-Jones, *Trans. Faraday Soc.*, **50**, 972 (1954).
269. Jones, P., and H. R. Thirsk, *Trans. Faraday Soc.*, **50**, 732 (1954).

270. Jones, P., H. R. Thirsk, and W. F. K. Wynne-Jones, *Trans. Faraday Soc.*, **52**, 1003 (1956).
271. Jones, R. O., and S. Rothschild, *J. Electrochem. Soc.*, **105**, 206 (1958).
271a. Jost, E., and F. Rufenacht, *J. Electrochem. Soc.*, **113**, 97 (1966).
272. Jungner, E. W., U. S. Pat. 692,298 (1902).
273. Jungner, W., Swedish Pat. 10,177 (1899); 11487 (1899); Brit. Pat. 7768 (1900).
274. Kabanov, B. N., I. G. Kiseleva, and D. I. Leikis, *Dokl. Adak. Nauk SSSR*, **99**, 805 (1954).
275. Kabanov, B. N., and D. I. Leikis, *Z. Elektrochem.*, **62**, 660 (1958).
276. Kabanov, B. N., D. I. Leikis, and E. I. Krepakova, *Dokl. Akad. Nauk SSSR*, **98**, 989 (1954).
277. Kameyama, N., and T. Fukumoto, *J. Soc. Chem. Ind. Japan*, **46**, 1022, (1943); Chem. Abstr., **42**, 6674 (1948).
278. Kameyama, N., and T. Fukumoto, *J. Soc. Chem. Ind. Japan*, **46**, 1022 (1943); *Chem. Abstr.*, **42**, 6675 (1948).
279. Kameyama, N., and T. Fukumoto, *J. Soc. Chem. Ind. Japan*, **49**, 155 (1946); *Chem. Abstr.*, **42**, 6689 (1948).
280. Katz, T., *Ann. Chim.*, **5**, 5 (1950).
281. Katz, T., and R. LeFaivre, *Bull. Soc. Chim. France*, **16**, D124 (1949).
282. Katz, T., M. Siramy, and R. Faivre, *Compt. Rend.*, **227**, 282 (1948).
283. Kay, M. I., *Acta Cryst.*, **14**, 80 (1961).
284. Kendrick, A., *Z. Elektrochem.*, **7**, 53 (1900).
285. Kinoshita, K., *Bull. Chem. Soc. Japan*, **12**, 164, 366 (1937).
286. Kiseleva, I. G., and B. N. Kabanov, *Dokl. Akad. Nauk SSSR*, **108**, 864 (1956); **122**, 1042 (1958).
287. Kittel, A., Dissertation Technische Hochschule, Prague, Czechoslovakia, 1944.
288. Kivola, P., and V. Vuorio, *Suomen Kemistilehti*, **34**, 179 (1961); *Chem. Abstr.*, **56**, 12647 (1962).
289. Klemm, W., *Z. Anorg. Allgem. Chem.*, **201**, 32 (1931).
290. Kobayashi, K., *Sci. Rept. Tokuku Univ.*, Ser. I., **35**, 173 (1951); *Chem. Abstr.*, **46**, 10842 (1952).
291. Koch, D. F. A., *Elektrochim. Acta*, **1**, 32 (1959).
292. Kolotyrkin, Ya. M., *Z. Elektrochem.*, **62**, 664 (1958).
293. Kornfeil, F., *Proc. Ann. Power Sources Conf.*, **12**, 18 (1958).
294. Krasil'shchikov, A. I., *Zh. Fiz. Khim.*, **26**, 216 (1952).
295. Krasil'shchikov, A. I., *Zh. Fiz. Khim.*, **37**, 531 (1963).
295a. Krasil'shchikov, A. I., and V. A. Andreeva, *Zh. Fiz. Khim.*, **27**, 389 (1953).
296. Krivolapova, E. V., and B. N. Kabonov, *Trudy Sovesh. Elektrokhim. Akad. Nauk SSSR*, **1950**, 539; *Chem. Abstr.*, **49**, 12161 (1955); *Zh. Fiz. Khim.*, **14**, 335 (1941).
297. Kronenberg, M. L., *J. Electroanal. Chem.*, **13**, 120 (1967).
298. Kronenberg, M. L., J. C. Banter, E. Yeager, and F. Hovorka, *J. Electrochem. Soc.*, **110**, 1007 (1963).
299. Ksada, C. J., *Am. J. Sci.*, **22**, 131 (1931).
300. Kuchinskii, E. M., and B. V. Ershler, *Zh. Fiz. Khim.*, **20**, 539 (1946).
301. Kukoz, F. I. et al., *Chem. Abstr.*, **56**, 1278, 1280, (1962); **57**, 15825 (1962).
302. Kukoz, L. A., and M. F. Skalozubov, *Chem. Abstr.* **58**, 12163 (1963).
303. Kul'kova, N. V., and M. I. Temkin, *Zh. Fiz. Khim.*, **36**, 1731 (1962).
304. Lachaud, M., and C. Lepierre, *Bull. Soc. Chim.*, **7**, 600 (1892).

305. Lake, P., and E. J. Casey, *J. Electrochem. Soc.*, **105**, 52 (1958); 106, 913 (1959).
306. Lander, J. J., *Acta Cryst.*, **4**, 148 (1951).
306a. Lander, J. J., *J. Electrochem. Soc.*, **95**, 174 (1949).
307. Lander, J. J., *J. Electrochem. Soc.*, **98**, 213 (1951).
308. Lander, J. J., *J. Electrochem. Soc.*, **98**, 220, 522 (1951).
309. Lander, J. J., *J. Electrochem. Soc.*, **99**, 339 (1952).
310. Lander, J. J., *J. Electrochem. Soc.*, **99**, 467 (1952).
311. Lander, J. J., *J. Electrochem. Soc.*, **103**, 1 (1956).
312. Lander, J. J., *J. Electrochem. Soc.*, **105**, 289 (1958).
313. Landsberg, R., and M. Hollnagel, *Z. Elektrochem.*, **58**, 680 (1954); 60, 1098 (1956).
313a. Langer, A., and J. T. Patton, *J. Electrochem. Soc.*, **114**, 113 (1967).
314. Lappe, F., *Phys. Chem. Solids*, **23**, 1563 (1962).
315. Latimer, W. M., *Oxidation Potentials*, 2nd ed., Prentice-Hall, Englewood Cliffs, N. J., 1952, p. 197.
316. Laves, F. G. Bayer, and A. Panagos, *Schweiz. Mineral. Petrog. Mitt.*, **43**, 217 (1963); *Chem. Abstr.*, **59**, 7041 (1963).
317. Labat, J., *J. Chim. Phys.*, **60**, 1251 (1963).
318. Labat, J., *Ann. Chim.*, **9**, 399 (1964).
319. LeBlanc, M., and E. Eberius, *Z. Physik. Chem. (Leipzig)*, **A160**, 69 (1932).
320. LeBlanc, M., and E. Möbius, *Z. Elektrochem.*, **39**, 753 (1933).
321. LeBlanc, M., and R. Müller, *Z. Elektrochem.*, **39**, 204 (1933).
322. LeBlanc, M., and H. Sachse, *Z. Elektrochem.*, **32**, 204 (1926).
323. LeBlanc, M., and H. Sachse, *Z. Elektrochem.*, **32**, 58 (1926).
324. LeBlanc, M., and H. Sachse, *Z. Anorg. Allgem. Chem.*, **168**, 15 (1927).
325. LeBlanc, M., and H. Sachse, *Z. Physik*, **32**, 887 (1931).
326. Leciejewicz, J., *Acta Cryst.*, **14**, 66 (1961).
327. Leciejewicz, J., *Acta Cryst.*, **14**, 1304 (1961).
328. Leikis, D. I., *Dokl. Akad. Nauk SSSR*, **135**, 1429 (1960).
329. Leikis, D. I., G. L. Vidovitch, L. L. Knoz, and B. N. Kabanov, *Z. Physik. Chem. (Leipzig)*, **214**, 334 (1960).
330. Levi, G. R., and G. Natta, *Nuovo Cimento*, **1**, 335 (1924); 3, 114 (1926).
331. Levi, G. R., and A. Quilico, *Gazz. Chim. Ital.*, **54**, 598 (1924); *Chem. Abstr.*, **19**, 942 (1925).
332. Lewis, G. N., *J. Am. Chem. Soc.*, **28**, 158 (1906).
333. Lewis, G. N., *Z. Physik. Chem. (Leipzig)*, **55**, 449 (1906).
334. Lopez-Lopez, A., *Compt. Rend.*, **225**, 3170 (1962).
335. Lorenz, R., and E. Lauber, *Z. Elektrochem.*, **15**, 157 (1909).
336. Lukovtsev, P. D., and G. J. Slaidin, *Electrochim. Acta*, **6**, 17 (1962).
337. Lukovtsev, P. D., and G. J. Slaidin, *Zh. Fiz. Khim.*, **36**, 2268 (1962); 38, 556 (1964).
338. Lunde, G., *Z. Anorg. Allgem. Chem.*, **163**, 345, (1927); 169, 405 (1928).
339. Lurie, M., H. N. Seiger, and R. C., Shair, *Proc. Ann. Power Sources Conf.*, **17**, 110 (1963).
340. Luther, R., and F. Pokorny, *Z. Anorg. Allgem. Chem.*, **57**, 290 (1908).
341. McMillan, J. A., *Acta Cryst.*, **7**, 640 (1954).
342. McMillan, J. A., *J. Inorg. Nucl. Chem.*, **13**, 28 (1960); *Nature*, **195**, 594 (1962); *Chem. Rev.*, **62**, 65 (1962).
343. McMillan, J. A., and B. Smaller, *J. Chem. Phys.*, **35**, 1698 (1961).

344. McMillan, J. A., and N. Hackerman, *J. Electrochem. Soc.*, **106**, 341 (1959).
345. MacRae, A. U., *Appl. Phys. Letters*, **2**, 88 (1963).
346. Maeda, M., *Denki Kagaku*, **25**, 195 (1957).
347. Maeda, M., *Acta Met.*, **6**, 66 (1958).
348. Maeda, M., *J. Electrochem. Soc. Japan*, **26**, 183 (1958).
349. Makrides, A. C., *J. Electrochem. Soc.*, **113**, 1158 (1966).
350. Malaguti, A., *Ann., Chim.* **41**, 241 (1951).
351. Margulis, E. V., and N. I. Kopylov, *Zh. Neorgan. Khim.*, **9**, 763 (1964).
352. Mark, H. B., *J. Electrochem. Soc.*, **109**, 634 (1962); 110, 945 (1963).
353. Mark, H. B., and W. C. Vosburgh, *J. Electrochem. Soc.*, **108**, 615 (1961).
354. Mathur, P. B., and R. Gangadharan, *Indian J. Tech.*, **2**, 331 (1964).
355. Maxim, I., and T. Braun, *J. Phys. Chem. Solids*, **24**, 537 (1963).
356. Mazitov, Yu. A., K. I. Rozental', and V. I. Veselovskii, *Zh. Fiz. Khim.*, **38**, 697 (1964).
357. Mazitov, Yu. A., K. I. Rozental', and V. I. Veselovskii, *Zh. Fiz. Khim.*, **38**, 449, (1964); *Elektrokhim.*, **1**, 36 (1965).
358. Moore, W. J., and J. K. Lee, *Trans. Faraday Soc.*, **48**, 916 (1952).
359. Moore, W. J., and L. Pauling, *J. Am. Chem. Soc.*, **63**, 1392 (1941).
360. Morehouse, C. K., R. Glicksman, and G. S., Lozier, *Proc. IRE*, **46**, 1462 (1958).
361. Morrison, W., U. S. Pat. 976,092 (1910).
362. Mott, N. F., *Trans. Faraday Soc.*, **43**, 429 (1947).
363. Mrgudich, J. N., *Trans. Electrochem. Soc.*, **81**, 165 (1942).
363a. Müller, J., *J. Catalysis*, **6**, 50 (1966).
364. Müller, W. J., *Z. Elektrochem.*, **33**, 401 (1927).
365. Müller, W. J., *Z. Elektrochem.*, **36**, 365 (1930).
366. Müller, W. J., H. K. Cameron, and W. Machu, *Monatsh*, **59**, 73 (1932).
367. Müller, W. J., and W. Machu, *Korrosion Metallschutz*, **16**, 187 (1940); *Chem. Abstr.*, **35**, 3538 (1941).
368. Murata, K., *Bull. Chem. Soc. Japan*, **3**, 57 (1928).
369. Nagel, K., R. Ohse, and E. Lange, *Z. Elektrochem.*, **61**, 795 (1957).
370. Natta, G., *Gazz. Chim. Ital.*, **58**, 344 (1928).
371. Naugle, R. B., and A. W. Speyers, *Proc. Ann. Power Sources Conf.*, **13**, 97 (1959).
372. Nefedova, I. D., and A. I. Krasil'schchikov, *Zh, Fiz. Khim.*, **21**, 855 (1947).
373. Neiding, A. B., and I. A. Kazarnovskii, *Dokl. Akad. Nauk SSSR*, **78**, 713 (1951).
374. Nerst, W., and H. von Wartenberg, *Z. Physik. Chem.*, (*Leipzig*), **56**, 534 (1906).
375. Newton, R. F., *J. Am. Chem. Soc.*, **50**, 3258 (1928).
376. Nikulin, V. I., *Zh. Fiz. Khim.*, **35**, 84 (1961).
377. Noyes, A. A., C. D. Coryell, F. Stitt, and A. Kossiakoff, *J. Am. Chem. Soc.*, **59**, 1316 (1937).
378. Noyes, A. A., D. deVault, C. D. Coryell, and T. Deahl, *J. Am. Chem. Soc.*, **59**, 1326 (1937).
379. Noyes, A. A., J. L. Hoard, and K. S. Pitzer, *J. Am. Chem. Soc.*, **57**, 1221 (1935).
380. Noyes, A. A., and A. Kossiakoff, *J. Am. Chem. Soc.*, **57**, 1238 (1935).
381. Noyes, A. A., K. S. Pitzer, and C. L. Dunn, *J. Am. Chem. Soc.*, **57**, 1229 (1935).
382. Ohse, R. W., *Z. Elektrochem.*, **63**, 1063 (1959).
383. Ohse, R. W., *Z. Elektrochem.*, **64**, 1171 (1960).
384. Ohse, R., *Werkstoffe Korrosion*, **11**, 220 (1960).

385. Okada, S. et al., *J. Chem. Soc. Japan, Ind. Chem., Sect.* **51**, 129 (1948); **52**, 37 (1949); **53**, 5 (1950).
386. Okada, S., T. Shiraishi, and T. Yasuhara, *J. Chem. Soc. Japan*, **53**, 5, 378 (1950); **54**, 16 (1951).
387. Okamota, G., H. Kobayashi, M. Nagayama, and N. Sato, *Z. Elektrochem.*, **62**, 775 (1958).
388. Ostrovskii, V. E., I. R. Karpovich, N. V. Kul'kova, and M. I. Temkin, *Zh. Fiz. Khim.*, **37**, 2569 (1963).
389. Ostrovskii, V. E., and M. I. Temkin, *Kinetika i Kataliz*, **7**, 529 (1966).
390. Otto, E. M., *J. Electrochem. Soc.*, **113**, 643 (1966).
391. Pacault, P., and J. Labat, *Compt. Rend.*, **258**, 5421 (1964).
392. Palmaer, K. H., *Z. Elektrochem.*, **29**, 415 (1923).
393. Palous, S., and R. Buvet, *Bull. Soc. Chim. France* **1962**, 1606; **1963**, 2490.
394. Park, R. L., and H. E. Farnsworth, *Appl. Phys. Letters*, **3**, 167 (1963).
395. Pavlenko, I. G., *Dopovidi Akad. Nauk RSR*, **1961**, 353; *Chem. Abstr.* **56**, 261 (1962).
396. Pelouze, J., *Ann. Chim. Phys.*, **79**, 108 (1841).
397. Petersen, M., *J. Am. Chem. Soc.* **63**, 2617 (1941).
398. Pettit, F. S., *J. Electrochem. Soc.*, **113**, 1249 (1966).
399. Piette, L. H., and H. E. Weaver, *J. Chem. Phys.*, **28**, 735 (1958).
400. Pinsker, G. Z., and G. I. Farmakovskaya, *Kristallografiya*, **6**, 269 (1961).
401. Piontelli, R., and G. Poli, *Z. Elektrochem.*, **62**, 320 (1958).
402. Piontelli, R., and G. Serravalle, *Z. Elektrochem.*, **62**, 759 (1958).
403. Pitman, A. L., and G. W. Work, *U. S. Naval Res. Lab. Rept.* 4845 (1956).
404. Pitzer, K. S., and W. V. Smith, *J. Am. Chem. Soc.*, **59**, 2633 (1937).
405. Planté, G., *Compt. Rend.*, **50**, 640 (1860).
406. Pleskov, Yu. V., *Dokl. Akad. Nauk SSSR*, **117**, 645 (1957).
407. Pospelova, I. N., A. A. Rokov, and S. Ya. Pehezhetskii, *Zh. Fiz. Khim.*, **30**, 1433 (1956).
408. Pourbaix, M., N. deZoubov, and E. Deltombe, *CITCE*, **7**, 193 (1955).
409. Prasad, N., and M. G. Tendulkar, *J. Chem. Soc.*, **1931**, 1403.
410. Preston, G. D., and L. L. Bircumshaw, *Phil. Mag.*, **19**, 160 (1935).
411. Pucker, L. E., *J. Electrochem. Soc.*, **99**, 204C (1952).
412. Quinn, C. M., and M. W. Roberts, *Nature*, **200**, 648 (1963).
413. Rädlein, G., *Z. Elektrochem.*, **61**, 727 (1957).
414. Ramaley, L., and C. G. Enke., *J. Electrochem. Soc.*, **112**, 947 (1965).
415. Randall, M., and J. Y. Cann, *J. Am. Chem. Soc.*, **52**, 589 (1930).
416. Reddy, A. K. N., M. G. B. Rao, and J. O'M. Bockris, *J. Chem. Phys.*, **42**, 2246 (1964).
417. Rencker, E., *Bull. Soc. Chim.*, **3**, 981 (1936).
418. Rencker, E., and M. Bassiere, *Compt. Rend.*, **202**, 765 (1936).
419. Reuter, B., and W. Töpler, *Angew. Chem.*, **71**, 137 (1959).
420. Richards, T. W., and R. C. Wells, *Z. Anorg. Allgem. Chem.*, **47**, 79 (1906).
421. Riesenfeld, E. H., and H. Sass, *Z. Elektrochem.*, **39**, 219 (1933).
422. Riley, H. L., and H. B. Baker, *J. Chem. Soc.*, **128**, 2501 (1926).
423. Rocard, J. M., M. Bloom, and L. B. Robinson, *Can. J. Phys.*, **37**, 522 (1959).
424. Rordam, H. N. K., *Z. Physik. Chem. (Leipzig)*, **99**, 474 (1921).
425. Rollet, A. P., *Ann. Chim.*, **13**, 202 (1930).
426. Rooksby, H. P., *Nature*, **152**, 304 (1943).

426a. Rouse, T. O., and J. L. Weininger, *J. Electrochem. Soc.*, **113**, 184 (1966).
427. Rozental', K. I., and V. I. Veselovskii, *Zh. Fiz. Khim.*, **35**, 2670 (1961).
428. Rüetschi, P., and R. T. Angstadt, *J. Electrochem. Soc.*, **105**, 555 (1958).
429. Rüetschi, P., and R. T. Angstadt, *J. Electrochem. Soc.*, **111**, 1323 (1964).
430. Rüetschi, P., R. T. Angstadt, and B. D. Cahan, *J. Electrochem. Soc.*, **106**, 547 (1959).
431. Rüetschi, P., and B. D. Cahan, *J. Electrochem. Soc.*, **105**, 369 (1958).
432. Rüetschi, P., and B. D. Cahan, *J. Electrochem. Soc.*, **104**, 406 (1957).
433. Rüetschi, P., J. B. Ockerman, and R. Amlie, *J. Electrochem. Soc.*, **107**, 325 (1960).
434. Rüetschi, P., J. Sklarchuk, and R. T. Angstadt, *Electrochim. Acta*, **8**, 333 (1963).
435. Rüetschi, P., and R. F. Amlie, *J. Electrochem. Soc.*, **108**, 813 (1961).
436. Ruer, R., *Z. Anorg. Allgem. Chem.*, **50**, 265 (1906).
437. Rumsh, M. A., and T. M. Zimkina, *Kristallografiya*, **6**, 56 (1961).
438. Sagoyan, L. N., *Izv. Akad. Nauk Arm. SSR Khim. Nauki*, **17**, 3 (1964); *Chem. Abstr.*, **61**, 3914 (1964).
439. Salkind, A. J., and P. F. Bruins, *J. Electrochem. Soc.*, **109**, 356 (1962).
440. Sandler, Y. L., and D. D. Durigon, *J. Phys. Chem.*, **69**, 4201 (1965).
441. Sato, N., and G. Okamoto, *J. Electrochem. Soc.*, **110**, 605, 897 (1963).
442. Sato, N., and G. Okamoto, *Electrochim. Acta*, **10**, 495 (1965).
443. Sawyer, D. T., and R. J. Day, *Electrochim. Acta*, **8**, 589 (1963).
444. Sawyer, D. T., and L. V. Interrante, *J. Electroanal. Chem.*, **2**, 310 (1961).
445. Scatturin, V., P. Bellon, and R. Zannetti, *Ric. Sci.*, **27**, 2163 (1957); *J. Inorg. Nucl. Chem.*, **8**, 462 (1958).
446. Scatturin, V., P. L. Bellon, and A. J. Salkind, *Ric. Sci.*, **30**, 1034 (1960); *Chem. Abstr.*, **55**, 4094 (1961).
447. Schaffer, L. H., *J. Electrochem. Soc.*, **113**, 1 (1966).
448. Schenck, R., *Metall u. Erz.*, **23**, 408 (1926); *Chem. Abstr.*, **21**, 1082 (1947).
449. Schwabe, K., and G. Dietz, *Z. Elektrochem.*, **62**, 751 (1958).
450. Sekido, S., and T. Yoko, *Denki Kagaku*, **31**, 15 (1963); *Chem. Abstr.*, **61**, 2731 (1964).
451. Shankman, S., and A. R. Gordon, *J. Am. Chem. Soc.*, **61**, 2370 (1939).
452. Sharpe, T. F., private communication.
452a. Shepard, C. M., and H. C. Langdon, *J. Electrochem. Soc.*, **109**, 657, 661 (1962); **114**, 8 (1967).
453. Shibasaki, Y. *J. Electrochem. Soc.*, **105**, 624 (1958).
454. Shrawder, J., and I. A. Cowperthwaite, *J. Am. Chem. Soc.*, **56**, 2340, 2345 (1934).
455. Shumilova, N. A., G. V. Zhutaeva, and M. P. Tarasevich, *Electrochim. Acta*, **11**, 967 (1965).
456. Sidgwick, N. V., *The Chemical Elements and Their Compounds*, Oxford Univ. Press, London, 1950, p. 118.
457. Sieverts, A., and J. Hagensacker, *Z. Physik. Chem. (Leipzig)*, **68**, 115 (1909).
458. Simon, A. C., *Electrochem. Tech.*, **1**, 82 (1963); **3**, 307 (1965).
458a. Simon, A. C., *J. Electrochem. Soc.*, **114**, 1 (1967).
459. Simon, A. C., and E. L. Jones, *J. Electrochem. Soc.*, **109**, 760 (1962).
460. Simons, J. H., *J. Phys. Chem.*, **36**, 652 (1932).
461. Sklyarenko, S. I., and O. S. Druzhinina, *Zh. Prikl. Khim.*, **13**, 1794 (1940).
462. Slaidin, G. J., and P. D. Lukovtsev, *Dokl. Akad. Nauk SSSR*, **142**, 1130 (1962).

463. Smeltzer, W. W., E. L. Tollefson, and A. Cambron, *Can. J. Chem.*, **34**, 1046 (1956).
464. Spencer, L., *J. Chem. Soc.*, **123**, 2124 (1923).
465. Steacie, E. W. R., and F. M. G. Johnson, *Proc. Roy. Soc. (London)*, **A112**, 542 (1926).
466. Stokes, R. H., *J. Am. Chem. Soc.*, **69**, 1291 (1947).
467. Strutt, R. J., *Proc. Roy. Soc. (London)*, **A87**, 302 (1912).
468. Sugden, S., *J. Chem. Soc.*, **1932**, 161.
469. Sugino, K., T. Tomonari, and M. Takahasi, *J. Chem. Soc. Japan*, **52**, 75 (1949).
470. Sugita, K., *Denki Kagaku*, **29**, 86 (1961); *Chem. Abstr.*, **62**, 3656 (1965).
471. Shwabe, G.-M., and G. Hartmann, *Z. Anorg. Allgem. Chem.*, **281**, 183 (1955).
472. Swanson, H. E., and R. K. Fuyat, *Natl. Bur. Std. Circ.*, **539**, 2, 30, 32 (1953).
473. Swanson, H. E., N. T. Gilfrich, M. I. Cook, R. Stinchfield, and P. C. Park, *Natl. Bur. Std. Circ.*, **539**, 8, 32 (1959).
474. Tarasevich, M. R., N. A. Shumilova, and R. Kh. Burhstein, *Izv. Akad. Nauk SSSR, Ser. Khim.*, **1964**, 17.
475. Temkin, M. I. and N. V. Kul'kova, *Dokl. Akad. Nauk SSSR*, **105**, 1021 (1955).
475a. Thacker, R., *Nature*, **198**, 179 (1963).
476. Thibaut, R., *Z. Elektrochem.*, **19**, 881 (1913).
477. Thomas, U. B., *Trans. Electrochem. Soc.*, **94**, 42 (1948).
478. Thomas, U. B., F. T. Forster, and H. E. Haring, *Trans. Electrochem. Soc.*, **92**, 313 (1947).
479. Thompson, M. deK. et al., *Trans. Electrochem. Soc.*, **7**, 95 (1905); **31**, 339 (1917).
480. Thompson, M. deK., and A. L. Kaye, *Trans. Electrochem. Soc.*, **60**, 229 (1931).
481. Tichenor, R. L., *Ind. Eng. Chem.*, **44**, 973 (1952).
482. Tronstad, L., *Z. Physik. Chem. (Leipzig)*, **142**, 241 (1929).
483. Troost, L., *Compt. Rend.*, **98**, 1427 (1884).
484. Trueb, L. F., G. Truempler, and N. Ibl, *Helv, Chim. Acta*, **44**, 1433 (1961).
485. Truempler, G., and H. E. Hintermann, *Helv. Chim. Acta*, **40**, 1947 (1957).
486. Tsinman, A. I., *Zh. Fiz. Khim.*, **37**, 1343 (1963).
487. Tuomi, D., *J. Electrochem. Soc.*, **112**, 1 (1965).
488. Turner, D. R., *J. Electrochem. Soc.*, **98**, 434 (1951).
489. Tur'yan, Ya, I., and A. I. Tsinman, *Zh. Fiz. Khim.*, **36**, 406, 659 (1962); *Dokl. Akad. Nauk SSSR*, **136**, 1154 (1961).
490. Uhlig, H., *Ann. N. Y. Acad. Sci.*, **1954**, 843; *Z. Elektrochem.*, **62**, 626, 701 (1958).
491. Van Arkel, A. E., *Rec. Trav. Chim. Pay-Bas*, **44**, 652 (1925).
492. Vaubel, W., *Chem. Z.*, **46**, 978 (1922).
493. Veselovskii, V. I., *Acta Physicochim. URSS*, **11**, 815 (1939).
494. Veselovskii, V. I., *Acta Physicochim. URSS*, **14**, 483 (1941); *Zh. Fiz. Khim.*, **15**, 145 (1941); **22**, 1302 (1948).
495. Vetter, K. J., *Z. Elektrochem.*, **62**, 642 (1958).
496. Vetter, K. J., *J. Electrochem. Soc.*, **110**, 597 (1963).
497. Vetter, K. J., and K. Arnold, *Z. Elektrochem.*, **64**, 244 (1960).
498. Vidovich, G. L., D. I. Leikis, and B. N. Kabanov, *Dokl. Akad. Nauk SSSR*, **124**, 855 (1959); **142**, 109 (1962).
499. Vielstich, W., *Z. Physik. Chem. (Frankfurt)*, **15**, 409 (1958).
500. Vinal, G. W., *Storage Batteries*, 4th ed., Wiley, New York, 1955.

501. Vinal, G. W., D. N. Craig, and C. L. Snyder, *J. Res. Natl. Bur. St.*, **25**, 417 (1940).
502. Volchkova, L. M., L. G. Antonova, and A. I. Krasil'shchikov, *Zh. Fiz. Khim.*, **23**, 714 (1949).
503. Volchkova, L. M., and A. I. Krasil'shchikov, *Zh. Fiz. Khim.*, **23**, 441 (1949).
504. Von Hanffstengel, K., and H. Hanemann, *Z. Metallk.*, **30**, 50 (1938).
505. Vosburg, W. C., and D. N. Craig, *J. Am. Chem. Soc.*, **51**, 2009 (1929).
506. Wagner, C., *J. Electrochem. Soc.*, **97**, 71 (1950).
507. Wagner, C., and K. Grünewald, *Z. Physik. Chem.* (*Leipzig*), **B40**, 455 (1938).
508. Wales, C. P., *J. Electrochem. Soc.*, **108**, 395 (1961).
509. Wales, C. P., *J. Electrochem. Soc.*, **109**, 1119 (1962).
510. Wales, C. P., *J. Electrochem. Soc.*, **111**, 131 (1964).
511. Wales, C. P., *J. Electrochem. Soc.*, **113**, 757 (1966).
512. Wales, C. P., and J. Burbank, *J. Electrochem. Soc.*, **106**, 885 (1959).
513. Wales, C. P., and J. Burbank, *J. Electrochem. Soc.*, **112**, 13 (1965).
514. Watson, E. R., *J. Chem. Soc.*, **89**, 578 (1906).
515. Weber, H. C. P., *Trans. Am. Electrochem. Soc.*, **32**, 391 (1917).
516. Weil, K. G., *Z. Elektrochem.*, **62**, 638 (1958).
517. Weininger, J. L., and M. W. Breiter, *J. Electrochem. Soc.*, **110**, 484 (1963).
518. Weininger, J. L., and M. W. Breiter, *J. Electrochem. Soc.*, **111**, 707 (1964).
519. Weiss, R., and R. Faivre, *Compt. Rend.*, **245**, 1629 (1957).
520. White, J. C., R. T. Pierce, and T. P. Dirkse, *Trans. Electrochem. Soc.*, **90**, 467 (1946).
521. White, J. C. et al., *J. Electrochem. Soc.*, **99**, 233C, 234C, 236C, 238C, 241C, and 243C (1952).
522. White, W. B., F. Dachille, and R. Roy, *J. Am. Ceram. Soc.*, **44**, 170 (1961).
523. White, W. B., and R. Roy, *J. Am. Ceram. Soc.*, **47**, 242 (1964).
524. Winkler, H., *CITCE*, **8**, 383 (1956).
524a. Winter, E. R. S., *J. Catalysis*, **6**, 35 (1966).
525. Wöhler, L., ahd O. Balz, *Z. Elektrochem.*, **27**, 406 (1921).
526. Wolf, E. F., and C. F. Bonilla, *Trans. Electrochem. Soc.*, **79**, 307 (1941).
527. Work, G. W., and C. P. Wales, *J. Electrochem. Soc.*, **104**, 67 (1957).
528. Yakovleva, A. A., T. I. Borisova, and V. I. Veselvoskii, *Zh. Fiz. Khim.*, **36**, 1426 (1962).
529. Yoshizawa, S., and Z. Takehara, *Electrochim. Acta*, **5**, 240 (1961).
530. Yoshizawa, S., and Z. Takehara, *Denki Kagaku*, **32**, 27, 35, 197 (1964); *Chem. Abstr.*, **61**, 11607 (1964).
531. Yost, D. M., *J. Am. Chem. Soc.*, **48**, 152 (1926).
532. Young, L., *Anodic Oxide Films*, Academic Press, New York, 1961, p. 13 *et seq.*
533. Zachlin, A. C., *Trans. Electrochem. Soc.*, **82**, 365 (1942); 92, 259 (1947).
534. Zaslavskii, A. I., Y. D. Kondrashov, and S. S. Tolkachev, *Dokl. Akad. Nauk SSSR*, **75**, 559 (1950).
535. Zaslavskii, A. I., and S. S. Tolkachev, *Zh. Fiz. Khim.*, **26**, 743 (1952).
536. Zender, J., *Z. Elektrochem.*, **11**, 809 (1905); 12, 463 (1906); 13, 752 (1907).
537. Zimmerman, J. G., *Trans. Electrochem. Soc.*, **68**, 231 (1935).
538. Zisman, W. A., *Rev. Sci. Instr.*, **3**, 367 (1932).
539. Zvonkova, Z. V., and G. S. Zhdanov, *Zh. Fiz. Khim.*, **22**, 1284 (1948).

chapter VIII

Porous Oxygen Electrodes

The accelerated interest of recent years in the construction and behavior of porous gas diffusion electrodes has been generated by the current activity in the development of the fuel cell. A fuel cell may be defined as an electrochemical cell in which a fuel (usually a fossil fuel, such as H_2, C_3H_8, CH_3OH, etc.) is continuously oxidized at the negative electrode, and an oxidant (usually O_2 or air) is continuously reduced at the positive electrode in a suitable electrolyte. Ideally, the electrodes and electrolytes remain unchanged during the operation of a fuel cell. In practice, however, corrosion of the structural materials, nonuniformity of the electrode catalyst structures, degradation of the electrolyte, and mass and heat transfer problems have slowed the development of the fuel cell as an economically competitive power source. A number of fuel cells connected in series or parallel or both are called a fuel cell battery.

For the efficient operation of an electrochemical cell, the oxidation process at the anode must be separated physically from the reduction process at the cathode so that a separation of charge may be maintained. The most success has been achieved with H_2–O_2 fuel cells. If H_2 and O_2 were bubbled over inert electrodes (Pt gauzes) immersed in a strong acid or alkaline electrolyte in a manner similar to the experiments performed by Grove (108), electrical current could be drawn from the cell, but very inefficiently. This is true because some H_2 or O_2 dissolved in the electrolyte could diffuse to the opposite electrode where it could react chemically without contributing current to the external circuit. With the very soluble fuels, such as methanol, the non-Faradaic chemical oxidation of the fuel at the oxygen electrode is a serious problem (122, 131).

To prevent the nonproductive chemical oxidation of the fuel in a fuel cell, porous gas diffusion electrodes (240) are employed. A simplified sketch is shown in Fig. 8.1. In such an arrangement, the gas is fed to the back of the porous electrode through which it diffuses to the front which is in contact with the electrolyte. At the electrolyte side of the diffusion electrode, some solution fills the pores in the surface layers of the electrode, and the gas dissolves in the electrolyte at this place. After the dissolved gas diffuses to a reaction site on the electrode surface, it participates in the electrode reaction by losing or accepting electrons.

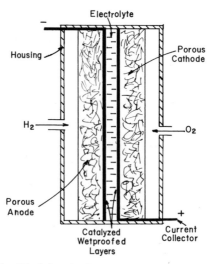

Fig. 8.1. Simplified sketch of fuel cell.   Gas exit ports not shown.

Ordinarily, the structural metal of the porous electrode does not act as an active catalyst for the electrode reaction, as a result, a catalytically more active material, such as platinum or silver, must be deposited on the electrolyte side of the porous electrode.

It is important to prevent the electrolyte from penetrating beyond the surface layers of the porous structure which causes a flooding of the pores of the electrode.   When flooding of the porous electrode occurs, there is a loss of the high surface area of the catalytically active layer, producing a loss in current density, and the diffusion parth of the gaseous reactant to the catalyzed reaction sites is lengthened, producing an increase in polarization and a lowering of the rate of the electrode reaction. Several methods may be used to obtain a stable gas–liquid interface. The simplest concept involves maintaining the gas under enough pressure to keep the electrolyte in the surface layers of the porous electrode, but such a procedure is very difficult to control because of the large range of values over which the forces to be balanced may vary.

A much more successful method of obtaining a stable gas–liquid interface is the construction of a dual pore electrode containing relatively large pores in the body of the structure with a thin layer of the porous material having much smaller pores bonded to the electrolyte side. Bacon (8) used porous Ni electrodes with a pore size of about 30 μ on the gas side with the thin layer of porous Ni having much smaller pores on the liquid side.   A small pressure difference set up across each electrode expelled the solution from the large pores, but gas could not bubble through the small pores since they were flooded owing to the surface tension of the liquid.

Another method of obtaining a stable gas–liquid interface is the use of the so-called double skeleton catalyzer (DSK) electrodes developed by Justi and Winsel (153). It was desired to combine the excellent catalytic activity and the low sensitivity to impurities of Raney nickel with the mechanical strength and good electrical conductivity of a porous metal electrode. After pulverizing the Raney alloy of 50% Ni and 50% Al, it was mixed with powdered Ni, and the mixture was pressed and sintered in the desired electrode form. Most of the Al was leached out with KOH, and the last of the Al was removed with anodic polarization, yielding a porous Ni matrix in which the pores contained the highly active Raney catalyst. Nickel electrodes were used for H_2, but Raney Ag electrodes were made for O_2. Because Ag is so ductile, more elaborate preparations and careful control of conditions were required. By careful choice of the particle size, a homoporous structure consisting chiefly of equilibrium pores may be obtained (if smaller than equilibrium size they flood: if larger, gas bubbles into the electrolyte, Fig. 8.2), and a stable gas–liquid interface may exist for moderate gas consumption rates. For higher gas consumption rates, a two-pore structure similar to that used by Bacon with a thin layer of fine pores on the electrolyte side of the electrodes must be used. The oxygen electrodes are always made with a two-pore structure.

The gas–liquid interface of homoporous electrodes may be stabilized by immobilizing the electrolyte in a solid matrix. Wynveen and Kirkland (297) describe the construction in which the KOH electrolyte

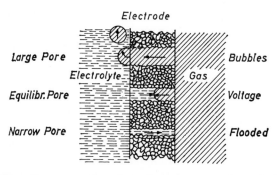

Fig. 8.2. Schematic cross section through a porous gas electrode (153) showing the different behavior of large, equilibrium, and narrow pores. (By permission of The Electrochemical Society.)

is held fixed in an asbestos disk placed between the H_2 and O_2 porous electrodes. Such fuel cells are not influenced by gravity and should be suitable for satellite applications.

By wetproofing the solution side of a diffusion electrode with a hydrophobic film, flooding of a porous electrode may be prevented. This is accomplished by applying a solution of paraffin wax in petroleum ether (260) to the solution side of the electrode and evaporating off the solvent. At the oxygen electrode, particularly at high current densities, the wax is oxidized, and the electrode floods (293). Polyethylene films are more resistant to oxidation (296), but the greatest success has been obtained with polytetrafluoroethylene (PTFE) films (260), which exhibit outstanding resistance to oxidation without inhibiting the electrochemical performance of the electrode.

Finally, the liquid electrolyte may be replaced by an ion-exchange membrane, as first proposed by Grubb (109) with the H_2 and O_2 diffusion electrodes clamped against the opposite faces of the membrane. A general review of the construction and behavior of ion-exchange membrane fuel cells has been presented by Niedrach and Grubb (208). To prevent the drying out of the membrane which may form cracks and pinholes, wicking from a reservoir of water is required.

At the present time, it appears that no adequate theory exists which can predict the exact behavior of porous electrodes (ref. 293, p. 44). Probably, the first significant attempt to formulate such a theory was presented by Daniel-Beck (64), and since that time many authors (20, 29,31,44,45,50,60,90,92,95,105,106,113–115,139,152,173,183,188,205,229, 271,285,292,295) have constructed models of the pore structure of porous electrodes to account for the experimentally observed behavior of such electrodes. At one time (e.g., 152) it was thought that the electrochemical reaction occurred only at the three-phase solid–liquid–gas boundary, a line which is described by the locus of points where the gas, diffusing through the pores, contacts the top of the liquid meniscus at the pore wall. Gas molecules absorbed on the metal wall of the pore diffuse along the surface to the electrolyte meniscus where reaction takes place.

According to Bennion and Tobias (20), Wagner first suggested that an invisible thin film of electrolyte might exist above the visible meniscus on the wall of a partially filled pore. In this case, an appreciable contribution to the total electrode reaction may take place on the electrode surface covered by the thin invisible electrolyte film. Previous investigators (e.g., 108,127) had observed that the current for a given gas flow was increased when the Pt electrodes were only partially immersed in the electrolyte. From studies of the variation of current with the length of Pt extending above the surface of the electrolyte at a partially immersed, plane, Pt electrode at constant potential, Will (292) was the first to demonstrate conclusively the existence of the thin invisible film of electrolyte above the visible meniscus. He concluded that most of the electrode reaction takes place in a narrow region next to the top of the visible meniscus, and surface migration of adsorbed H_2 molecules along the Pt wall to the meniscus contributed insignificantly to the total reaction.   Consequently, the three-phase boundary concept (152) must be rejected.

Although Will (292) studied the electrode reaction of H_2 in H_2SO_4 at Pt, Maget and Roethlein (183) studied the reduction of O_2 in both H_2SO_4 and H_3PO_4 solution and obtained essentially the same results.   The distribution of current as a function of the distance from the electrolyte level is shown in Fig. 8.3. Most of the reaction occurs in a narrow region with the maximum rate occurring at about 0.28 cm above the electrolyte level.   In regions above this, the resistance of the thin film increases rapidly with distance above the liquid, and the rate falls off.   In regions below, the diffusion length required for the gas molecules to reach the electrode surface increases rapidly as the liquid surface is approached, and the rate falls off again.   At about 0.28 cm above the liquid surface, the opposing quantities of resistance and diffusion length have the most favorable values, and the maximum current is observed

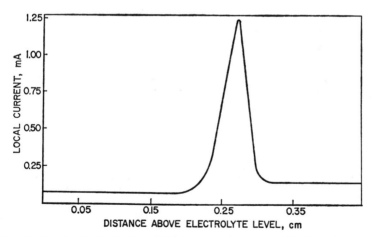

Fig. 8.3. Local cell distribution (183) in the meniscus and thin film at a constant potential of 0.400 V vs. Pt/H_2 in $1N$ H_2SO_4.   (By permission of The Electrochemical Society.)

at this point.   Using sectioned cylindrical Ni or Ag electrodes, Bennion and Tobias (20) found a thin film mechanism accounts for the behavior of the reduction of oxygen in alkaline solutions, but the thin film extended several millimeters above the visible meniscus.   They reported that Muller determined the thickness of the films as about 1 μ.   Mazitov and co-workers (187,187a,188) obtained similar results on partially immersed Ag wires for the reduction of O_2 in 10.6N KOH solution.

Although the evidence for the existence of the thin film is obtained from partially immersed plane electrodes, further evidence was required to show that this mechanism prevailed in the pores of a porous electrode. Katan and co-workers (156) studied the reduction of O_2 on a bed of Ag spheres (55 μ in diameter) held by a fine Ag screen in the mouth of a Pyrex tube.   The pores of the bed of Ag beads were much larger than the holes in the Ag screen so that the electrolyte could be prevented from flooding the Ag-bead bed.   During polarization measurements, beads could be added to the bed, and the steady-state potential (galvanostatic) or current density (potentiostatic) was determined as a function of the depth of the Ag-bead bed.   The overvoltage decreased as the bed thickness increased up to bout 0.5 mm, after which no change occurred with further increases in bed thickness.   They concluded that the results were in agreement with a thin film mechanism, and the film thickness ranged between 10^{-5} and 10^{-3} cm.

For a discussion of the theory of the galvanostatic and potentiostatic charging of the double layer at porous electrodes, the reader is referred to the literature (e.g., 30,69,172,220).

If the gas pressure on both the H_2 and O_2 diffusion electrodes is adjusted so that gas bubbles through the electrode (80,219), a diaphragm must be placed in the cell between the electrodes to prevent mixing of the gases (ref. 283, p. 95).   This is an undesirable design since the internal resistance of the cell must be kept as low as possible, but the presence of a diaphragm greatly increases the cell internal resistance.   For this reason the electrolyte chambers are made as thin as possible, and the use of diaphragms is discouraged.

Another problem encountered in fuel cell technology is the removal of the products of the cell reaction.   In the case of the H_2–O_2 fuel cell, the product is H_2O which must be removed, or the performance of the cell will be lowered owing to a dilution of the electrolyte (83).   By proper engineering of the design of electrodes, water vapor may be carried off in the gas stream (83).   If the fuel is a hydrocarbon, the problem of carbonate formation produced by the presence of the reaction product CO_2 in the electrolyte is encountered.   This same problem to a lesser extent may be present if air is used instead of O_2. The most ideal solution is the use of a CO_2-rejecting electrolyte such as H_2SO_4 or H_3PO_4 solutions.

The reader is referred to books edited by Williams (293), Mitchell (195), Young (301), Gould (104), and Baker (10) for detailed discussions of fuel cell theory and technology.

Since many diffusion electrodes are made with porous carbon or graphite as the structural support material, a discussion of the interaction of carbon with oxygen is taken up before the behavior of the porous oxygen electrode is considered.

A. THE OXIDES OF CARBON

1. The Forms of Carbon

Carbon exists in two modifications, diamond and graphite. In diamond, the carbon atoms form covalent bonds at the corners of a tetrahedron with a C—C bond distance of 1.54 Å, as determined by Bragg and Bragg (33) from x-ray studies. The three-dimensional covalent structure gives diamond its extreme hardness. Diamond belongs to a cubic crystal structure.

Graphite has a layered structure first proposed by Bernal (23). The carbon atoms of the layers form condensed hexagonal rings with a C—C bond distance of 1.415 Å. Resonating π electrons give the C—C bond a partial double bond character so that the hexagons have a sort of quinoid structure. In well crystallized graphite, the layers are held together 3.3538 Å apart by van der Waals forces, as determined by x-rays (94,204,280). Because the layers in graphite are so weakly held together, graphite is the only known solid lubricant for electrical (motor brushes) and high temperature applications (247). In normal graphite, the hexagons of consecutive layers overlap one another half way in an $ABABAB$ sequence forming a hexagonal crystal structure. The unit c axis length is twice the distance between layers. However, Taylor and Laidler (261), observed lines in the x-ray diffraction pattern of naturally occurring graphite which could not be accounted for by the hexagonal structure. Lipson and Stokes (179) reported that a rhombohedral form of graphite, which is metastable to the hexagonal form and occurs to the extent of about 10% of a given sample, has a layered structure with a sequence of $ABCABC$, giving rise to the extra x-ray lines observed by Taylor and Laidler. In this case, the unit length of the c axis is three times the distance between layers.

When graphite is ground to produce smaller particle sizes, there is a continuous deterioration of the crystal perfection as reported by Bacon (9). The hexagonal form may be converted to the rhombohedral form with grinding, and the reverse process may be carried out by heating the rhombohedral graphite to about 1300°C (9). When the crystallite size falls below 150 Å with further grinding (9,94,218), the layers become disordered with respect to the vertical sequence but remain parallel to one another. Biscoe and Warren (26) called this disordered carbon "turbostratic carbon."

When complete disorder is reached (26), the layers are separated by 3.44 Å, and Bacon (9), Franklin (94) and Pinnick (218) found that the interlayer spacing increased as the crystallite size decreased. Amorphous carbon, carbon black, or soot are turbostratic carbon. It has been known for some time (1) that turbostratic carbon can be converted to graphite by heating it above 2000°C. The x-ray diffraction patterns for amorphous carbon are different from those obtained for graphite (247). Electron micrographs show that the particles of carbon black are spherical (279), and when heated between 2000 and 3000°C, an ordering of the layers or graphitization sets in. Under these conditions the spheres becomes polygons with angular sides.

The density of diamond is 3.51 g/cm³ (251); of graphite, 2.2 g/cm³; and of turbostratic carbon, about 1.9 g/cm³ (247). When pressure was applied to graphite at elevated temperatures (4,43,272,290), artificial diamonds were produced. Bundy (43) compressed graphite at 120 kbars and 3300°K to form diamonds in a few milliseconds. By using a strong shock wave to compress graphite to 350 kbars at 1000°K, Alder and Christian (4) produced diamonds in a few microseconds. Diamonds have been obtained not only from various forms of carbon, but also from a number of organic compounds in the pressure range 95–150 kbars and over a temperature range from 1300 to 3000°C by Wentorf (290). Most carbonaceous materials were converted to diamond in periods of time ranging from 0.2 to 55 min depending on the nature of the carbonaceous material used. At atmospheric pressure diamond is converted to graphite at temperatures above 1500°C (246).

Although carbon black is produced by the thermal decomposition of hydrocarbons, such as methane, where the carbon particles are collected on a cool surface, a dense polycrystalline graphite known as pyrolytic graphite (PC) is obtained by the thermal decomposition of hydrocarbon gas directed against a heated (between 1700 and 2500°C) substrate usually composed of graphite (107). From x-ray diffraction studies, Guentert and co-workers (111,112) concluded that the crystallites of PC consist of groups of parallel and equidistant layers with a random layer order along the c axis. As pointed out by Pappis and Blum (214), the structure of PC is similar to that of turbostratic carbon. With modern technology, thick deposits of PC may be grown (99,111, 112,214), and the density increases with the temperature of formation, from 1.50 g/cm^3 at 1700°C to 2.20 g/cm^3 at 2500°C (111,112). When massive PC is heat-treated at 3600°C, evidence is obtained (162) that single crystals may be formed.

A discussion of the properties of PC may be found in the literature (28,99,107,111,112,161,214,258). Because of the layered structure of graphite and PC, the thermal, electrical, and mechanical properties are anisotropic. The electrical resistance of graphite parallel to the c axis (perpendicular to the layers) is about 100 times greater than in a direction perpendicular to the c axis (222) (about 4×10^{-5} ohm-cm parallel to the layers, 10^{-2} ohm-cm perpendicular to the layers). Pyrolytic graphite has a lower resistance than graphite (214) with a higher degree of anisotropy (about 0.2×10^{-3} ohm-cm parallel and 0.2 ohm-cm perpendicular to the layers). However, these values depend on the temperature at which they are determined. Also, the thermal conductivity is about 1000 times greater parallel to the layers in PC than perpendicular to them (214). Such properties as these, together with lightness and mechanical strength at very high temperatures, make PC an important structural material for rocket and satellite applications, such as reentry heat shields.

The reader may consult Ubbelohde and Lewis (270), Blackman (27), and the *Proceedings of the Biennial Carbon Conferences* (5 volumes) for extensive discussion of the properties and uses of carbon.

Of particular interest to fuel-cell research is the production of porous carbon structures. These are made by mixing ground coke with melted pitch as a binder at about 150°C. After filling the desired mold with this mixture, it is baked in the absence of air to about 1000°C. Such a treatment causes the pitch to be pyrolyzed with a loss of about 40% of the binder (279) and the production of a highly porous structure. The porosity of the structure may be lowered by impregnating the porous carbon with a coal tar followed by pyrolysis at 1000°C. Additional carbon deposits in the pores produced by the pryolysis of the coal tar reduces the porosity. The procedure may be repeated until the desired porosity is obtained.

When the porous carbon is baked in the absence of air between 2500 and 3000°C, the carbon is converted to graphite (graphitized) (279).

Porous graphite is a more desirable product because the thermal, electrical, and mechanical properties are improved over porous carbon. In addition, a purer product is obtained since much more of the adsorbed hydrogen, sulfur, and metallic impurities are removed. The suitability of various porous carbon structures for fuel-cell electrode application has been discussed by Rusinko, Parker, and Marek (235), by Werking (291), and by Paxton et al. (216).

2. Gaseous Oxides of Carbon

Carbon forms at least three stable oxides. When carbon or an organic compound is completely oxidized in air or oxygen, the product is the colorless and odorless gas CO_2. It has been established (201) that carbon dioxide is a linear molecule, and since it is diamagnetic, has a configuration of closed shells. It has been used to maintain an inert atmosphere in polarographic investigations. The properties of CO_2 may be found in reports recorded by Meyers and Van Dusen (192) and Giauque and Egan (101).

When carbon or a hydrocarbon is burnt in an atmosphere deficient in oxygen, a lower oxide, carbon monoxide, is produced. The laboratory preparation consists of the interaction of H_2SO_4 on formic acid, but the thermal decomposition of nickel carbonyl at 200°C (191) yields the purest samples of CO. This colorless, odorless gas is extremely poisonous since it forms a more stable compound with hemoglobin than oxygen does, and thus prevents the blood from carrying oxygen to the vital organs of the body (185,200,227). Langmuir (178) proposed that the triple bonded structure $\overset{-}{C}\!\equiv\!\overset{+}{O}$ explains the observed properties of CO better than the structure C=O, although the latter makes some contribution to the overall resonant structure since the observed bond distance lies between the pure double and triple bond values. In agreement with the triple bond structure is the fact that CO has an electric moment (119,283,302). Hall (118) has presented a review of some of the properties of CO.

Carbon can interact with steam at elevated temperatures to produce CO and CO_2 according to the following equations.

$$C + H_2O \rightarrow CO + H_2 \tag{8.1}$$

$$C + 2H_2O \rightarrow CO_2 + 2H_2 \tag{8.2}$$

$$CO + H_2O \rightarrow CO_2 + H_2 \tag{8.3}$$

$$C + CO_2 \rightarrow 2CO \tag{8.4}$$

The first of these reactions is the basis for the production of artificial gas.

The details of these reactions have been discussed in the literature (e.g., 96,257,282).   Hydrocarbons react with steam at temperatures between 750 and 1000°C over a catalyst such as Ni to form CO and H_2.

$$C_3H_8 + 3H_2O \rightarrow 3CO + 7H_2 \tag{8.5}$$

This process is known as steam reforming of hydrocarbons and is used as a source of H_2 for fuel at a fuel cell anode as discussed by Rogers and Crooks (230), Meek and Baker (189), and Williams (ref. 293, p. 217). Additional H_2 is obtained by the reaction of CO with steam, according to Eq. 8.3.   Either the CO_2 must be removed from the H_2 stream for an alkaline fuel cell system, or else the alkaline electrolyte must be replaced with an acid one.   CO may be used directly as a fuel in high temperature fuel cells, e.g., see Gorin and Recht (103).

A third stable oxide, carbon suboxide C_3O_2, is known.   It was first made by Diels and co-workers (72,73) by dehydrating malonic acid with P_2O_5 at reduced pressures.   However, the best method reported (138, 212) for making C_3O_2 is by the pyrolysis of diacetyltartaric anhydride. From electron diffraction studies, it was determined (36), that C_3O_2 is a linear molecule, $O{=}C{=}C{=}C{=}O$, but the C—C bonds have some triple bond character due to resonance with structures such as

$$\overset{+}{O}{\equiv}C{-}C{\equiv}C{-}\overset{-}{O}$$

The C—O bond distance is 1.20 Å and the C—C, 1.30 Å.   This linear structure is supported by evidence from the Raman spectrum (88), infrared absorption spectra, (180) and the ultraviolet absorption spectra (264) of C_3O_2.   Carbon suboxide can polymerize (164) to form a reddish yellow product and can decompose to CO_2 and C.   In fact, C_3O_2 is a good material from which pyrolytic carbon can be obtained. The properties of this evil-smelling gas which has many uses in organic synthesis are discussed by Klemenc and co-workers (164), Hagelloch and Feess (116), and Vol'kenshtein (275).

Klemenc and Wagner (163) reported that an oxide, C_5O_2, was obtained from the thermal decomposition of C_3O_2 in the gas phase to the extent of about 3% of the products.   Most of the reaction takes place by Eq. 8.6,

$$C_3O_2 \rightarrow CO_2 + C_2 \tag{8.6}$$

but some C_3O_2 can react with C_2 to form C_5O_2

$$C_3O_2 + C_2 \rightarrow C_5O_2 \tag{8.7}$$

However, Diels (71) casts doubts on the existence of C_5O_2.

Other oxides have been reported.   In the spectrum of the combustion of CO in O_2, a band appears at 1615 cm^{-1} which is attributed (259) to the presence of CO_3.   When CO was ionized by streaming the gas through an ac field of 1 kV/cm and 250 cycles/sec at pressures from 200 to 690 mm Hg, a product was obtained (182) having an empirical formula, $C_5O_3 \cdot xH_2O$.

3. Absorbed Oxygen Films

When carbon is in contact with oxygen, two types of adsorption may be detected, a very rapid physical adsorption and a slow chemisorption of oxygen (15,87,237).   At low temperatures, the adsorption is physical adsorption of O_2 molecules as determined by magnetic susceptibility measurements at $-183°C$ (154).   As the temperature is raised, chemisorption of O_2 sets in with film formation (15) up to about 144°C. The physically absorbed O_2 can be desorbed reversibly under vacuum even at room temperatures (237), but the chemisorbed oxygen cannot be removed in this way.   By heating the sample, O_2 is not evolved, but only CO_2 and CO are detected (15,42,87,237) which indicates that the thermal breakdown of the absorbed oxygen layer takes place rather than the desorption of oxygen.   According to Barrer, the adsorbed layer begins to break down at temperatures above about 370°C.

The presence of surface carbon–oxygen complexes has been reported by a number of investigators (e.g., 34,167,177,202,215,226,239,241,256). Lambert (177) concludes that carbon is oxidized by two processes; first, by the direct impact of O_2 molecules with a clean carbon surface forming CO_2, and second, by the breakdown of the surface carbon–oxygen complexes either spontaneously or by the impact of an O_2 molecule with the complex forming CO_2 and CO.   The presence of certain metallic salts on the carbon surface promotes the breakdown of the adsorbed complexes with a resultant increase in the amount of CO_2 and CO evolved at a given temperature (87,215,237).   Apparently, the foreign metal sites act as catalytic centers for the decomposition of the complex.   At temperatures above 850°C (237), the surface is purged of absorbed oxygen.

4. Solid Oxides of Carbon

Because graphite has a layered structure, various kinds of compounds can be made in which a layer of atoms or ions is inserted between the hexagonal carbon layers.   Swelling of the graphite accompanies the reaction.   Any number of compounds with a given sequence of layers are known, and Rüdorff (231) has given a description of several of these. Rüdorff and Hofmann (232) studied the salts of graphite obtained by the reaction with strong acids. The hexagonal structure of the carbon layer is maintained in these compounds. A H_2SO_4 salt of graphite obtained by reaction of graphite with H_2SO_4 is described as blue graphite by Hofmann and co-workers (132,134). Blue graphite is unstable in air (186) because it adsorbs moisture from the air and turns black, showing that it has decomposed to graphite.

As early as 1859, Brodie (37) was the first to record the preparation of a solid oxide of carbon obtained by the action of strong oxidizing agents on graphite. It was called graphitic oxide or graphitic acid because suspensions of this material turn litmus red. The best known method of preparation is that given by Staudenmaier (255) in which a gram of powdered graphite is suspended in a mixture of 15 cc of 66% HNO_3 and 30 cc of concentrated H_2SO_4. Over a period of 4–7 days, 10 g of $KClO_3$ is added slowly so that the temperature is kept below 50°C. It requires about 14 days to complete the reaction, and the ClO_2 produced is swept away with an inert gas such as CO_2 or N_2. The procedure may be repeated two or three times, and a final treatment with $KMnO_4$ solution may be required to ensure complete oxidation (63). The color of the product ranges from yellow to brown (37,49,63,67,125,132,133,135,186, 231,232,255). Hofmann and co-workers (57) obtained white graphitic oxide by washing the oxide with an HCl solution of ClO_2 in the absence of light. If the crystallites of the powdered graphite are too small, the structurally imperfect carbon is oxidized mostly to gaseous oxides and a poor yield of graphitic oxide is obtained (63). With macrocrystalline graphite, yields up to 95% of the theoretical may be realized (132).

Because the Staudenmaier preparation of graphitic oxide is tediously long with a potential danger of an explosion, Hummers and Offerman (135) recently described a safer and quicker method of preparation. Graphite is treated with a water-free mixture of concentrated H_2SO_4, $NaNO_3$, and $KMnO_4$ which requires 2 hr at temperatures below 45°C. Water is added to the brownish grey paste, and the excess $KMnO_4$ is reduced with H_2O_2. After washing and filtering, dry graphitic oxide is obtained after heating. Carbon-to-oxygen ratios of 2.3/1 are found which agree well with a ratio of 2.9/1 obtained by Staudenmaier's method. An empirical formula of $C_7H_2O_4$ may be assigned (67) to graphitic oxide. According to Hummers and Offeman, the higher the oxygen content, the lighter colored is the sample; the highest are yellow, and as the samples become poorer in oxygen, the color ranges through brown to green and finally black. Bottomley and Blackman (32) oxidized a wide variety of aromatic compounds with boric acid buffered solutions of H_2O_2 in the presence of cupric or ferric ions and obtained a brown precipitate exhibiting the properties of graphitic oxide. It is concluded that the formation of a graphitic structure from material containing single benzene rings must occur through polymerization mechanisms.

The structure of graphitic oxide has been the subject of several reports (5,53,57,67,68,133,232,233). Infrared analysis of graphitic oxide (8,57, 68) shows the presence of OH, CO, and ether-like COC groups, and as noted by Alexanian (5), the OH and CO bands increased in intensity as H_2O molecules were added to the structure. The OH groups may be attached to a tertiary carbon or to a double-bonded carbon (57), and it is the OH groups near the double bond that give the acidic character to the oxide. Hofmann et al. (57) reported that the white graphitic oxide was darkened by exposure to light, by heating to 80°C, or by treating with NaOH but the reverse process could take place again by acidic oxidation in the dark. Both forms have the same structure, and the reversible change between the light and dark forms is explained by a keto–enol shift of the OH groups positioned near the double bond. The white graphitic oxide is the keto form. Some knowledge about the distribution of the hydrogen between OH groups and interstitial H_2O can be obtained from methylation studies.

From x-ray diffraction patterns taken on graphitic oxide (5,53), it is concluded that the hexagonal layers of carbon atoms found in graphite are still present, but there is complete disorder in the sequence in which they are stacked. In this respect, graphitic oxide is more closely related to carbon black (turbostratic carbon) than to graphite. Since the carbon layers are spaced much farther apart in graphitic oxide than in graphite, the unit cell dimensions are much larger.

More information is available from electron diffraction studies (5,68) which show that graphitic oxide has a structure of puckered hexagonal layers spaced 2.43 Å apart. No order is found in the stacking of the layers, and the OH and ether-like COC groups are attached to the six-membered carbon rings in a disordered array. As noted by Alexanian (5), the unit cell length increases as the interstitial water content increases. This effect is explained in terms of hydrogen bonding. From this viewpoint, the carbon layers are held apart by hydrogen bonding through COOH groups. When H_2O molecules enter the structure, they form even longer bridges, and the graphitic oxide structure swells.

The thermal decomposition of graphitic oxide has been studied (67, 126,166,186). When graphitic oxide is heated rapidly, it explodes with a small flash of light (186) forming soot, but when heated very slowly it decomposes to graphite.   During the pyrolysis of graphitic oxide (186), the weight decrease was followed at a constant rate of heating.   Sharp changes in the decomposition rate were observed to occur at the points where decreases in the interlayer spacing were detected with x-rays. Between 120 and 190°C, De Boer and Van Doorn (67) detected CO and O_2 along with H_2O and CO_2 in the rapid thermal decomposition products. As the temperature was increased, the amount of O_2 present decreased. With slow oxidation rates, no O_2 was detected in the N_2 stream in which the graphitic oxide was thermally decomposed.

Some heats of immersion and heats of adsorption of water vapor were carried out by Slabaugh and Hatch (252) on the graphitic oxide–water system.

Membranes of graphitic oxide 0.03–0.05 mm thick have been made by Clauss and Hoffmann (56) by evaporating a colloidal solution of graphite to dryness.   These membranes are permeable to H_2O vapor but not to O_2 or N_2.   Since the interstitial water determines the equilibrium between the vapor pressure on the two sides of the membrane, these membranes were used to determine the vapor pressure of various liquids. Clauss, Hofmann, and Weiss (58) have measured the membrane potential across graphitic oxide membranes in various electrolytes.   Graphitic oxide is a nonconductor of electricity (40,41,132,134,161).

Reviews of the properties and structure of graphitic oxides and interstitial compounds have been written by Croft (63), Henning (125), and Ubbelohde and Lewis (270).

B. THE OXYGEN ELECTRODE ON CARBON

Because graphite is a good electronic conductor and because interfering oxide films such as those found on metals are not present on carbon surfaces (22), such materials make good inert electrodes for the construction of an oxygen electrode.

1. The O_2 Anode in Acid Solutions

In strong acids, particularly in the presence of oxidizing agents, porous carbon or graphite anodes swell (41,263) during the electrode process with the production of graphitic oxide.   Finally, the electrode breaks up and falls to the bottom of the cell in the form of a black sludge.   As pointed out by Brown and Storey (41), the product is impure graphitic oxide since it is composed of grains of graphite around which a shell of nonconducting graphitic oxide is formed.   The highest yields of graphitic oxide are obtained with dilute HNO_3 (41) where 5–6% of the anodic product is entrapped oxygen. As the temperature is increased, less oxygen is occluded until above 60°C negligible corrosion of the carbon electrode is observed in dilute HNO_3. The explanation offered rests on polarization effects. It appears that a high potential is required for the formation of graphitic oxide, and as the temperature is increased, the polarization of the electrode process decreases until a point is reached where the electrode potential falls below a threshold value.

When the pores in the graphite are filled with paraffin, the rate of the electrode process is slowed down, and the electrode is broken into finer particles. In this way, more oxygen may be occluded, and up to 10 or 11% oxygen content has been reported (41). At high current densities, the electrode is broken into large chunks, each surrounded by a shell of nonconducting graphitic oxide. Consequently, the overall oxygen content of the product is lower than the one obtained at the lower current densities.

This occluded oxygen is held relatively loosely as demonstrated by the ability of a slurry of graphitic oxide to oxidize $FeSO_4$. Brown (40) proposed that MnO_2 be replaced with graphitic oxide as a depolarizer in Leclanché type primary cells. Since the graphitic oxide is a nonconductor, the carbon cathode must be made from a mixture of powdered graphite and graphitic oxide. Electrochemically produced graphitic oxide is already a mixture, and the crushed, dry material is fabricated (40) into workable Leclanché type cells. The operating voltage is lower than that of cells using MnO_2.

2. The Carbon Indicator Electrode

In electroanalytical investigations, there is need for a wider range of operating potentials than obtained with the DME, and although Pt and Au fill this need, the presence of adsorbed films causes variations in performance with the history of electrode preparation. Several desirable properties made carbon a material worthy of consideration as an indicator electrode for electroanalytical applications. Both the oxygen and hydrogen overvoltages are high, the surface is easily renewed by machining, and surface oxides are not a problem.

Lord and Rogers (181) first introduced graphite as an indicator electrode, and since then many workers (14,84–86,97,98,198,277,281) have reported on its behavior in various electroanalytical systems. As pointed out by Elving and co-workers (83,84), exceptionally pure graphite must be used as unwanted spurious polarographic waves will appear, and the use of spectroscopic grade graphite was recommended by them. However, the chief disadvantage to the uses of graphite indicator electrodes is the large residual currents obtained on them.   The residual current may be reduced to small values by impregnating the graphite with a wax, such as ceresin wax.   To ensure the maximum filling of the pores, the impregnation with wax is carried out under reduced pressure. The sides of the wax-impregnated graphite rod are protected from contact with solution by coating them with glyptol or an epoxy resin. The tip of the electrode is sanded to expose the graphite surface to the electrolyte.   By sawing, machining, or breaking the tip, a renewed surface may be obtained.   Yet, erratic results were obtained with the impregnated graphite electrodes.   Elving and Smith (84) suggested that the electrolyte could not wet the hydrophobic surface of the wax impregnated electrode, thus producing irreproducible results.

If the graphite electrode was treated with a wetting agent by immersing it in a dilute solution of Triton X-100 for 1 min (84), good reproducible results were obtained for a number of systems.   The effect of Triton X-100 on the properties of the graphite electrode was investigated by Elving and Smith (86).   Using very dilute solutions of Triton X-100, no anomalies in the polarograms were observed between 0 and 1.2 V.   When greater amounts ($>0.003\%$) of Triton X-100 were used, a current peak which increased with increasing concentration of Triton X-100 appeared at 1.35 V.   According to Elving and Smith (86), the surfactant lowers the oxygen overvoltage, and the peak is due to the evolution of O_2 from H_2O.

Organic oxidation processes are best studied on graphite electrodes since excellent results have been reported (84,85,277,281) for a number of such systems. In such cases the interfering effects of adsorbed films are absent.   Graphite indicator electrodes give poorer results than Pt or Au indicator electrodes for inorganic systems.   Possibly, the poor results obtained in inorganic systems arises from the penetration of the pores with electrolyte in mineral acid solutions (84).

To avoid the problem of the highly irregular surface of the porous carbon indicator electrode or the necessity of adding foreign substances such as waxes and wetting agents to the electrode material, Adams and co-workers (2,210) used a pasted carbon electrode.   A paste was made of powdered graphite and an organic binder, such as bromoform or bromonaphthalene.   In a well , hollowed out of one end of a Teflon rod through which a Pt–lead wire was passed, the paste was tamped so that a smooth surface flush with the end of the rod was obtained.   Only this smooth graphite surface was in contact with the electrolyte.   Adams reports good results for several systems with the pasted carbon indicator electrode and Brezina (35) has investigated pasted electrodes made from a variety of carbons for their activity towards the reduction of oxygen.

In recent years, pyrolytic graphite indicator electrodes have been the object of much attention. The first reported use of PC indicator electrode was published by Laitinen and Rhodes (176) who studied the vanadium redox system in the molten LiCl–KCl eutectic with a PC indicator electrode composed of 6–11 mils of PC deposited on a ¼-in. graphite rod. As reported by Pappis and Blum (214), no pores exist in PC, and this material is impermeable to gases and liquids. Beilby and co-workers (19) studied the ferri–ferrocyanide system with a PC indicator electrode made by decomposing natural gas at 1025°C on an alumina rod which was mounted in a Teflon holder. The PC electrode gave results similar to those obtained with Pt electrodes and was more reversible than the wax-impregnated graphite electrode.

Miller and co-workers (193,194) warn that the plane of the c axis of PC must not be exposed to the surroundings, because in this direction, PC is permeable to aqueous solutions and molten salts. As a result, the PC electrode is covered with an epoxy resin, and the plain of the a axis is exposed by sanding off the resin. A PC electrode has a useful potential range of from -0.8 V to $+1.0$ V vs. SCE, and no treatment beyond a preliminary cleaning is required (194). Several advantages, such as freedom of trapped gases and metal impurities, chemical inertness, and a high degree of impermeability to liquids and gases, are realized with PC electrodes. Miller (193) was obtained good results with PC electrodes in acid–base titrations as well as in redox systems.

3. Steady-State Polarization Studies

Yeager and Kozawa (299) noted that it is difficult to interpret steady-state polarization measurements obtained (296,298) at porous electrodes because of the complications associated with mass transport and distributed IR drops within the pores of the electrode. To avoid these complications, he studied the anodic evolution and cathodic reduction of oxygen on paraffin-filled porous graphite and the reduction of O_2 on PC (299,300). Some measurements were also made (300) with rotating disks of these materials. Yeager and co-workers (296, 298–300) could not obtain reproducible polarization curves for the reduction of oxygen on carbon electrodes in H_2O_2-free electrolytes. Since carbon is a poor catalyst for the decomposition of H_2O_2 (64a,296), the H_2O_2 concentration in the electrolyte builds up continuously during the period of cathodic polarization. As the composition of the system changes, so do the polarization characteristics.

In oxygen saturated solutions, the observed rest potential on a graphite electrode is about 0.3 V less noble than the reversible oxygen potential (202,263). As H_2O_2 is added to the electrolyte, the rest potential shifts to increasingly less noble values as shown in Fig. 8.4. From the variation of the potential both with the log of the HO_2^- ion concentration in alkaline solution containing peroxide ($dE/d \log [HO_2^-]$ ~ 0.03) and with the P_{O_2} ($dE/d \log P_{O_2} \sim 0.03$), Yeager et al. (300) concluded that the potential-determining reaction is the O_2/H_2O_2 couple,

$$O_2 + H_2O + 2e \rightleftarrows HO_2^- + OH^- \qquad (8.8)$$

As demonstrated first by Berl (22) and later by Yeager and co-workers (296,298–300), the O_2 electrode on carbon electrodes in electrolytes containing added peroxide (peroxide-stabilized systems) is reversible to the HO_2^- ion in alkaline solutions.   Similar results have been obtained by Weisz and Jaffe (289) and by Kordesch and co-workers (168,169).

Oxygen tracer studies using ^{18}O showed (64a) that in the reduction of O_2 on carbon electrodes all of the peroxide produced comes from O_2 molecules, and that the O_2 was reduced at about 100% current efficiency to peroxide. Taube and co-workers (46,137) showed that exchange of oxygen between the O—O linkages and H_2O is negligible. Other workers (21,24,155,199) have reported the quantitative production of H_2O_2 by the reduction of O_2 on carbon electrodes.   Consequently, the O—O bond is not broken in the reduction of O_2 on carbon electrodes, and reproducible results cannot be obtained on carbon cathodes until

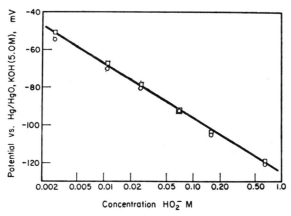

Fig. 8.4. Dependence of the rest potential on HO_2^- concentration (300) for porous carbon, (○), and graphite (□) electrodes in $5M$ KOH at 25°C. The P_{O_2} is 0.97 atm.   (By permission of Pergamon.)

the peroxide concentration in the electrolyte builds up to a stable value. The oxygen overvoltage is much lower in peroxide-stabilized systems than in ones not containing peroxide (296,298–300). As the carbon cathode in H_2O_2-free KOH solution is cycled between increasing and decreasing polarization, the overvoltage decreases with each cycle as the H_2O_2 concentration builds up in the electrolyte. As pointed out by Yeager (296,298–300) this hysteresis is avoided by adding H_2O_2 directly to the electrolyte to stabilize it.

In the pH range from 9 to 12, a Tafel region is observed for the cathodic reduction of O_2 in KOH solutions stabilized with additions of H_2O_2 to the electrolyte (24,266,299,300). A slope ranging between 0.11 and 0.15 is found on wax-impregnated graphite and on PC cathodes (300). Yeager (300) suggests the following series of steps as a possible mechanism for the reduction of O_2 on carbon cathodes.

$$O_2 + e \rightarrow (O_2^-)_{ads} \tag{8.9}$$

$$(O_2^-)_{ads} + H_2O \rightarrow (HO_2)_{ads} + OH^- \tag{8.10}$$

$$(HO_2)_{ads} + e \rightarrow HO_2^- \tag{8.11}$$

With this scheme, the experimentally observed Tafel slope of about 0.12 is consistent with a mechanism in which Eq. 8.9 is rate limiting. In support of this mechanism is the experimentally observed fact that the overvoltage depends on P_{O_2} (300) but is independent of pH (24). At high enough potentials, a limiting current is observed which depends on the P_{O_2}. Under these conditions the system is mass transfer-controlled by the diffusion of dissolved O_2 to the electrode (199,300).

Polarization studies of the evolution of O_2 on carbon anodes in H_2O_2 stabilized alkaline solutions have been made (13,128,296,298–300), and a Tafel slope of about 0.12 is observed. However, the anodic data were found (300) to be poorly reproducible so that a mechanism was not proposed. Possibly, such irreproducible data may be explained in terms of the anodic oxidation of peroxide with the resultant loss of H_2O_2 to the system as found for noble metal anodes (130) in H_2O_2-stabilized solutions discussed in Chapter IV. For each cycle of increasing and decreasing anodic polarization, peroxide is consumed continuously in the electrode process. As a consequence, the rest potential drifts to more noble values and the overvoltage increases.

Yeager (300) has found that the anodic limiting current depends on the HO_2^- concentration but is independent of the OH^- concentration. The Tafel slope of 0.12 is consistent with a mechanism in which a one-electron transfer step is rate controlling. In this case, a possible mechanism consists of the reverse, of Eq. 8.11 followed in succession by the reverse of Eqs. 8.10 and 8.9 with Eq. 8.11 as rate determining. From this viewpoint, the O_2/H_2O_2 couple is reversible on carbon electrodes because the path of the anodic reaction is the reverse of the path of cathodic reaction, although the rate-determining step is different in the two cases. On carbon the O—O bond is not broken.

Mechanisms in which carbon–oxygen surface complexes, such as reducible organic groups, may participate in the rate-controlling step are considered less favorable because there is negligible variation in the apparent exchange current density found on cathodes made from several types of carbon (299).

Very recently, Morcos and Yeager (197) have studied the reduction of O_2 on the surfaces of high temperature annealed (162) pyrolytic graphite and on naturally occurring single crystals of graphite. As a point of interest, the reduction of O_2 to H_2O_2 and the oxidation of H_2O_2 to O_2 occurred readily on all surfaces except the cleavage surface where these reactions were relatively inhibited. Since such an effect was not observed in the ferri–ferrocyanide reactions on these graphite surfaces, an explanation in terms of a variation in electronic conduction of electron transfer from one surface to another cannot apply. Rather, they (197) suggest that the effect is produced by a lack of favorable chemisorption sites for O_2 or some reaction intermediate on the cleavage surface.

C. THE OXYGEN DIFFUSION ELECTRODE

At an early date, Rideal and Evans (228) pointed out that an outstanding problem in the development of the fuel cell is the construction of an efficiently performing O_2 electrode. This problem is still with us at the present time. To prevent the mixing of oxygen by diffusion from the cathode to the anode chambers, a porous diffusion electrode must be used. The oxygen molecule is relatively stable at ambient temperatures, and the O—O bond is difficult to break. Consequently, a catalyst must be used at the fuel cell cathode to make the electrochemical reduction of O_2 proceed at a useful rate.

Any catalyst on the O_2 cathode surface must perform two functions if an efficient O_2 electrode is to be realized. It must promote the transfer of electrons from the electrode metal to the adsorbed oxygen, and it must catalyze the decomposition of the intermediate peroxide to water. In H_2–O_2 fuel cells, the anodic catalysts are so active that the electrode process is diffusion controlled. Because the distribution of the diffusion paths in the pores of a porous fuel cell anode ranges over a

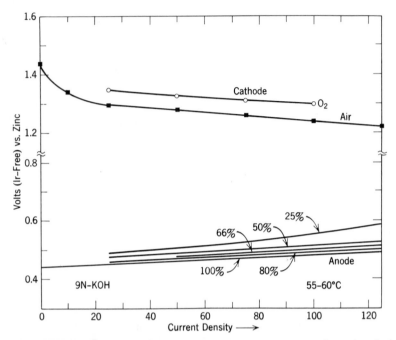

Fig. 8.5. Polarization curves (54) for catalyzed porous carbon anodes and cathodes showing linear behavior of H_2 anode and logarithmic for the O_2 or air cathode. The percentage of H_2 in H_2–N_2 gas mixtures is indicated. (By permission of The Electrochemical Society.)

considerable spread of values, a linear polarization (E–i) curve is obtained at the mass transfer-controlled fuel electrode as shown in Fig. 8.5. At the fuel cell cathode, the catalysts available for the reduction of O_2 are so poor that the electrode kinetic processes are slower than the diffusion processes. In this case, the electrode process is under activation control, and the logarithmic polarization curve of Fig. 8.5 is obtained. Bennion and Tobias (20) concluded that the reduction of O_2 at the porous cathode is charge transfer controlled on the catalytic surface below the thin film of electrolyte rather than limited by the diffusion of O_2 to the active sites through the thin film. Much of the research carried out on fuel-cell O_2 electrodes is concerned with a search for more active catalysts for the reduction of O_2.

1. Porous Carbon Electrodes

Because of its chemical inertness and its good electrical conductivity, porous carbon is a good material from which oxygen electrodes may be fashioned (22,64a,91,169,216,224,240,276,296). To increase the active surface, and hence, the current density, finely powdered graphite is rubbed or pasted in the pores of the porous carbon body. Such activated porous carbon electrodes can support much larger current densities than the untreated graphite electrodes (22,296). The active carbon layer may be sprayed or painted on the porous electrode by using a solution of active carbon in a volatile solvent such as benzene containing about 5% polyethylene or wax as a binder (296). After evaporating the solvent a porous active layer (high surface area) of carbon is obtained.

Since carbon is a poor peroxide-decomposing catalyst (22,64a,276), the end product of the reduction of O_2 is H_2O_2 with a transfer of only two electrons per oxygen molecule. The potential of the O_2/H_2O_2 reaction against a hydrogen electrode in the same solution is 0.682 V which is considerably less than the four-electron O_2/H_2O potential (1.229 V vs. Pt/H_2 in the same solution). The more the four-electron O_2/H_2O reaction contributes to the total reduction current, the higher is the potential for a given current density, and hence the greater is the energy realized from the electrochemical processes occurring in the cell. As a result, it is important to have an active peroxide-decomposing catalyst present on the cathode surface.

Náray-Szabó (202) electrochemically deposited Pt black on porous carbon electrodes and got a rest potential of 1.064 V vs. Pt/H_2 in the same solution ($2N$ H_2SO_4) at the oxygen electrode. For a given potential the platinized carbon supported twice the current that the uncatalyzed carbon did. Metals other than platinum were electrodeposited on porous carbons by Knobel (276) in O_2 electrode studies. To prevent the flooding of electrodes, Tobler (265) wetproofed a porous carbon plug with paraffin before applying a layer of catalyst. He tried a number of metals also. In general, it was found that carbon in which metallic impurities such as iron were present made better O_2 cathodes than the very pure carbon.

The development of porous carbon diffusion electrodes was accelerated by the need for long-lasting air electrodes for the primary Zn–air cell. Early attempts were disappointing in general because, among other things, there as a lack of good carbon materials. A review of the early literature pertaining to the Zn–air cell is given by Nasarischwily (203). Heise and co-workers (123,124,243) developed activated carbon electrodes having long life in strong alkaline solutions for use in the Zn–air cell.

One of the attractive features of the primary Zn–air cell, which is composed of an amalgamated Zn anode and a porous carbon cathode (exposed to the air) in strong alkaline solution, is the ability of this cell to deliver a given current at a very steady potential (123). The open-circuit potential ranges between 1.4 and 1.45 V, and under load, current is delivered along a flat discharge curve in a range of potentials from 1.1 to 1.2 V. The current is limited by the rate at which O_2 can diffuse from the surrounding atmosphere to the active sites on the air cathode. Heise and Schumacher (123,243) estimated that current loads of 0.5–0.65 A require the use of 9–12 cc of air/min. Such air cell batteries designed for railroad applications have up to 600 amp-hr capacity.

In the cell reaction, the Zn behaves as a consumable anode by dissolving during discharge as zincate ions. Early workers (120,129,242) identified the form of this ion in solution as $HZnO_2^-$, but later work by Dirkse (74) indicates that the zincate ion exists as $Zn(OH)_4^{2-}$. As the cell reaction proceeds, Zn goes into solution by reaction with the electrolyte,

$$Zn + 4OH^- \rightarrow Zn(OH)_4^{2-} + 2e \qquad (8.12)$$

until the solution becomes saturated with zincate at which point zinc oxide precipitates from solution,

$$Zn(OH)_4^{2-} \rightarrow ZnO + H_2O + 2OH^- \qquad (8.13)$$

In some cases, the precipitated zinc oxide deposits on the active sites of the O_2 cathode and quickly blocks the cathodic reduction of O_2. Crystallization of the oxide can occur so quickly (123) that as soon as the solution becomes saturated with zincate the output of the cell falls rapidly to zero. As a result, very little of the cell capacity is delivered at potentials below a value of 1.1 or 1.2 V. In any event, the cell is discharged when the alkaline electrolyte is spent. To extend the life of a Zn–air cell, lime is placed in the cell in the form of briquets (123,124) made by mixing the lime with an extender, such as cellulose floc. Calcium zincate separates out, regenerating the NaOH electrolyte,

$$Ca(OH)_2 + Zn(OH)_4^{2-} \rightarrow CaZn(OH)_4 + 2OH^- \qquad (8.14)$$

It is important to prevent the porous carbon electrode from wetting, which is accomplished by treating the carbon electrode with a solution of paraffin in a volatile solvent. The active carbon (very small particle size) is worked into the pores of the porous graphite body. If it can be assumed that the O_2 of the air is reduced to water in the cell process, then the overall cell reaction is

$$2Zn + O_2 + 2H_2O + 4OH^- \rightarrow 2Zn(OH)_4^{2-} \qquad (8.15)$$

Since the standard potential for Eq. 8.12 is -1.215 V (66) and for the O_2/H_2O reaction is 0.401 V (66) in strong alkaline solutions, the cell potential for Eq. 8.15 is calculated to be 1.616 V.   This value is much closer to the experimentally observed open-circuit potential of 1.45 V (124) than the one calculated (1.139 V) with the O_2/H_2O_2 reaction in place of the O_2/H_2O reaction.   Apparently, metallic impurities were present in the active carbon cathodes used in the Zn–air cells which acted as effective peroxide-decomposing catalysts.

Present-day Zn–air cells (34,48,51,89) use the greatly improved oxygen diffusion cathodes developed for fuel cells, and the migration of Zn to the O_2 cathode can be blocked with ion-exchange membrane separators (238).   Magnesium may also be used as an anode material in place of Zn (157).   Low cost and light weight make the air cells attractive.

Very recently, the Zn–air system has been studied as a secondary cell (248) because the theoretical energy density of the metal–air secondary cells is very high.   In the secondary cell, the anode consists of a layer of Zn plated on a thin metal plate so that on complete discharge all the Zn is converted to zincate.   The cathode is a catalyzed porous nickel diffusion electrode operating on pressurized air.   The cycling characteristics of this cell are not ideal.   Excess ZnO precipitates out of the electrolyte, and on the charging cycle Zn tends to deposit as dendrites instead of a sponge.   Circulation of the electrolyte relieves some of these problems and is required to remove the inactive N_2 from the system.

Kordesch and co-workers (168,169) described methods of catalyzing porous carbon electrodes.   The porous electrode was soaked in the salt (such as the nitrate) of the catalyst metal, after which the electrode was baked in an oven in a reducing atmosphere (so that the carbon was not oxidized) for a specified period of time at a particular temperature to decompose the salt thermally to the free metal or metal oxide inside the pores.   Metals such as Co, Al, Fe, Mn, and Ag were effective catalysts.   Flooding of the catalyzed electrode was prevented by spraying a solution of paraffin in a volatile solvent, such as toluene, on the electrode surface.   The solvent was removed by warming the electrode in an oven.   Special precautions had to be taken with regard to the concentration of paraffin or a nonconducting film of wax would cover the catalyzed sites completely and the electrode performance would deteriorate.

Yeager and co-workers (296) noted that paraffin and polystyrene formed films on the electrodes but that polyethylene did not.   When the electrode was sprayed with a 5% benzene solution of polyethylene and the benzene was evaporated away, the polyethylene formed small particles interspersed with the active carbon particles. The highly hydrophobic nature of polyethylene imported good wetproofing characteristics to the electrode surface while at the same time exposing a large uncovered area of the electrode surface to the electrolyte. Teflon dispersions make particularly effective wetproofing layers (260) because the Teflon is also deposited as hydrophobic particles interspersed with the catalyst particles.

Studies of the nature and properties of the surfaces of catalyzed porous-oxygen electrodes appear in the literature (e.g., 3,25,59,165,211, 244,245,274,296). In alkaline solutions Co, Mn and Ag are effective catalysts on porous carbon cathodes for the reduction of O_2 (168,169,296). Oswin (211) reports that Pd is more active than Pt which, in turn, is more active than Ag for the reduction of O_2 in KOH solutions. According to Cohn (59) Ru and alloys of Ru and Pt and Pd are active catalysts for the oxidation of H_2 or methanol at fuel cell anodes (3) but seem to be inferior to Pd, Pt, and Ag for the reduction of O_2. Studies of the reduction of O_2 on Ru electrodes in alkaline solution have recently been reported (203a). Platinum is by far the best catalyst for the reduction of O_2 in acid solutions with the other noble metals exhibiting less activity. Gold has virtually no activity. No alloy of a base metal not containing expensive noble metals has been found with greater activity than Pt for the reduction of O_2 in acid solution.

As pointed out by Williams (ref. 293, p. 92), the optimum use of the precious metal catalysts is made by dispersing the metal in finely divided form on substrates of very high surface area such as active carbon. Any method of increasing the surface area is desirable since this not only lowers the electrode polarization but also reduces the amount of catalyst required to give a demanded electrode performance. Catalyst preparation as well as its chemical composition surely play a role in determining electrode performance (3). Schwabe and co-workers (244,245) report that special pretreatment of carbon electrodes at a very high temperature produces a porous graphite structure which is highly resistant to both acid and alkaline solutions even at relatively high temperatures.

2. Porous Metal Electrodes

One of the first applications of a porous metal electrode as an oxygen diffusion electrode was made by Bacon (8) in a high temperature (200°C) high pressure (400 psi) fuel cell. The electrodes were fashioned from porous nickel by sintering a Ni powder at elevated temperatures and pressures.   To obtain a stable liquid–gas interface in the pores of the electrode, a dual porosity construction was made by forming a thin layer of porous Ni having smaller pores on the surface of the body of the porous Ni electrode.   Since the surface area of these electrodes was low, the electrodes were impregnated with $Ni(NO_3)_2$ after which the salt was decomposed thermally and reduced to finely divided Ni in the pores with hydrogen.   At the temperature and pressure of operation, the Ni itself had sufficient catalytic activity so that other catalysts were not required.   Because the oxygen electrode surface was oxidized to a nonconducting green NiO, lithium was introduced into the surface of the Ni (lithiated) to form the semiconducting black double oxide of Li and Ni. This was done by impregnation with LiOH followed by thermal decomposition of the lithium salt.

Justi and co-workers (153) have combined the high activity and resistance to poisoning by impurities of Raney type catalysts with the mechanical strength and electrical conductivity of sintered metal electrodes, by forming the double skeleton structure (DSK) from a mixture of the metal-Al alloy and pure metal powders, as described at the beginning of this chapter.   Because Ni is not an active catalyst for the reduction of O_2 at ambient temperatures, Raney-Ag electrodes were made (75) as DSK fuel cell cathodes.

Silver-aluminum alloys are not brittle like the Ni-Al alloys; and instead of crumbling in a press, they flatten into sheets.   However, if great care is taken in choosing the proper weight per cent of Ag and in applying the proper heat treatment (75), a suitable brittle alloy is obtained from which the Raney-Ag powder may be made.   The Raney-Ag powder is mixed with pure Ni powder, and the mixture is hot pressed between 300 and 500°C under a pressure of 1 ton/cm².   By leaching out the Al with KOH, the structure is activated by producing finely divided Ag in the Ni pores.   The Ag DSK electrodes make every high performance alkaline fuel cell cathodes.

The great advantages of using metal electrodes over porous carbon ones are high mechanical strength, good electrical conductivity, and the ease with which good electrical contact may be made to the electrode. Oxygen cathodes may be made from porous nickel plaque (267,297) which are thin (0.03–0.06 in. thick) and mechanically rugged.   Such electrodes are catalyzed by impregnating them with solutions of the catalyst metal salt following by thermal decomposition or chemical reduction similar to the catalyzing procedures used for carbon electrodes. Wetproofing is accomplished by spraying or painting the surface with a Teflon emulsion.

Porous metal electrodes made of Ni can only be used in alkaline solutions so that some other material must be used in acid solutions. Krupp et al. (171) made Raney-type structures from sintered powders of alloys of Al with Pt, Pd, and Rh activated in KOH solutions. As one might expect, the Raney-Pt electrodes give excellent results, but they are frightfully expensive. A less expensive material is tantalum. Although Ta corrodes in acid solutions, alloys of Ta containing small amounts of Pt or Pd are highly passive in acid media (184). Some studies of porous Ta diffusion electrodes catalyzed with noble metal catalysts have been carried out (55). Usually the porous metal structure does not have a high enough surface area to give acceptable performance. High surface areas are obtained by applying a paint of active carbon to the porous metal electrode, and upon this surface the active metal catalysts are deposited.

3. Composite Porous Electrodes

Carbon electrodes cannot be made thinner than $\frac{1}{4}$ in. thick without becoming too weak mechanically, and when operated on air, inert nitrogen collects in the pores of the cathode, increasing the concentration polarization caused by the hindered diffusion of oxygen through the blanket of nitrogen (54). The effects of N_2-blanketing are seen in Fig. 8.6. On the other hand, porous metal electrodes can be made very

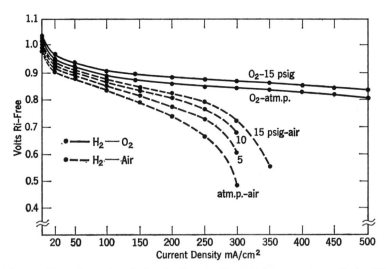

Fig. 8.6. The effect of variation in P_{O_2} on O_2 and air-composite cathodes (54). The limiting current observed at air electrodes near 300 mA/cm² shows the effect of N_2 blanketing (By permission of the Electrochemical Society.)

thin but do not have the high surface area of the active carbon electrodes. Kordesch and co-workers (54) combined the mechanical strength of porous metal and the high area of active carbon to get a high area thin-diffusion electrode. The backing consists of a layer of porous Ni (about 0.035 in. thick) sintered on a Ni screen which is wetproofed by spraying with a Teflon emulsion followed by drying at about 100°C. A thin layer of unactive carbon (large particle size) is applied to the Ni surface by spraying a suspension of powdered carbon and Teflon on the Ni. The Teflon-bonded carbon layer is sintered at 400°C in a N_2 atmosphere. Several layers of activated carbon (fine particle size) are sprayed onto the electrode surface using a hot solution of polyethylene containing the active carbon. Finally, the structure is pressed at 1000 psi and 130°C to increase the density and the bonding of the layers. These electrodes are catalyzed by reduction of impregnated salts of the metal catalyst. Excellent results have been obtained with these electrodes (500 mA/cm^2 at 0.8 V on O_2).

4. Pasted Diffusion Electrodes

To keep the diffusion path as short as possible and to decrease the size and weight of the cell, it is advantageous to make the diffusion electrodes as thin as possible. Very thin electrodes were made by Grubb and Michalske (110) by pasting a slurry of Pt black in water in a smooth layer on a Pt-wire gauze. The electrode was thoroughly dried in a vacuum desiccator. Since such electrodes are very thin, they are easily flooded, and so the electrolyte must be immobilized by holding it in a matrix such as an asbestos pad. The thin electrodes are placed on each side of the electrolyte-saturated asbestos pad, and the O_2 and fuel gases are fed to the electrodes through glass frits. The entire assembly is held in place with a clamp as described by Thacker (262). One of the most serious defects of such an electrode is its extremely fragile nature which seriously limits the amount of handling that the electrode can take without causing the shedding of the Pt black. To overcome this drawback, a binder must be used.

As early as 1938, Spiridonov (254) made a diffusion cathode for use in alkaline electrolytes by pressing activated carbon on a wire grid with rubber cement as a binder. Such electrodes were activated by impregnating them with catalysts. The first to use Teflon as a binder were Elmore and Tanner (83), who constructed an O_2-diffusion electrode by enclosing a Teflon-bonded Ag powder between two Ag screens. To immobilize the electrolyte, a paste was used consisting of an NaOH–KOH eutectic mixture containing 20% $Ca(OH)_2$ for alkaline systems and a thick slurry of H_3PO_4 and SiO_2 powder for acid systems.

Teflon-bonded thin electrodes may be made about as thin as the wire mesh on which the material is bonded (52,62,117). According to Niedrach and Alford (207), a thin film of Teflon is formed on an Al foil by spraying a Teflon emulsion on the heated foil ($\sim$130°C) placed on a hot plate. On the Teflon film is spread a thin slurry of Pt black in a Teflon emulsion, and the solvent is completely removed by heating to about 250°C. A metal screen is placed between two pieces of this material with the Pt-black layers next to the screen. This sandwich is sintered at 350°C for 2 min at pressures between 1800 and 3000 psi, and the final electrode is obtained by dissolving the thin Al backing in warm 20% KOH. For alkaline electrolytes, the metal screens are made of Ni or Ag, but for acid electrolytes, a Pt screen is used. They (207) found that the thinner the electrode the less was the polarization of the air electrode caused by the blanketing of the pores by N_2.

Haldeman and co-workers (62,117) made very thin electrodes (0.004–0.008 in. thick) by pasting a slurry of Pt black in a Teflon emulsion on a metal screen. The final product is obtained by sintering the Pt-black–Teflon layer at elevated temperatures and pressure. Nickel screen is used for alkaline systems and tantalum screen for acid systems. Electron microscopic examination of this material showed that the Pt-black formed loosely packed aggregates interspersed with Teflon particles and threads. They (117) concluded that the Teflon threads bind the material into a mechanically secure structure.

Carbon black was also mixed with the Pt black as an extender, and in this way, catalyst loading from 4 down to 0.5 mg Pt/cm² can be obtained. Without the extender, loadings ranged from 7 to 10 mg Pt/cm². The effect of catalyst loading on the polarization characteristics of the O_2 diffusion electrode is similar to that shown in Fig. 8.7.

Christopher (52) made a Teflon-bonded carbon black electrode by sintering a layer of carbon black pasted on a Ta screen from a slurry of carbon black in a Teflon emulsion at 250°C and 2000 psi. Such electrodes were catalyzed by brushing a solution of H_2PtCl_6 or $PdCl_2$ on the carbon surface and decomposing the salt thermally or by reducing the salt chemically with sodium borohydride. Equally suitable, the catalyst solution could be added to the carbon–Teflon slurry and reduced followed by the sintering procedure.

To reduce the weight still further, Barber and Woodberry (12) replaced the metal screen with a paper made from acrylic fibers. Chloroplatinic acid was reduced to Pt black with sodium borohydride directly on the paper pulp. With loading higher than 7 mg Pt/cm², no conducting filler was required to give the electrode the proper conductivity, but

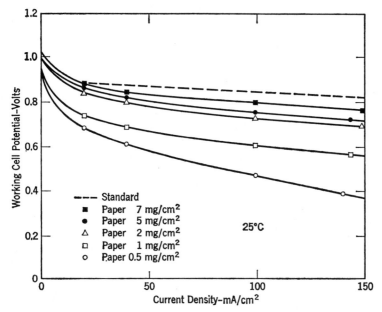

Fig. 8.7. Oxygen electrode performance for acrylic paper electrodes (12) as a function
   by Pt catalyst loadings.   (By permission of The Electrochemical Society.)

with lower loading, powdered graphite was added to lower the electrical
resistance.   By adding 18% Teflon emulsion to the pulp, sufficient wetproofing of the final electrode structure was obtained.   The paper was
made on a standard papermaking machine.   When an oxygen electrode
containing 7 mg Pt/cm² was used in a matrix cell with 5N H_2SO_4 held
in a glass fiber paper matrix, a current of 600 mA/cm² could be drawn
at 0.57 V at 70°C.   They showed that the paper electrodes were stable
in 5N H_2SO_4 solution at 80°C for at least a month.

Gregory and Williams (294) evaporated a metal (usually Au or Ag)
on a microporous plastic (polyvinyl chloride) to give a light, conducting
substrate.   On this surface, the catalyst was deposited either electrochemically or as a Teflon-bonded layer.   Free electrolyte systems could
be used, since flooding of the electrolyte was prevented by maintaining
a pressure differential (1–3 psi) across the electrode–electrolyte interface.
Such electrodes could be used in either acid or alkaline electrolytes, and
good performance was obtained both on O_2 and on air (ref. 293, p. 100).

Thin electrodes are used with ion-exchange membrane fuel cells (150).
Joachim and Vielstich (148) held an Ag powder or a 10% mixture of
Pt black in carbon black against the exchange membrane with Ni gauze.

In the first membrane cells used by Niedrach (206), platinized porous sintered nickel electrodes were used, but the surface area was small, and only 5.7 mA/cm² at 0.5 V could be drawn from the cell. Much higher area electrodes were obtained with platinized porous carbon electrodes (217), 20 mA/cm² at 0.5 V, using the same membrane. Because of poor contact between the electrode and the membrane, high resistance circuits developed, causing hot spots and the resultant drying out of the membrane. This problem was overcome by developing the Pt-imprinted membrane (217). A dried sample of membrane was placed on a smooth stainless steel plate. Platinum black was sprinkled on the surface of the membrane to give loadings of 5–10 mg of Pt/cm². Then the membrane–catalyst system was subjected to 60,000 psi at 115.5°C for 5 min to form the platinum-imprinted membrane. The performance was increased to 33 mA/cm² at 0.5 V (217).

Dravnieks and co-workers (78) studied the use of inorganic ion-exchange membranes, such as zirconyl phosphate. A layer of Teflon-bonded zirconyl phosphate was formed first, on top of which a layer of Pt black and powdered zirconyl phosphate was placed. Over these layers was sprinkled a layer of Pt black, and the entire structure was sintered at 380°C and 20,000 psi. The zirconyl-phosphate layer was saturated with phosphoric acid. The results (78) were not as good as those obtained with organic exchange membranes (150).

During the sintering process of Teflon-bonded Pt black electrode structures, the high surface area of the Pt powder is lowered. Despic and co-workers (70,79) have studied powdered O_2 electrodes to avoid the destruction of the high surface area of the Pt black by sintering. This electrode structure is constructed so that the catalyst powder is held between a porous Ni plate and a fine Ni gauze all clamped in place in a steel housing. Some of the catalysts studied were powdered Raney Ag, silvered glass-wool, silvered ion-exchange resin powder, and silvered porous ceramic powder. The particles were silvered by the chemical reduction of the silver–ammonium complex in a slurry of the particles. With the Raney Ag, the performance of the powdered O_2-electrode was reported to approach that of Justi's DSK electrode. Good results were found with the silvered resins but not with the silvered glass wool because the adhesion of the Ag was poor in this case.

D. REDOX OXYGEN ELECTRODES

Since the O_2 molecule is relatively stable and the electrochemical reduction of O_2 occurs with difficulty (with large overvoltage), it is possible to take advantage of a more favorably reducible system added as an intermediate or electron carrier. A redox system such as Fe^{2+}/Fe^{3+} is added to the O_2-electrode compartment. For this example, the electrochemical reduction of Fe^{3+} ion to Fe^{2+} ion takes place readily at the Pt cathode in acid solutions. The Fe^{2+} ion comes into contact with O_2 or air where the Fe^{3+} ion is regenerated by the oxidation of Fe^{2+} ion by O_2. In this way, the electrochemical reduction of the redox system is coupled with the chemical reduction of O_2 by the ferrous ions acting as electron carriers.

To realize the greatest efficiency from such a system, the reversible potential of the redox system should be as close as possible to the reversible O_2 potential (1.229 V), and the reaction of O_2 with the reduced form of the redox system should be rapid. As pointed out by Posner (221), neither of these conditions is met with the Fe^{2+}/Fe^{3+} system, but a better system is the Br_2/Br^- couple with a reversible potential of 1.09 V (66). The Br_2/Br^- couple has been studied in detail by Reneke (225) as a possible redox system for fuel cells. It was shown that a practical fuel-cell cathode could not be made using the Br_2/Br^- couple; for among other things, Br_2 escaped continuously from the system, and a membrane which increased the internal resistance of the cell was necessary to separate the anode and cathode compartments.

The HNO_3 acid couple (82,196,273) has been extensively studied by Shropshire and Tarmy (249) as the O_2 redox electrode for fuel-cell application. Open-circuit potentials up to 1.16 V have been recorded (249) at platinized carbon electrodes in $3.7M$ H_2SO_4 which is also $0.2M$ in HNO_3. Because HNO_3 is highly oxidizing, the more dilute the solution is in HNO_3 the less is the corrosion. It was determined that $0.2M$ HNO_3 was the optimum concentration of HNO_3. They (249) observed that the polarization of a catalyzed carbon–oxygen cathode was reduced by the addition of HNO_3 to the acid electrolyte in single electrode studies. According to Shropshire and Tarmy, the active intermediate of the HNO_3/HNO_2 couple is NO, which is oxidized to NO_2 by O_2. The HNO_2 is regenerated by the interaction of NO_2 and water.

Although good results were obtained with the nitric acid oxygen electrode in single electrode studies, the electrode suffered from the same disadvantages that plague all redox systems when used in a fuel cell with a fuel anode. A diaphragm must be used to prevent the HNO_3 or some reaction intermediate from reaching the fuel electrode where it can be reduced by reactions at the anode, producing an adverse effect on the potential. The presence of the diaphragm increases the internal resistance of the cell. In addition, both HNO_3 and HNO_2 are relatively volatile and are gradually lost from the system with time. As noted by Williams (ref. 293, p. 102), the HNO_3 system is neither invariant nor simple. Redox systems can be devised for fuel cell anodes, but success here as at the cathode is limited.

E. HIGH TEMPERATURE OXYGEN ELECTRODES

To increase the rate of a given reaction, one may either lower the activation energy barrier by using a catalyst, or raise the average energy of the reactants with respect to the energy barrier by raising the temperature of the system. In previous sections, a discussion of the various catalyst structures employed in the construction of O_2 electrodes for fuel cells has been presented. The alternative approach of operating the fuel cell at elevated temperatures is considered in the following sections. Good reviews of the status of high temperature fuel cells may be found in the literature by Gorin and Recht (103), Jones (151), and Smith (253) and reviews of the preparation and use of O_2 electrodes in molten salts have been presented by Laitinen and Bhatia (174) and Reddy (223).

1. Solid Oxide Cells

At elevated temperatures, many ceramic materials, such as the various glasses, become conducting. If a ceramic could be found in which the conducting mechanism consisted of the transport of O^{2-} ions, a very simple system, indeed, could be made. As in all electrochemical cells, the electrolyte must be an ionic conductor so that a separation of charge may be maintained between the anode and cathode, and it must be impermeable to gases so that the non-Faradaic chemical combuston of the fuel and oxidant gases which do not contribute current to the external circuit may be prevented.

Zirconium oxide is a refractory nonconducting material which exists in two modifications. The transition from the room temperature monoclinic form to the high temperature tetragonal form is accompanied by a large change in volume (81), making it sensitive to thermal shock. When CaO or MgO is added to ZrO_2, it is transformed to a cubic fluorite structure, and the polymorphic transition is suppressed above a threshold concentration of CaO or MgO. Such a mixed oxide system is said to be stabilized. It is known (61,76,81,100,136,149,209,213,234) that other oxides, particularly the rare earth oxides, will stabilize ZrO_2.

The stabilized zirconia systems are highly conducting at elevated temperatures, and x-rays (81,159) show the presence of O^{2-} ion vacancies in the crystal lattice. When zirconia or thoria are doped with oxides of metals having a valence less than 4 (136), the high conductivity of these ceramic materials at temperatures above 600°C is attributed to the presence of anionic vacancies (149). Wagner (278) proposed the presently accepted mechanism of conduction by the diffusion of O^{2-} ions through the defect lattice of the stabilized zirconia. A number of investigations of the conductivity of calcia and yttria-stabilized zirconia over the wide range of temperature and concentration have been carried out (76,159,160,269,278). A minimum in the resistance–composition curve occurs at the minimum concentration required to stabilize the zirconia and occurs at about 8% yttria (76) and at about 12% calcia (76,149,159,269). At this point, there are about 7.5% of the possible anion sites vacant (269,288).

Kingery and co-workers (159) studied the diffusion of oxygen through $(ZrO_2)_{0.85} \cdot (CaO)_{0.15}$ using ^{18}O isotope and concluded that the diffusion coefficient of O^{2-} ions is sufficiently large to account for the entire observed conductivity. In addition, the resistance was found (159) to be independent of the partial pressure of O_2, a fact which supports the O^{2-} ion conduction thesis. The resistance of $(ZrO_2)_{0.85} \cdot (CaO)_{0.15}$ is about 15 ohm-cm and of $(ZrO_2)_{0.9} \cdot (Y_2O_3)_{0.1}$, 9 ohm-cm.

If porous Pt films are deposited on each side of a thin wafer of calcia-stabilized zirconia, it is found (287) that the system acts like an O_2-concentration cell when O_2 is fed to one side of the disk.   At the cathode, O_2 is reduced to O^{2-} ions,

$$O_2 + 4e \rightarrow 4O^{2-} \tag{8.16}$$

which migrate through the ceramic lattice to be oxidized at the anode,

$$4O^{2-} \rightarrow O_2 + 4e \tag{8.17}$$

Weissbart and Ruka (287) suggested that such a system could be used as an oxygen partial-pressure gauge for vacuum systems. A tube of $(ZrO_2)_{0.85} \cdot (CaO)_{0.15}$ was coated on the inside and the outside with a thin porous film of Pt, and the system operated between 600° and 750°C. They found that the current was proportional to the P_{O_2} gradient, and the transference number of the O^{2-} ion was virtually unity.

If fuel were fed to the anode side of the solid electrolyte cell, it would react with the O^{2-} ions,

$$H_2 + O^{2-} \rightarrow H_2O + 2e \tag{8.18}$$

to reduce the P_{O_2} to a very small value.   This is the basis of the solid oxide electrolyte fuel cell.

A large number of systems have been investigated for fuel cell application, and Baur and Tobler (17) have given an extensive review of these early efforts. The early cells were characterized by an inability to operate for long periods of time because the electrodes and electrolytes were not invariant. Also, the internal conductivity of these cells was too low, thereby limiting the current to impractical values. Most of the early systems were not truly solid electrolyte cells (38), but at the operating temperature were really molten salt systems held in a solid matrix.

The first to use a solid electrolyte in a fuel cell were Baur and Preis (16). Solid electrolytes composed of rate earth oxides, clay, and tungstic oxide were cast in the form of tubes. The inside of the tube was filled with Fe_3O_4 as the O_2 cathode catalyst, and the tube was imbedded in a bed of granular iron as the anode. Air was fed to the inside of the tube, and the fuel gases (H_2 and CO) were pumped through the bed of granular iron. The resistance of the cell was 40 ohm/cm at 1100°C. In a relatively short time the Fe_3O_4 was oxidized to Fe_2O_3, and the output was reduced drastically. They also used yttria-stabilized zirconia.

At the present time, attention is directed to the production of very thin wafers of the stabilized zirconia (ref. 293, p. 204), since the current obtained from these cells is limited mainly by the resistance of the cell (6,286,288). Activation polarization in these systems cannot be detected. Disks only about 0.15 cm thick have been used (286), and Pt films were vapor deposited on the disks as the O_2 electrode. Archer and co-workers (7) have discussed this system in terms of a fuel cell battery.

Carter and co-workers (ref. 293, p. 209) have operated a solid electrolyte fuel cell at 1100°C by bubbling a stream of air into a pool of molten silver (as the cathode) held in a tube made of stabilized zirconia. Natural gas as a fuel was directed over the outside of the zirconia tube, and a film of porous carbon was formed on the tube's surface by the pyrolysis of the natural gas and served as the anode. Current densities up to 150 mA/cm² at 0.7 V have been recorded.

Such cells operate at very high temperatures ($\sim$1000°C), are limited by the resistance of the electrolyte, and employ fragile, thin ceramic disks which limit the size to which the cells may be scaled up because of their sensitivity to mechanical and thermal shock. Cells using molten salt electrolytes may be operated at lower temperatures and are considered next.

2. Molten Salt Cells

The work of Baur and Preis prompted Davtyan (65) to study fuel cells operating at about 700°C and using an electrolyte of 43% Na_2CO_3, 27% monozite sand, 20% WO_3, and 10% soda glass which was fused at 1200°C and cast into disks.   A paste of Fe and Fe_3O_4 powders in clay was applied to one side of the disk as the air electrode and a paste of Fe and Fe_2O_3 powders in clay on the other side as the fuel anode.   The pasted electrodes were held in place by perforated iron plates, and the sandwich cell was clamped in a steel housing.   Not only did the electrolyte change composition with time, but also it could not stand the thermal shock in heating up and cooling down the system so that cracks allowed the fuel to mix non-Faradaically with the air inside the cell.

Broers and Ketelaar (38) have thoroughly investigated the electrolytes used by Davtyan and have come to the conclusion that they are composed of a solid matrix of rare earth oxides containing a melt of carborates, phosphates, tungstates, and silicates.   If a carbon-containing fuel is used, such as CO or a hydrocarbon, the CO_2 formed at the anode interacts with any molten salt composed of chlorides, sulfates, nitrates, etc. and, in time, converts them to carbonates.   Consequently, carbonate molten salts should be used to obtain an invariant electrolyte.

Using tubes of sintered MgO filled with Fe powder (anode) and Fe_3O_4 powder (cathode), Baur and co-workers (18) described a fuel cell using an electrolyte composed of a mixture of sodium and potassium carbonates which operated at 800–900°C.   The internal resistance was very high below 1000°C, and the electrodes were not invariant.

Because of the rapid development of fuel cell technology during the last decade, a number of fuel cell systems based on molten carbonate electrolytes and having good performance characteristics have been described (11,38,47,93,103,158,170,236,250,268,284) although none have operated longer than about 6 months.   The molten carbonates are extremely corrosive limiting the number of usable construction materials, and problems with gaskets and seals have been a continuing source of trouble.   Most cells lose performance because of corrosion of the structural materials of the cell or because of a loss of electrolyte producing an increase in the resistance of the cell.

The physical properties of molten carbonate melts have been investigated by Janz and co-workers (142,145), and the solid–liquid phase diagram of Li, Na, K carbonate melts has been constructed (143). Melting points of carbonates are: Li_2CO_3, 726°C; Na_2CO_3, 858°C; and K_2CO_3, 899°C.   Mixtures of carbonates lower the melting points: equal mixtures of Li and Na carbonates melt at about 500°C; mixtures of the carbonates of Li, Na, and K in a ratio $4:3:3$ melt at $397°C$.   The lower the temperature the less severe is the corrosive nature of the melt. Janz and co-workers (141) investigated the stability of Pt, Au, Ag, and MgO in (Li, Na, K)$_2$CO$_3$ melt at $900°C$.   Gold showed little if any attack, Pt was definitely attacked, and silver actually dissolved to a small but finite extent.   The most stable material is an alloy of 80% Au with 20% Pd (145).

According to Janz and Lorenz (142), the conductivity of molten carbonates is attributed to a Grotthuss type of conduction.   The principal charge carriers are CO_3^{2-} ions, M^+ ions, and the weakly bonded ion pairs MCO_3^-.   Metal ions jump from one CO_3^{2-} ion to another.

$$MCO_3^- + CO_3^{2-} \rightarrow CO_3^{2-} + MCO_3^- \qquad (8.19)$$

Janz and co-workers (140,144) also made galvanostatic steady-state polarization measurements of the evolution of oxygen on Pt, Au, and Ag foil anodes in (Li, Na, K)$_2$CO$_3$ melts between 500 and $900°C$.   An unpolarized Pt wire was used as a reference electrode.   Laitinen and Liu (175) have discussed the use of an anodized Pt wire as a reference electrode.   The overall reaction for the evolution of O_2 is given by Janz as

$$2CO_3^{2-} \rightarrow 2CO_2 + O_2 + 4e \qquad (8.20)$$

Tafel regions with a slope of about 0.06 were observed on the polarization curves taken on Pt and Au anodes at lower temperatures.   From an analysis of the electrode kinetics, Janz concluded that the rate-limiting step involved the interaction of an oxide on the anode surface with an oxide ion in the melt, such as

$$MO + O^{2-} \rightarrow M + O_2 + 2e \qquad (8.21)$$

It is assumed that an equilibrium between CO_3^{2-} and CO_2 exists.

$$CO_3^{2-} \rightleftharpoons CO_2 + O^{2-} \qquad (8.22)$$

The two electrons in Eq. 8.21 are probably discharged sequentially but rapidly enough that one cannot distinguish two one-electron steps.   As the temperature is raised, the overvoltage is lowered until at high enough temperatures (near $900°C$) a Tafel region can no longer be detected. Under these conditions corrosion processes set in to complicate the electrode kinetics.   On Ag anodes a Tafel region is not detected even at lower temperatures because Ag dissolves in these melts.

Porous O_2 electrodes used as cathodes in high temperature carbonate fuel cells cannot be wetproofed with any known substance so that other methods must be used to prevent the electrode from being flooded with the melt. In one method (38,47,103,158,170,236,250,251a,268,284) the electrolyte is immobilized by impregnating a solid matrix of MgO with the melt. The MgO matrix is made by sintering pure MgO powder at 1200°C (158) and may be cast in the form of disks or wafers (158,268) or in the form of a tube (170). When high purity MgO was not used to form the matrix, it was found by Gorin and Recht (103) that the matrix was attacked by the melt.

It is generally agreed that for high temperature cells Ag is the best metal for the fabrication of porous O_2 cathodes, even though it may dissolve in the melt very slowly. Baker and co-workers (11) have operated cells with Ag electrodes for 4000 hr at 600°C. Ketelaar and Broers (38,158) held Ag powder against the fixed electrolyte with a fine Ag gauze, backed up by perforated steel plates and clamped in a steel housing. Similarly, Sandler (236) used a Ag mesh filled with particles of Ag and backed up with perforated metal plates. Baker (11) and Trachtenberg (268) bonded the porous Ag electrode directly to the MgO matrix. The Ag electrode can be formed on the ceramic by vapor deposition or by the thermal decomposition of a painted layer of a silver dispersion suspended in an organic binder (11). Thin films can be formed by impinging a beam of high energy electrons on a target material and condensing the vapor on a cool substrate (250). Fine particles of material may be injected in the O_2 stream of an oxyacetylene or oxyhydrogen torch, and the molten particles may be deposited on a substrate (flame spraying). A thin porous sintered Ag plaque with imbedded Ag screen may be used (250) as an O_2 cathode held against the MgO matrix.

Kronenberg (170) could not solve the gasketing and sealing problems with the disk type of cell so he devised a cell in the form of a tube. The porous MgO tube was impregnated with the melt, and electrodes were deposited on the inside (anode) and outside (cathode) of the tube. The electrodes were formed by applying a slurry of the metal powders to the MgO surface and baking at 500°C. Perforated metal foils were used as current collectors and to hold the porous electrodes in place.

Trachtenberg (268) studied the polarization characteristics of the disk cells by using an isolated unpolarized piece of the diffusion electrode as a reference electrode. In agreement with others (103,158), he found that activation polarization was virtually nonexistent. Ohmic polarization is most important at the Ag cathode where some oxide films may form. Because of the back diffusion of products, concentration polarization is most important at the anode.

The use of a dual pore electrode structure is a second method of preventing flooding of the porous electrode. Douglas (77) and Chambers and Tantrum (47) used a free molten carbonate system in which the electrodes were dual-pore, Bacon-type metal electrodes. Nickel was found to corrode severely (77,103) so that Ag was used as the cathode since it best resisted corrosion. An advantage of the free electrolyte is the ability to use a reference electrode. Douglas used a Au/CO_2-O_2 electrode as a reference system.

Because the MgO matrix type cell could not be made gas or electrolyte tight, and also, because the fragile MgO wafers are sensitive to mechanical shock and deteriorate with time under operation, Broers and Schenke (39) decided to avoid the ceramic problem by the elimination of the MgO matrix. Although this is accomplished by the free electrolyte cells of Chambers and Tantrum (47) and Douglas (77), such cells have all their strength in the electrodes which must be dual-pore structures backed up with perforated metal plates. By the use of paste electrolytes, a lighter, more sturdy system which avoids the ceramic problem can be made. Broers and Schenke made pellets by cold pressing a mixture of the alkali carbonates and MgO powders. Above the eutectic melting point [510°C for $NaLiCO_3$ and 390°C for equal parts of $(Li, Na, K)_2CO_3$] these pellets form a stiff paste which can be molded into many shapes and is gas tight at 700°C. The O_2 cathode is a Ag powder imbedded in the meshes of a Ag gauze. In another design, Ag was flame sprayed on the pellet. Similarly, because of sealing and fabrication problems, Baker and co-workers (11) have abandoned the matrix and the free electrolyte cells in favor of the paste cells.

If a CO_3^{2-} electrolyte cell is operated with a hydrocarbon as fuel and with pure O_2 or air, a layer of oxide forms at the cathode (103) which diffuses through the CO_3^{2-} electrolyte. Eventually both CO_3^{2-} ion and O^{2-} ion concentration gradients which increase the cell polarization and hence the cell output are set up in the electrolyte because the reaction of the fuel at the anode uses up carbonate,

$$CH_4 + 4CO_3^{2-} \rightarrow 2H_2O + 5CO_2 + 8e \qquad (8.23)$$

To maintain the electrolyte invariant, CO_2 must be added to the air or O_2 stream so that now the overall cathode is reaction is

$$O_2 + 2CO_2 + 4e \rightarrow 2CO_3^{2-} \qquad (8.24)$$

instead of Eq. 8.16. Concentration polarization is thus eliminated by replacing the carbonate ions at the cathode. Chambers and Tantrum (47) consider the carbonate cell as a CO_2-O_2 concentration cell, but

Gorin and Recht (103) have criticized this interpretation on the basis that these electrodes operate irreversibly in fuel cells.

When O_2 is adsorbed on the cathode surface, the oxygen, according to Trachtenberg (268), is dissociated to form a surface oxide,

$$O_2 + 2M \rightarrow 2MO \qquad (8.25)$$

followed by the electron transfer steps,

$$MO + 2e \rightarrow M + O^{2-} \qquad (8.26)$$

Finally, the O^{2-} ion interacts with an adsorbed CO_2 molecule to form the carbonate ion,

$$O^{2-} + CO_2 \rightarrow CO_3^{2-} \qquad (8.27)$$

In this way, the transfer of oxygen from the cathode to the fuel anode takes place through the migration of CO_3^{2-} ions.

F. OTHER OXYGEN ELECTRODES

The most active catalysts for low temperature fuel-cell cathodes, particularly in acid solutions, is based on the expensive noble metals. By using very high surface area structures, the catalyst loading may be decreased.   In recent years, there has been a concerted effort made to find other economically cheaper materials to replace the expensive platinum catalysts without a loss of electrode performance.

Jasinski (146) has suggested the use of nickel boride as a catalyst for fuel-cell anodes, but it has been found that the performance of these anodes is no better than a Pt catalyst.   In addition, nickel boride is rather expensive to make.   It has been reported (121) that nickel boride as well as cobalt boride are suitable catalysts for fuel-cell cathodes.

Chelate structures, such as cobalt phthalocyanine (CoPc), have been considered (147) as possible catalysts for the reduction of oxygen.   At first, electrodes were made by spreading a slurry of CoPc on a porous Ni plaque, but impregnation of the pores of the Ni plaque was difficult because of the low solubility of CoPc.   A better method involved the sintering of a paste of CoPc and carbon black in a Teflon emulsion on a Ni screen.   Although the Teflon-bonded CoPc electrode required activation by prepolarizing to H_2 evolution potentials, the results obtained with CoPc were poorer than with Pt or Ag.

According to Giner and Swette (102), titanium nitride is an electrical conductor.   Disks of TiN were made and rotated at 600 rpm.   The catalytic activity for the reduction of O_2 was studied by obtaining potential-sweep traces on the rotated TiN disk in KOH.   Although TiN is more active than Ti alone, it is not as active as a Pt or Ag catalyst. Teflon-bonded boron carbide supported on a sintered Ni plaque was investigated by Meibuhr (190) as an O_2 cathode catalyst, but the results were disappointing, although Sawyer and Seo (237a) reported that boron carbide was a good material from which an electrode for the reduction of O_2 can be made. To date, a suitable substitute for the noble metal catalysts has not been found.

References

1. Acheson, E. G., *J. Franklin Inst.*, **147**, 175 (1899).
2. Adams, R. N., *Anal. Chem.*, **30**, 1576 (1958).
3. Adlhart, O. J., and A. J. Hartner, *Proc. Ann. Power Sources Conf.*, **20**, 11 (1966); **19**, 1 (1965).
4. Alder, B. J., and R. H. Christian, *Phys. Rev. Letters*, **7**, 307 (1961).
5. Alexanian, C., *J. Chim. Phys.* **58**, 133 (1961).
6. Archer, D. H., et al., *Advan. Chem.*, **47**, 332 (1965).
   Archer, D. H., et al., *Proc. Ann. Power Sources Conf.*, **16**, 34 (1962); **18**, 36
7. Archer, D. H., et al., *Proc. Ann. Power Sources Conf.*, **16**, 34 (1962); **18**, 36 (1964).
8. Bacon, F. T., *Fuel Cells*, Vol. 1, G. J. Young, Ed., Reinhold, New York, 1960, p. 51.
9. Bacon, G. E., *Acta Cryst.*, **3**, 320 (1950); **4**, 558 (1951); **5**, 392 (1952).
10. Baker, B. S., *Hydrocarbon Fuel Technology*, Academic Press, New York, 1965.
11. Baker, B. S., et al., in *Advances in Chemistry*, No. 47, Am. Chem. Soc., Washington, D.C., 1965, p. 247.
12. Barber, W. A., and N. T. Woodberry, *Electrochem. Tech.*, **3**, 194 (1965).
13. Bardina, N. G., and L. I. Krishtalik, *Elektrokhim.*, **2**, 216, 334, 357 (1966).
14. Barendrecht, E., *Anal. Chim. Acta*, **24**, 498 (1961).
15. Barrer, R. M., *J. Chem. Soc.*, **138**, 1261 (1936).
16. Baur, E., and H. Preis, *Z. Elektrochem.*, **43**, 727 (1937); **34**, 695 (1938).
17. Baur, E., and J. Tobler, *Z. Elektrochem.*, **39**, 169 (1933).
18. Baur, E., W. D. Treadwell, and G. Trümpler, *Z. Elektrochem.*, **27**, 199 (1921).
19. Beilby, A. L., W. Brooks, and G. L. Lawrence, *Anal. Chem.*, **36**, 22 (1964).
20. Bennion, D. N., and C. W. Tobias, *J. Electrochem. Soc.*, **113**, 589, 593 (1966).
21. Berl, E., *Trans. Elektrochem. Soc.*, **76**, 359 (1939).
22. Berl, W. G., *Trans. Electrochem. Soc.*, **83**, 253 (1943).
23. Bernal, J. D., *Proc. Roy. Soc. (London)*, **A106**, 749 (1924).
24. Bianchi, G., *Corrosion Anti-corrosion*, **5**, 146 (1957).
25. Bianchi, G., *J. Electrochem. Soc.*, **112**, 233 (1965).
26. Biscoe, J., and B. E. Warren, *J. Appl. Phys.*, **13**, 364 (1942).
27. Blackman, L. C. F., *Research*, **13**, 390, 441 (492 (1960).
28. Bockros, J. C., *Carbon*, **3**, 17 (1965).
29. Bonnemay, M., G. Bronoël, and E. Levart, *Compt. Rend.*, **257**, 3394, 3885 (1963); *J. Electrochem. Soc.*, **111**, 265 (1964); *Electrochim. Acta*, **9**, 727 (1964).
30. Bonnemay, M., et al., *Compt. Rend.*, **258**, 4256, 6139 (1964).
31. Borucka, A., and J. N. Agar, *Electrochim. Acta*, **11**, 603 (1966).
32. Bottomley, G. A., and J. G. Blackman, *Nature*, **171**, 620 (1953).

33. Bragg, W. H., and W. L. Bragg, *Proc. Roy. Soc. (London)*, **A89**, 277 (1913).
34. Bratzler, K., *Z. Elektrochem.*, **54**, 81 (1950).
35. Brezina, M., *Nature*, **212**, 283 (1966).
36. Brockway, L. O., and L. Pauling, *Proc. Natl. Acad. Sci. U.S.*, **19**, 860 (1933).
37. Brodie, B. C., *Phil. Trans. Roy. Soc. London*, **149**, 249 (1859); *Ann.*, **114**, 6 (1860).
38. Broers, G. H. J., and J. A. A. Ketelaar, in *Fuel Cells*, Vol. 1, G. J. Young, Ed., Reinhold, New York, 1960, p. 78; *Ind. Eng. Chem.*, **52**, 303 (1960).
39. Broers, G. H. J., and M. Schenke, in *Fuel Cells*, Vol. 2, G. J. Young, Ed., Reinhold, New York, 1963, p. 6.
40. Brown, B. K., *Trans. Electrochem. Soc.*, **53**, 113 (1928).
41. Brown, B. K., and O. W. Storey, *Trans. Electrochem. Soc.*, **53**, 129 (1928).
42. Bull, H. I., M. H. Hall, and W. E. Garner, *J. Chem. Soc.*, **133**, 837 (1931).
43. Bundy, F. P., *J. Chem. Phys.*, **38**, 631 (1963).
44. Burnshtein, R. Kh., et al., *Electrochim. Acta*, **9**, 773 (1964).
45. Buvet, R., M. Guillou, and B. Warszawski, *Electrochim. Acta*, **6**, 113 (1962).
46. Cahill, A. E., and H. Taube, *J. Am. Chem. Soc.*, **74**, 2312 (1952).
47. Chambers, H. H., and A. D. S. Tantrum, in *Fuel Cells*, Vol. 1, G. J. Young, Ed., Reinhold, New York, 1960, p. 94; *Ind. Eng. Chem.*, **52**, 295 (1960).
48. Charkey, A., and G. A. Dalin, *Proc. Ann. Power Sources Conf.*, **20**, 79 (1966).
49. Charpy, G., *Compt. Rend.*, **148**, 920 (1909).
50. Chizmadzhev, Yu. A., *Elektrokhim.*, **2**, 1 (1966).
51. Chodosh, S. M., E. G. Katsoulis, and M. G. Rosansky, *Proc. Ann. Power Sources Conf.*, **20**, 83 (1966).
52. Christopher, H. A., *Proc. Ann. Power Sources Conf.*, **20**, 18 (1966).
53. Clark, G. L., A. C. Eckert, and R. L. Burton, *Ind. Eng. Chem.*, **41**, 201 (1949).
54. Clark, M. B., W. G. Darland, and K. V. Kordesch, *Electrochem. Tech.*, **3**, 166 (1965); *Proc. Ann. Power Sources Conf.*, **18**, 11 (1964).
55. Clarkin, P., *ASTIA Rept.* AD 446 629, General Electric (June 1964).
56. Clauss, A., and U. Hoffmann, *Angew. Chem.*, **68**, 522 (1956).
57. Clauss, A., R. Plass, H. P. Boehm, and U. Hofmann, *Z. Anorg. Allgem. Chem.*, **291**, 205 (1957).
58. Clauss, A., U. Hofmann, and A. Weiss, *Z. Elektrochem.*, **61**, 1284 (1957).
59. Cohn, G., *Proc. Ann. Power Sources Conf.*, **15**, 12 (1961).
60. Coleman, J. J., *J. Electrochem. Soc.*, **90**, 545 (1946); **98**, 26 (1951).
61. Collongues, R., et al., *Bull. Soc. Chim. France*, **1961**, 70.
62. Colman, W. P., et al., *Proc. Ann. Power Sources Conf.*, **19**, 14 (1965).
63. Croft, R. C., *Quart. Rev.*, **14**, 1 (1960).
64. Daniel-Beck, V. S., *Zh. Fiz. Khim.*, **22**, 697 (1948).
64a. Davies, M. O., M. Clark, E. Yeager, and F. Hovokra, *J. Electrochem. Soc.*, **106**, 56 (1969).
65. Davtyan, O. K., *Bull. Acad. Sci. USSR, Classe Sci. Tech.*, **1**, 107 (1946); **2**, 215 (1946).
66. de Bethune, A. J., and N. A. S. Loud, *Standard Aqueous Electrode Potentials and Temperature Coefficients*, Clifford A. Hampel, Skokie, Ill., 1964.
67. DeBoer, J. H., and A. B. C. Van Doorn, *Koninkl Ned. Akad. Wetenschap. Proc.*, **B57**, 181 (1954); **B61**, 12, 17, 160, 242 (1958); **B64**, 34 (1960); **B66**, 165 (1963).
68. De la Cruz, F. A., and J. M. Cowley, *Nature*, **196**, 468 (1962); *Acta Cryst.*, **16**, 531 (1963).

69. De Levie, R., *Electrochim. Acta*, **8**, 751 (1963).
70. Despic, A. R., D. M. Drazic, and I. V. Kadija, *Electrochem. Tech.*, **4**, 451 (1966).
71. Diels, O., *Ber.*, **B71**, 1197 (1938).
72. Diels, O., and P. Blumberg, *Ber.*, **41**, 82 (1908).
73. Diels, O., and B. Wolf, *Ber.*, **39**, 689 (1906).
74. Dirkse, T. P., *J. Electrochem. Soc.*, **101**, 328 (1954); **102**, 497 (1955); **106**, 155 (1959).
75. Dittmann, H. M., E. W. Justi, and A. W. Winsel, in *Fuel Cells*, Vol. 2, G. J. Young, Ed., Reinhold, New York, 1963, p. 133.
76. Dixon, J. M., et al., *J. Electrochem. Soc.*, **110**, 276 (1963).
77. Douglas, D. L., in *Fuel Cells*, Vol. 1, G. J. Young, Ed., Reinhold, New York, 1960, p. 129; *Ind. Eng. Chem.*, **52**, 308 (1960).
78. Dravnieks, A., D. B. Boies, and J. I. Bregman, *Proc. Ann. Power Sources Conf.*, **16**, 4 (1962).
79. Drazic, D. M., A. R. Despic, and G. A. S. Maglic, *Electrochem. Tech.*, **4**, 453 (1966).
80. Duddy, J. C., D. T. Farrell, and P. Rüetschi, *Proc. Ann. Power Sources Conf.*, **16**, 9 (1962).
81. Duwez, P., F. Odell, and F. H. Brown, *J. Am. Ceram. Soc.*, **35**, 107 (1952); *J. Electrochem. Soc.*, **98**, 356 (1951).
82. Ellingham, H. J. T., *J. Chem. Soc.*, **134**, 1565 (1932).
83. Elmore, G. V., and H. A. Tanner, *J. Electrochem. Soc.*, **108**, 669 (1961).
84. Elving, P. J., and A. F. Krivis, *Anal. Chem.*, **30**, 1645, 1648 (1958).
85. Elving, P. J., and D. L. Smith, *Anal. Chem.*, **32**, 1849 (1960).
86. Elving, P. J., and D. L. Smith, in *Microchemical Techniques*, N. D. Cheronis, Ed., Interscience, New York, 1962, p. 829.
87. Emmett, P. H. *Chem. Rev.*, **43**, 69 (1948).
88. Engler, W., and K. W. F. Kohlrausch, *Z. Physik. Chem. (Leipzig)*, **B34**, 214 (1963).
89. Euler, J., *Z. Elektrochem.*, **55**, 316 (1951).
90. Euler, J., *Electrochim. Acta*, **4**, 27 (1961); **7**, 205 (1962).
91. Euler, J., *Electrochim. Acta*, **11**, 1667 (1966).
92. Euler, J., and W. Nonnenmacher, *Electrochim. Acta*, **2**, 268 (1960).
93. Flood, H., T. Forland, and K. Motzfeldt, *Acta Chem. Scand.*, **6**, 257 (1952).
94. Franklin, R. E., *Proc. Roy. Soc.*, **209**, 196 (1950); *Acta Cryst.* **4**, 253 (1951).
95. Frumkin, A. N., *Zh. Fiz. Khim.*, **23**, 1477 (1949).
96. Gadsley, J., C. N. Hinshelwood, and K. W. Sykes, *Proc. Roy. Soc. (London)*, **A187**, 129 (1946).
97. Gaylor, V. F., A. L. Conrad, and J. H. Landerl, *Anal. Chem.*, **29**, 224, 228 (1957).
98. Gaylor, V. F., P. J. Elving, and A. L. Conrad., *Anal. Chem.*, **25**, 1078 (1953).
99. Gebhardt, J. J., and J. M. Berry, *AIAAJ*, **3**, 302 (1965).
100. Geller, R. F., and P. J. Yavorsky, *J. Res. Natl. Bur. Std.*, **35**, 87 (1945).
101. Giauque, W. F., and C. J. Egan, *J. Chem. Phys.*, **5**, 45 (1937).
102. Giner, J., and L. Swette, *Nature*, **211**, 1291 (1961).
103. Gorin, E., and H. L. Recht, in *Fuel Cells*, W. Mitchell, Ed., Academic Press, New York, 1963, p. 193.
104. Gould, R. F., *Advances in Chemistry*, Am. Chem. Soc., Vol. 47, 1965.
105. Grens, E. A., and C. W. Tobias, *Ber. Buns. Ges. Physik. Chem.*, **68**, 236 (1964).

106. Grens, E. A., R. M. Turner, and T. Katan, *Advan. Energy Conversion*, **4**, 109 (1964).

107. Grisdale, R. O., A. C. Pfister, and W. van Roosbroeck, *Bell System Tech. J.*, **30**, 271 (1951).

108. Grove, W. R., *Phil. Mag.*, **14**, 127 (1839); **21**, 417 (1842); *Proc. Roy. Soc. (London)*, **4**, 463 (1843); **5**, 557 (1845).

109. Grubb, W. T., U.S. Pat. 2,913,511 (1959); *Proc. Ann. Power Sources Conf.*, **11**, 5 (1957).

110. Grubb, W. T., and C. J. Michalske, *J. Electrochem. Soc.*, **111**, 477 (1964).

111. Guentent, O. J., *J. Chem. Phys.*, **37**, 884 (1962); *J. Appl. Phys.*, **35**, 1841 (1964).

112. Guentert, O. J., and S. Cvikevich, *Acta Cryst.*, **13**, 1059 (1960).

113. Guillou, M., and R. Buvet, *Electrochim. Acta*, **8**, 489 (1963).

114. Gurevich, I. G., *Inzhener Fiz. Zh. Akad. Nauk Belorous SSR*, **2**, 78 (1959).

115. Gurevich, I. G., and V. S. Bagotsky, *Electrochim. Acta*, **9**, 1151 (1964).

116. Hagelloch, G., and E. Feess, *Chem. Ber.*, **84**, 730 (1951).

117. Haldeman, R. G., et al., *Advan. Chem.*, **47**, 106 (1965).

118. Hall, C. C., *Research*, **9**, 7 (1965).

119. Hammick, D. L., R. C. A. New, N. V. Sidgwick, and L. E. Sutton, *J. Chem. Soc.*, **132**, 1876 (1930).

120. Hantzsch, A., *Z. Anorg. Allgem. Chem.*, **30**, 298 (1902).

121. Hayworth, D. T., U. S. Pat. 3,183,123 (1965).

122. Heath, C. E., *Proc. Ann. Power Sources Conf.*, **18**, 33 (1964).

123. Heise, G. W., and E. A. Schumacher, *Trans. Electrochem. Soc.*, **62**, 383 (1932).

124. Heise, G. W., E. A. Schumacher, and C. R. Fischer, *Trans. Electrochem. Soc.*, **92**, 173 (1947).

125. Hennig, G. R., *Prog. Inorg. Chem.*, **1**, 125 (1959).

126. Herold, A., *Compt. Rend.*, **232**, 1484 (1915).

127. Hersch, P., *Nature*, **169**, 792 (1952).

128. Hickling, A., and W. H. Wilson, *J. Electrochem. Soc.*, **98**, 425 (1951).

129. Hildebrand, J. H., and W. G. Bowers, *J. Am. Chem. Soc.*, **38**, 785 (1916).

130. Hoare, J. P., *J. Electrochem. Soc.*, **112**, 608 (1965).

131. Hoare, J. P., *J. Electrochem. Soc.*, **113**, 846 (1966).

132. Hofmann, U., and A. Frenzel, *Ber.*, **63**, 1248 (1930); *Z. Electrochem.*, **37**, 613 (1931); **40**, 511 (1934).

133. Hofmann, U., and E. König, *Z. Anorg. Allgem. Chem.*, **234**, 311 (1937).

134. Hofmann, U., and W. Rüdorff, *Trans. Faraday Soc.*, **34**, 1017 (1938).

135. Hummers, W. S., and R. E. Offerman, *J. Am. Chem. Soc.*, **80**, 1339 (1958).

136. Hund, F., et al., *Z. Physik. Chem. (Leipzig)*, **199**, 142 (1952); **201**, 268 (1952); *Z. Anorg. Allgem. Chem.*, 265, 67 (1951); *Z. Elektrochem.*, **55**, 363 (1951).

137. Hunt, J. P., and H. Taube, *J. Am. Chem. Soc.*, **74**, 5999 (1952).

138. Hurd, C. D., and F. D. Pilgrim, *J. Chem. Soc.*, **55**, 757 (1933).

139. Iczkowski, R. P., *J. Electrochem. Soc.*, **111**, 605 (1964).

140. Janz, G. J., F. Colom, and F. Saegusa, *J. Electrochem. Soc.*, **107**, 581 (1960).

141. Janz, G. J., A. Conte, and E. Neuenschwander, *Corrosion*, **19**, 292t (1963).

142. Janz, G. J., and M. R. Lorenz, *J. Electrochem. Soc.*, **108**, 1052 (1961).

143. Janz, G. J., and M. R. Lorenz, *J. Chem. Eng. Data*, **6**, 321 (1961).

144. Janz, G. J., and F. Saegusa, *J. Electrochem. Soc.*, **108**, 663 (1961).

145. Janz, G. J., and F. Saegusa, *J. Electrochem. Soc.*, **110**, 452 (1963).

146. Jasinski, R., *Advan. Chem.*, **47**, 95 (1965).

147. Jasinski, R., *Nature*, **201**, 1212 (1964); *J. Electrochem. Soc.*, **112**, 526 (1965).
148. Joachim, E., and W. Vielstich, *Electrochim. Acta*, **2**, 341 (1960); **3**, 244 (1960).
149. Johansen, H. A., and J. G. Cleary, *J. Electrochem. Soc.*, **109**, 1076 (1962); **111**, 100 (1964).
150. Johnson, T. K., *Proc. Ann. Power Sources Conf.*, **18**, **25** (1964).
151. Jones, F., in *An Introduction to Fuel Cells*, K. R. Williams, Ed., Elsevier, New York, 1966, p. 156.
152. Justi, E., M. Pilkuhn, W. Scheibe, and A. Winsel, *Hochbelastbare Wasserstoff-Diffusions-Elektroden für Betrieb bei Umgebungstemperature und Niederdruck*, Franz Steiner, Wiesbaden, 1959, p. 33.
153. Justi, E. W., and A. W. Winsel, *Naturwiss.*, **47**, 298 (1960); *J. Electrochem. Soc.*, **108**, 1073 (1961).
154. Juza, R., and R. Langheim-Heidelberg, *Z. Elektrochem.*, **45**, 689 (1939).
155. Kamüke, O., E. Naruko, and I. Sawada, *J. Electrochem. Soc. Japan*, **22**, 608 (1964).
156. Katan, T., S. Szpak, and E. A. Grens, *J. Electrochem. Soc.*, **112**, 1166 (1965).
157. Kent, C. E., and W. N. Carson, *Proc. Ann. Power Sources Conf.*, **20**, 76 (1966).
158. Ketelaar, J. A. A., and G. H. J. Broers, *Ind. Eng. Chem.*, **52**, 303 (1960).
159. Kingery, W. D., et al., *J. Am. Ceram. Soc.*, **42**, 393 (1959).
160. Kiukkola, K., and C. Wagner, *J. Electrochem. Soc.*, **104**, 379 (1954).
161. Klein, C. A., *Rev. Mod. Phys.*, **34**, 56 (1962).
162. Klein, C. A., D. Straub, and R. J. Diefendorf, *Phys. Rev.*, **125**, 468 (1962).
163. Klemenc, A., and G. Wagner, *Ber.*, **B70**, 1880 (1937); **B71**, 1625 (1938).
164. Klemenc, A., R. Wechsberg, and G. Wagner, *Z. Physik. Chem. (Leipzig)*, **170**, 97 (1934); *Monatsh.*, **66**, 337 (1935).
165. Knobel, M., *Ind. Eng. Chem.*, **17**, 826 (1925).
166. Kohlschütter, V., and P. Haenni, *Z. Anorg. Allgem. Chem.*, **105**, 121 (1919).
167. Kolthoff, I. M., *J. Am. Chem. Soc.*, **54**, 4473 (1932).
168. Kordesch, K., and A. Marko, *Oester. Chem. Ztg.*, **53**, 125 (1951).
169. Kordesch, K., and F. Martinola, *Monatsh.*, **84**, 39 (1953).
170. Kronenberg, M. L., *J. Electrochem. Soc.*, **109**, 753 (1962).
171. Krupp, H., et al., *J. Electrochem. Soc.*, **109**, 553 (1962).
172. Ksenzhek, O. S., *Zh. Fiz. Khim.*, **37**, 2007 (1963); **38**, 1846 (1964).
173. Ksenzhek, O. S., and V. V. Stender, *Dokl. Akad. Nauk SSSR*, **106**, 487 (1956); 107, 280 (1956); *Zh. Prikl. Khim.*, **32**, 110 (1959); *Electrochim. Acta*, **9**, 629 (1964).
174. Laitinen, H. A., and B. B. Bhatia, *J. Electrochem. Soc.*, **107**, 705 (1959).
175. Laitinen, H. A., and C. H. Liu, *J. Am. Chem. Soc.*, **80**, 1015 (1958).
176. Laitinen, H. A., and D. R. Rhodes, *J. Electrochem. Soc.*, **109**, 413 (1962).
177. Lambert, J. D., *Trans. Faraday Soc.*, **32**, 452 (1936); **34**, 1080 (1938).
178. Langmuir, I., *J. Am. Chem. Soc.*, **41**, 1543 (1919).
179. Lipson, H., and A. R. Stokes, *Nature*, **149**, 328 (1942).
180. Lord, R. C., and N. Wright, *J. Chem. Phys.*, **5**, 642 (1937).
181. Lord, S. S. and L. B. Rogers, *Anal. Chem.*, **26**, 284 (1954).
182. Lunt, R. W., and R. Venkateswaran, *J. Chem. Soc.*, **129**, 857 (1927).
183. Maget, H. J. R., and R. Roethlein, *J. Electrochem. Soc.*, **112**, 1034 (1965); **113**, 581 (1966).
184. Maget, H. J. R., and G. Wheeler, *Electrochem. Tech.*, **4**, 412 (1966).
185. Martin, H. A., *Ohio State Med. J.*, **34**, 1251 (1938).

186. Matuyama, E., *J. Phys. Chem.*, **58**, 215 (1954).

187. Mazitov, Yu. A., *Elektrokhim*, **1**, 218, 340 (1965).

187a. Mazitov, Yu. A., N. A. Fedotov, and N. A. Aladzhalova, *Zh. Fiz. Khim.*, **39**, 218 (1965).

188. Mazitov, Yu. A., K. I. Rosental', and V. I. Veselovskii, *Zh. Fiz. Khim.*, **38**, 449, 697 (1964).

189. Meek, J., and B. S. Baker, *Advan. Chem.*, **47**, 221 (1965).

190. Meibuhr, S. G., *Nature*, **210**, 409 (1966).

191. Meyer, G., R. A. Henkes, and A. Slooff, *Rec. Trav. Chim. Pay-Bas*, **54**, 797 (1935).

192. Meyers, C. H., and M. S. Van Dusen, *J. Res. Natl. Bur. Std.*, **10**, 381 (1933).

193. Miller, F. J., *Anal. Chem.*, **35**, 929 (1963).

194. Miller, F. J., and H. E. Zittel, *Anal. Chem.*, **35**, 1866 (1963).

195. Mitchell, W., Ed., *Fuel Cells*, Academic Press, New York, 1963.

196. Monk, R. G., and H. J. T. Ellingham, *J. Chem. Soc.*, **137**, 125 (1935).

197. Morcos, I., and E. Yeager, *Ext. Abst. Electrochem. Soc., Cleveland*, **4**, 13 (May 1966).

198. Morris, J. B., and J. M. Schempf, *Anal. Chem.*, **31**, 286 (1959).

199. Moussa, A. A., and H. K. Embaby, *Egypt. J. Chem.*, **1**, 175 (1958).

200. Mueher, B., *Med. Klinik*, **34**, 1487, 1523 (1938).

201. Mulliken, R. S., *Phys. Rev.*, **40**, 55 (1932); *J. Chem. Phys.*, **3**, 720 (1935); *Rev. Mod. Phys.*, 14, 204 (1942).

202. Náray-Szabó, St. V., *Z. Elektrochem.*, **33**, 15 (1927).

203. Nasarischwily, A., *Z. Elektrochem.*, **29**, 320 (1923).

203a. Nekrasov, L. N., and E. I. Krushcheva, *Elektrokhim.*, **3**, 166 (1967).

204. Nelson, J. B., and D. P. Riley, *Phys. Soc. London*, **57**, 477 (1945).

205. Newman, J. S., and C. W. Tobias, *J. Electrochem. Soc.*, **109**, 1183 (1962).

206. Niedrach, L. W., *Proc. Ann. Power Sources Conf.*, **13**, 120, (1959).

207. Niedrach, L. W., and H. R. Alford, *J. Electrochem. Soc.*, **112**, 117 (1965).

208. Niedrach, L. W., and W. T. Grubb, in *Fuel Cells*, W. Mitchell, Ed., Academic Press, New York, 1963, p. 253.

209. Noddach, W., H. Walch, and W. Dobner, *Z. Physik. Chem. (Leipzig)*, **211**, 180 (1959).

210. Olson, C., and R. N. Adams, *Anal. Chim. Acta*, **22**, 582 (1960).

211. Oswin, H. G., *Plat. Met. Rev.*, **8**, 42 (1964).

212. Ott, E., and K. Schmidt, *Ber.*, **47**, 239 (1914); **55**, 2126 (1922).

213. Pal'guev, S. F., et al., *Zh. Fiz. Khim.*, **34**, 452 (1960); *Trudy Inst. Khim. Akad. Nauk SSSR*, **1958**, 183; *Chem. Abstr.*, **54**, 9542 (1960); *Zh. Neorg. Khim.*, **4**, 2571 (1959); *Chem. Abstr.*, **54**, 15057 (1960); *Tr. Inst. Elektrokhim. Akad. Nauk SSSR*, **1960**, 111 ; *Chem. Abstr.*, **55**, 19556 (1961).

214. Pappis, J., and S. L. Blum, *J. Am. Ceram. Soc.*, **44**, 592 (1961).

215. Patai, S., E. Hoffmann, and L. Rajbenbach, *J. Appl. Chem.*, **2**, 306, 311 (1952).

216. Paxton, R. R., J. F. Demendi, G. J. Young, and R. B. Rozelle, *J. Electrochem. Soc.*, **110**, 933 (1963).

217. Perry, J., *Proc. Ann. Power Sources Conf.*, **16**, 1 (1962).

218. Pinnick, H. T., *J. Phys. Chem.*, **20**, 756 (1952).

219. Plust, H. G., *Brown Boveri Rev.*, **49**, 3 (1962).

220. Posey, F. A., and T., Morozumi, *J. Electrochem. Soc.*, **113**, 176 (1966).

221. Posner, A. M., *Fuel*, **34**, 330 (1955).

222. Primak, W., and L. H. Fuchs, *Phys. Rev.*, **95**, 22 (1954).
223. Reddy, T. B., *Electrochem. Tech.*, **1**, 325 (1963).
224. Reed, M. W., and R. J. Brodd, *Carbon*, **3**, 241 (1965).
225. Reneke, W. E., ASTIA Rept. No. AD 273299 (1961).
226. Rhead, T. F. E., and R. V. Wheeler, *J. Chem. Soc.*, **101**, 831,846 (1912); **103**, 461 (1913).
227. Richmond, A. E., *J. Roy. Army Med. Corps*, **73**, 79, 145 (1939).
228. Rideal, E. K., and U. R. Evans, *Trans. Faraday Soc.*, **17**, 466 (1922).
229. Rockett, J. A., and R. Brown, *J. Electrochem. Soc.*, **113**, 207 (1966).
230. Rogers, M. C. F., and W. M. Crooks, *J. Appl. Chem.*, **16**, 253 (1966).
231. Rüdorff, W., *Z. Physik. Chem. (Leipzig)*, **B45**, 42 (1939).
232. Rüdorff, W., and U. Hofmann, *Z. Anorg. Allgem. Chem.*, **238**, 1 (1938).
233. Ruff, O., *Trans. Faraday Soc.*, **34**, 1022 (1938).
234. Ruff, O., and F. Ebert, *Z. Anorg. Allgem. Chem.*, **180**, 19 (1929).
235. Rusinko, F., R. W. Marek, and W. E. Parker, *Proc. Ann. Power Sources Conf.*, **15**, 9 (1961); **16**, 6 (1962).
236. Sandler, Y. L., *J. Electrochem. Soc.*, **109**, 1115 (1962).
237. Sato, H., and H. Akamatu, *Fuel*, **33**, 195 (1954).
237a. Sawyer, D. T., and E. T. Seo, *J. Electroanal. Chem.*, **3**, 410 (1962).
238. Schaffer, L. H., *J. Electrochem. Soc.*, **113**, 1 (1966).
239. Schilov, N., H. Schatunovkaya, and K. Tschmutov, *Z. Physik. Chem. (Leipzig)*, **148**, 233 (1930); **149**, 211 (1930).
240. Schmid, A., *Helv. Chim. Acta*, **7**, 370 (1924).
241. Schoenfield, F. K., *Ind. Eng. Chem.*, **27**, 571 (1935).
242. Scholder, R., and G. Hendrich, *Z. Anorg. Allgem. Chem.*, **241**, 76 (1939).
243. Schumacher, E. A., and G. W. Heise, *J. Electrochem. Soc.*, **99**, 191C (1952).
244. Schwabe, K., R. Köpsel, K. Wiesener, and E. Winkler, *Electrochim. Acta*, **9**, 413 (1964).
245. Schwabe, K., et al., *Electrochem. Tech.*, **3**, 189 (1965).
246. Seal, M., *Nature*, **182**, 1265 (1958); **185**, 522 (1960).
247. Seeley, S. B., *Kirk-Othmer, Encyclopedia of Chemical Technology*, Vol. 4, 2nd ed., A. Standen et al. Eds., Interscience, New York, 1964, p. 304.
248. Shipps, P. R., *Proc. Ann. Power Sources Conf.*, **20**, 76 (1966).
249. Shropshire, J. A., and B. L. Tarmy, *Advan. Chem.*, **47**, 153 (1965).
250. Shultz, E. B., et al., in *Fuel Cells*, Vol. 2, G. J. Young, Ed., Reinhold, New York, 1963, p. 24.
251. Sidgeick, N. V., *The Chemical Elements and Their Compounds*, Oxford Univ. Press, London, 1950, p. 492.
251a. Silakov, A. V., G. S. Tyurikov, and N. P. Vailistov, *Elektrokhim.*, **2**, 205 (1966).
252. Slabaugh, W. H., and C. V. Hatch, *J. Chem. Eng. Data*, **5**, 453 (1960).
253. Smith, J. G., in *An Introduction to Fuel Cells*, K. R. Williams, Ed., Elsevier, New York, 1966, p. 183.
254. Spiridonov, P. M., *Novosti Techniki*, **7**, 42 (1938); *Chem. Abstr.*, **35**, 3537 (1941).
255. Staudenmaier, L., *Ber.*, **31**, 1481 (1899); **32**, 1394 (1899); **33**, 2824 (1900).
256. Strickland-Constable, R. F., *Trans. Faraday Soc.*, **34**, 1074 (1938).
257. Strickland-Constable, R. F., *Proc. Roy. Soc. (London)*, **A198**, 1 (1947).
258. Syskov, K. I., et al., *Koks i Khim.*, **1965**, 34; *Chem. Abstr.*, **63**, 2446 (1965).
259. Tagirov, R. B., and I. P. Shevchuk, *Dokl. Akad. Nauk SSSR*, **116**, 797 (1957).
260. Tascheck, W. G., *Advan. Chem.*, **47**, 9 (1965).

261. Taylor, A., and D. Laidler, *Nature*, **146**, 130 (1940).
262. Thacker, R., *Electrochem. Tech.*, **3**, 312 (1965).
263. Thiele, H., *Z. Elektrochem.*, **40**, 26 (1934); *Z. Anorg. Allgem. Chem.*, **206**, 407 (1932); **207**, 340 (1932).
264. Thompson, H. W., and N. Healey, *Proc. Roy. Soc. (London)*, 157, 331 (1963).
265. Tobler, J., *Z. Elektrochem.*, **39**, 148 (1933).
266. Tomashov, N. D., *Dokl. Akad. Nauk SSSR*, **52**, 601 (1946).
267. Tomter, S. S., and A. P. Antony, *Chem. Eng. Progr.*, **59**, 47 (1963).
268. Trachtenbert, I., *Advan. Chem.*, **47**, 232 (1965); *J. Electrochem Soc.*, **111**, 110 (1964).
269. Trombe, F., and M. Foëx, *Compt. Rend.*, **236**, 1783 (1953).
270. Ubbelohde, A. R., and F. A. Lewis, *Graphite and Its Crystal Compounds*, Clarenden Press, Oxford, 1960.
271. Urbach, H. B., in *Fuel Cells*, Vol. 2, G. J. Young, Ed., Reinhold, New York, 1963, p. 77.
272. Vereschagin, L. F., et al., *Dokl. Akad. Nauk SSSR*, **162**, 1027 (1965).
273. Vetter, K. J., *Z. Anorg. Allgem. Chem.*, **260**, 242 (1949); *Z. Physik. Chem. (Leipzig)*, **194**, 199 (1950).
274. Vielstich, W., *Z. Physik. Chem. (Frankfurt)*, **15**, 409 (1958).
275. Vol'kenshtein, M. V., *Uspekhi Khim.*, **4**, 610 (1935).
276. Von Doehren, H., and G. Wolf, *Electrochim. Acta*, **11**, 53 (1966).
277. Voorhies, J. D., and S. M. Davis, *Anal. Chem.*, **32**, 1855 (1960).
278. Wagner, C., *Naturwiss.*, **31**, 265 (1943).
279. Walker, P. L., *Am. Scientist*, **50**, 259 (1962).
280. Walker, P. L., and G. Imperial, *Nature*, **180**, 1184 (1957).
281. Ward, G. A., *Talanta*, **10**, 261 (1963).
282. Warner, B. R., *J. Am. Chem. Soc.*, **65**, 1447 (1943).
283. Watson, H. E., G. G. Rao, and K. L. Ramaswamy, *Proc. Roy. Soc. (London)*, **A143**, 558 (1934).
284. Webb, A. N., B. Mather, and R. M. Suggett, *J. Electrochem. Soc.*, **112**, 1061 (1965).
285. Weber, H. C., H. P. Meissner, and D. A. Sama, *J. Electrochem. Soc.*, **109**, 884 (1962).
286. Weissbart, J., and R. Ruka, *J. Electrochem. Soc.*, **109**, 723 (1962).
287. Weissbart, J., and R. Ruka, *Rev. Sci. Instr.*, **32**, 593 (1961).
288. Weissbart, J., and R. Ruka, in *Fuel Cells*, Vol. 2, G. J. Young, Ed., Reinhold, New York, 1963, p. 37.
289. Weisz, R., and S. Jaffe, *Trans. Electrochem. Soc.*, **93**, 128 (1948).
290. Wentorf, R. H., *J. Phys. Chem.*, **69**, 3063 (1965).
291. Werking, L. C., *Trans. Electrochem. Soc.*, **74**, 365 (1938).
292. Will, F. G., *J. Electrochem. Soc.*, **110**, 145, 152 (1963).
293. Williams, K. R., *An Introduction to Fuel Cells*, Elsevier, New York, 1966, p. 98.
294. Williams, K. R., and D. P. Gregory, Brit. Patent 874,283 (1961).
295. Winsel, A., *Z. Elektrochem.*, **66**, 287 (1962); *Advan. Energy Conversion*, **3**, 427 (1963).
296. Witherspoon, R. R., H. Urbach, E. Yeager, and F. Hovorka, Tech. Rept. No. 4, ONR Contract Nonr. 581(00), Western Reserve Univ. (1954).
297. Wynveen, R. A., and T. G. Kirkland, *Proc. Ann. Power Sources Conf.*, **16**, 24 (1962).

298. Yeager, E., *Ing. Vetenskaps Akad. Medd.*, **134**, 89 (1963).
299. Yeager, E., and A. Kozawa, ASTIA Rept. No. AD 429248 (Jan. 1964).
300. Yeager, E., P. Krouse, and K. V. Rao, *Electrochim. Acta*, **9**, 1057 (1964).
301. Young, G. J., *Fuel Cells*, Vol. 1, Reinhold, New York, 1960; Vol. 2, 1963.
302. Zahn, C. T., and J. B. Mills, *Phys. Rev.*, **32**, 497 (1928).

chapter IX

The Role of Oxygen in Corrosion Mechanisms

To complete the discussion of the electrochemistry of oxygen, something should be said about the part played by oxygen in the kinetics of corrosion mechanisms. Since the varied aspects of corrosion have been so clearly discussed in the remarkably comprehensive book by Evans (40), no extended discourse is taken up on the theories and mechanisms of corrosion processes other than that required as background material for an understanding of the role played by oxygen in such processes. This chapter is designed to give the reader only an outline of the many facets associated with the corrosion of metals and alloys.

The dry destruction of metals exposed to a highly oxidizing gas, particularly at elevated temperatures, is not considered because these processes most likely occur by a chemical reaction; the electron exchange in the oxidation–reduction process takes place at the same reaction site. As a result, our attention is directed to electrochemical or wet corrosion in which the destructive attack occurs by an electrochemical or local cell mechanism; oxidation occurs at a site physically remote from the site at which reduction takes place.

When a metal undergoes attack in an ionically conducting medium, certain sites on the electrode surface become centers at which the surface atoms of the metal lattice are oxidized to ions which go into solution, and electrons are donated to the metal. As this process continues, electrons accumulate on the metal surface until a point is reached when the energy required to add an electron to the charged surface is greater than the energy gained from the oxidation of an atom of the metal lattice to a solvated ion in solution. If no other process can be established to act as a sink for electrons, the metal dissolution process comes to a halt. Because the surfaces of metals are heterogeneous in nature, some sites are cathodic to others and can be come centers at which an available reduction half-reaction can occur. Under these conditions electrons flow from the anodic sites to the cathodic sites through the metal, and the circuit is completed by the migration of ions in the ionic conductor (an aqueous electrolyte or a molten salt). In this way, the electrochemical dissolution of the metal is maintained.

In contrast to the usual electrochemical cell where the anodic and the cathodic reactions take place on two different electrodes, the reactions

357

at the local anodes and cathodes in a corrosion cell take place at different sites on the same electrode. Such cells are called local cells, and all wet corrosion processes, in the absence of an applied external emf, proceed by local cell mechanisms. If the ionic conductor has a high conductivity, the rate of the corrosion process will be controlled by the local electrode polarization; if not, the rate may be controlled by the resistance of the electrolyte path. For highly conducting electrolytes, the rate may be controlled by activation or concentration polarization steps at either the anode or the cathode.

Because a metal surface is an equipotential surface and a separation of charge cannot be maintained between two points on it, the local cell anodic and cathodic half-reactions must occur at the same potential known as a mixed potential (150). This requirement demands that the anodic and cathodic half-reactions be polarized from their respective open-circuit equilibrium potentials to the mixed potential which must lie between the two equilibrium potentials as limiting values. As a result, the local cell is set up such that the anodic reaction is polarized toward more noble potentials while the cathodic reaction is polarized toward less noble potentials. As seen in the diagram of Fig. 9.1, at point C, the polarization curve A for the local cathode will cross curve B for the local anode. The potential at this point is the mixed potential E_m and the current is the corrosion current I_c.

For a given local cell current I_c, the local anodic and cathodic partial currents are the same, although the local partial current densities are

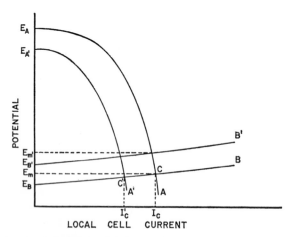

Fig. 9.1. A sketch of the local cell polarization curves for a corrosion process under cathodic control.

usually not identical, since the anodic and cathodic areas are not equal. The shape of the local polarization curves (A and B) depend on the local current densities.   Two limiting situations may be distinguished.

In one case, the cathodic reaction is strongly polarized by the local current, while the anodic reaction is only slightly polarized, so that the mixed potential E_m is very near the equilibrium potential E_B of the anodic reaction.   When the concentration of one of the reactants in the cathodic process is reduced, a polarization curve A', which crosses B at C', is obtained.   The local cell current (rate of corrosion) is strongly affected by changes in the rate-determining factors of the cathodic reaction, and the corrosion process is said to be under cathodic control. When the reactant concentration of the anodic reaction is changed so that curve B' is obtained, very little change is produced in the local cell current.   However, the mixed potential is influenced by the anodic reaction and not by the cathodic reaction.   In summary, then, the potential-determining reaction for the mixed potential is the anodic reaction, whereas the rate-determining process of the corroding system is the cathodic reaction.

The second case, of course, exists when the anodic reaction is polarized to potentials in the vicinity of the equilibrium potential E_A of the cathodic reaction as shown in Fig. 9.2.   Under these conditions, the potential-determining reaction for the mixed potential is the cathodic reaction, and the rate-determining reaction for the corrosion process is the anodic reaction.   Such a process is said to be under anodic control.   Various situations of mixed control occur when both anodic and cathodic reactions are strongly polarized by the local cell current.   The local cell polarization diagram would be similar to Fig. 2.18.

Figures 9.1 and 9.2 represent corrosion systems in which the electrolyte is highly conducting.   If the electrolyte path is very long or if the electrolyte conductivity is very low, an IR drop appears along the ionically conducting path.   In such a case, the corrosion current is lowered from the value of I_c in Figs. 9.1 and 9.2 as demonstrated in Fig. 9.3.   As the resistance of the path increases, the local cell current decreases, and the corrosion process is now under resistance control.

The corrosion process may be halted by causing the IR drop to increase to a point where zero local cell current is reached.   This may be accomplished by keeping the metal surface dry so that the ionic-conducting path is broken.   Similarly, when metal ions are prevented from entering solution by covering the surface of the metal with a protective film of oxide (produced chemically or electrochemically), paint, or resin, the IR is large enough to reduce the corrosion current to zero.

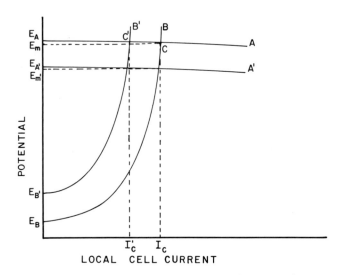

Fig. 9.2. A sketch of the local cell polarization curves for a corrosion process under anodic control.

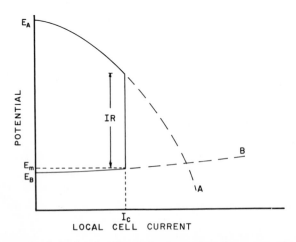

Fig. 9.3. A sketch of the local cell polarization curves for a corrosion process in which a large IR drop occurs along the electrical path through the electrolyte.

Another method of preventing the corrosion process from taking place is known as cathodic protection. If the potential of the metal is depressed to potentials less noble than the equilibrium potential of the anodic reaction E_B in Fig. 9.3 by means of an externally applied potential source with a suitable counterelectrode, the anodic oxidation process of metal dissolution cannot occur. Mears and Brown (106) and Hoar (78) have discussed the theory and the problems associated with cathodic protection.

Finally, a metal surface may be protected by using a sacrificial anode. When a metal with a dissolution potential occurring at a potential less noble than that (E_B in Fig. 9.3) of the metal to be protected is placed in electrical contact with the metal to be protected, the anodic reaction will be the dissolution of the more active metal. Since the protected metal sites rest at a more noble potential than those of the more active metal, the cathodic reduction of oxygen or evolution of hydrogen is established on these protected sites. Metals such as Zn, Mg, and alloys of these with Al have been used to protect steel and cast iron pipes and tanks (113,114), steel gas mains (141), and the steel hulls of ships (4,5,66).

A. DIFFERENTIAL AERATION PROCESSES

When a base metal such as Zn or Fe is placed in a corrosive medium in which the products of the metal dissolution process are soluble (acid solutions), a generalized attack of the metal takes place with the liberation of H_2 until the metal has dissolved. In such cases, the reaction proceeds by a local cell mechanism with the metal dissolving at the local anodes and hydrogen evolving at local cathodes. As the pH is increased toward neutral solutions and further toward alkaline solutions, this mechanism is greatly retarded because of the low activity of H^+ ion in such solutions. In addition, stable oxide films which impede the corrosion process are formed on the metal surface in high pH solutions.

In neutral salt solutions in the presence of oxygen or air, another local cell mechanism can take place with the reduction of O_2 occurring at the cathodic sites. Under these conditions, the process is usually controlled by the diffusion of O_2, as diagrammed in Fig. 9.1 for a process under cathodic control. Those sites easily accessible to O_2 will act as cathodes, and the O_2-starved sites will act as anodes where the metal will be attacked. Corrosion processes set up by O_2 concentration gradients have been designated as differential aeration processes by Evans (41,43).

Lynes (97) views corrosion processes in the light of O_2 concentration cells and has presented a review of the early literature concerned with this subject.   Since most systems corroding in solutions in contact with air are under cathodic control, it is useful to investigate the factors which affect the dissolution of O_2 in water.   This has been done by Downing and co-workers (35,36) and by Gameson and Robertson (64). It was found that the rate of solution of O_2 in distilled water and sea water is increased with increasing agitation of the water, and between 0 and 35°C the rate increases with increasing temperature.   In general, surface contamination decreases the rate of solution of O_2.   For example, in studies with oil films (35), not much effect is noticed until the oil film becomes thicker than 10^{-4} cm, after which the rate of solution of O_2 decreases with increasing film thickness.   The solubility of O_2 in water solutions increases with partial pressure of O_2 above the solution (95) and decreases with increasing concentration of dissolved salt (98).

1. Partially Immersed Metal Plates

When base metals such as Fe, Zn, Cu, or Al are exposed to oxygen or air, a thin film of oxide quickly forms on the metal surface (50,147).   At ambient temperatures, the diffusion of O_2 to the metal or of metal ions to the O_2 through the thin oxide film is very slow, and the oxidation process comes to a halt after a thin, invisible protective film of oxide is formed. Such films become visible and may be studied when transferred to a glass or plastic substrate (57).   This is done by cementing the surface of the oxidized metal sample to a glass plate and by removing the metal with anodic dissolution in a suitable electrolyte.

Borgmann and Evans (19) and Bianchi (14) have investigated the distribution of corrosion on a Zn plate partially immersed in a neutral electrolyte such as an aqueous solution of KCl, in which the anodic products are soluble.   At first, the areas of corrosion attack appear at random over the entire submerged surface (43) since corrosion is in-itiated at weak spots in the film (21,45).   These weak spots are associ-ated with sites on the metal surface where the film has covered impurity sites or lattice defect sites on the surface.   After a period of time, a steady-state distribution of corrosion sets in.   In this situation, the bottom part and the sides of the Zn plate are corroded, whereas a pro-tected, unattacked band exists just below the water line.   At the intersection of the corroding and protected regions, a membraneous layer of zinc oxides and hydroxides (59) extends into the solution.

An explanation for these observations follows.   When the zinc plate is first placed in the KCl solution, Zn goes into solution as a soluble salt at the anodic weak spots on the metal surface, and the reduction of dissolved oxygen occurs on the oxide film covering the protected sites. Soon the dissolved O_2 is consumed by the cathodic reaction, and only those sites which have access to O_2 dissolved by contact of the solution with the atmosphere are protected. Consequently, a band of sites exists just below the water line, where the solution is relatively rich in dissolved O_2 and behaves as the cathode. At sites further removed from the water line, metal dissolution takes place, since the oxide film is not stable in KCl at these O_2-free sites.

At the cathodic sites, the O_2 is reduced by the usual two-step process (82) with peroxide being formed as a stable intermediate. Churchill (27) and Bianchi (13) have detected the presence of H_2O_2 during the corrosion of a number of metals in contact with O_2 or air. When air was excluded from the system so that the cathodic evolution of H_2 was the cathodic process, no H_2O_2 was detected. Some of the peroxide formed is decomposed in the neutral salt solution to hydroxyl ions and O_2. As a result, the pH increases in the vicinity of the local cathodes. At places where the metal ions in solution meet the hydroxyl ions produced from the reduction of O_2, a loose precipitate of zinc hydroxides is formed. Because of its membraneous nature, this precipitate offers little protection to the metal sites beneath it. Corrosion continues unchecked in such a solution on the lower regions of the Zn plate remote from the water line and O_2 replenishment.

Similar experiments were carried out by Evans and co-workers (44,45, 52,54,55) on partially immersed foils of iron and steel. The situation is complicated compared to that of Zn because iron can exist in two valence states. In acid solutions free of oxidizing agents, Fe dissolves as Fe^{2+} ions (153), but in neutral salt solutions containing dissolved oxygen, a compound hydroxide or basic salt containing both Fe^{2+} and Fe^{3+} ions is formed (60). Those salts containing less than about 50% Fe^{3+} are green, and those above are brown. Feitknecht and Keller (60) have discussed the nature and properties of iron oxides and hydroxides. The same corrosion pattern is obtained on partially immersed iron or steel plates in KCl solutions with corrosion at the lower regions and sides of the plate, protected regions just below the water line, and a layer of porous iron hydroxides separating the corroded and protected regions where hydroxyl ions meet the metal ions in solutions.

By using O_2–N_2 mixtures, it was found (17,52,54,55) that the amount of corrosion and the area of the protected region decreased with P_{O_2}. When the O_2 atmosphere above the salt solution was replaced with N_2 or H_2, the region of attack extended to the water line, and there was general corrosion over the entire surface of the metal. In this case, the cathodic reaction is the evolution of H_2, which occurs very slowly so that the amount of corrosion is very small compared with that occurring when O_2 is present.

Evans and Hoar (54) studied the rate of corrosion of partially immersed iron plates in KCl solutions as a function of the concentration of KCl from 0.001 to $3M$. The corrosion rate increases with increasing concentration until a maximum rate in the neighborhood of about $0.5M$ KCl, after which the rate falls off again. The rate decreases with concentration at high KCl concentrations because the solubility of O_2 at these salt concentrations decreases with increasing concentrations of KCl. Since the local cell current for highly conducting solutions is limited by the diffusion of oxygen to the cathodic sites (Fig. 9.1), the rate of corrosion is expected to drop as the concentration of KCl is increased. For low KCl concentrations, the resistance of the electrolyte path becomes important, and the corrosion mechanism is resistance limited. As the KCl solution becomes more dilute, the IR loss becomes greater, and the local cell current becomes less (Fig. 9.3).

The existence of local cell currents has been demonstrated by Evans and Hoar (54). The corroding specimen was cut in two at the boundary between the corroding and protected regions of a partially immersed iron plate. The sections were clamped in position and connected electrically through a milliammeter. In this way the local cell current was measured. The amount of corrosion which took place during the experiment was determined by the loss in weight experienced by the lower segment of the Zn plate. They found that the number of coulombs passed in a given zone were accounted for (by nearly 100%) by the loss in weight of the Zn anode.

By using a calomel reference electrode and Luggin capillary, the local cell polarization curves for the cathodic and anodic segments were recorded when the system was driven by an external source of potential. The experimental setup is shown in Fig. 9.4, and the results in Fig. 9.5. As expected, the O_2 reduction process is polarized to the potential of the anodic dissolution process (rate of corrosion is under cathodic control). Such investigations as those carried out by Evans and Hoar provide experimental evidence for the differential aeration theory. Evans (ref. 40, p. 865) offers an explanation for the straight line polarization curves obtained experimentally for the cathodic reaction in Fig. 9.4 instead of the ideal logarithmic curve in Fig. 9.1 in terms of the wandering of the anodic and cathodic sites over the metal surface during the corrosion process.

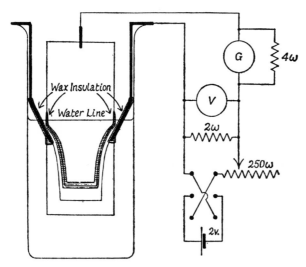

Fig. 9.4. Experimental arrangement for determining local cell current and polarization curves (54).   (By permission of The Royal Society.)

Additional evidence for this theory is obtained from agitation experiments.   As mentioned before, the rate of solution of O_2 is increased by stirring the solution (35).   Schaschl and Marsh (136) studied the corrosion of steel in buffered 3% NaCl solutions with and without agitation of the electrolyte.   The corrosion rate increased as the pH was lowered and as the rate of stirring was increased.   By increasing with stirring the supply of oxygen to the cathodic sites, the local cell current (corrosion rate) was increased since the process was under cathodic control.   For low O_2 partial pressures (below 2 ppm), stirring had little or no effect, showing that the principal role played by solution agitation is the increased transport of O_2 to the cathodic sites.

Evans (42) constructed a device for whirling a sample of metal in the corroding medium.   In KCl solutions, the amount of corrosion was increased with respect to the stationary experiment, but the attack was spread over the entire surface of the sample.   This indicates that a uniform distribution of oxygen over the faces of the metal was obtained so that the anodic and cathodic areas were small.   It is interesting that oxygen acts both as an inhibitor of corrosion by promoting the formation of a protective film and as an accelerator of corrosion by increasing the cathodic current of a process under cathodic control.   As noted by Bianchi (14), the attack of Zn in stirred solutions is slight, and a uniform film of corrosion product is formed which slows down the rate of corrosion.

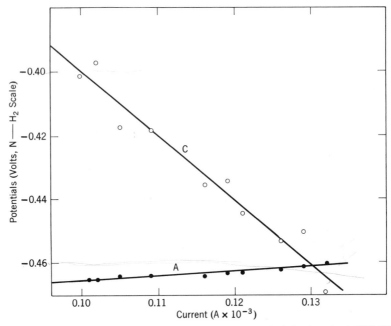

Fig. 9.5. Local anodic and cathodic polarization curves (54) for iron in 0.05N KCl. (By permission of The Royal Society.)

2. Totally Immersed Metal Plates

In a series of investigations reported by Bengough and co-workers (9–12), machined disks of Zn or Fe completely immersed in the corrosive medium were corroded in salt solutions and sea water. The metal specimens were supported horizontally in the corrosion cell, and the progress of the reaction was followed by the consumption of O_2, which was contained in a gas space above the liquid by measurement with a gas buret. They found that the corrosion rate was controlled by the diffusion of O_2 to the metal surface. When the metal was close to the surface of the liquid, the corrosion rate decreased as the metal was moved to greater depths because the diffusion path over which the dissolved oxygen must be transported to the metal surface was increased. For a totally immersed specimen, the total amount of corrosion for a given period of time increased with the area of the liquid surface exposed to the atmosphere. In unit time, more O_2 may be dissolved by the same amount of solution in a wide vessel than in a narrow vessel because of the larger liquid–gas interface, and hence, a greater local cell current can be supported by the cathodically controlled process.

When pure O_2 was used instead of air, the rate of corrosion was increased.   The corrosion rate was increased still further if the pressure of the oxygen above the submerged metal sample was increased above atmospheric pressure.   Any process which increases the rate of solution of O_2 in a given electrolyte accelerates the corrosive attack of the corroding metal.

For a totally submerged specimen, no pattern of corrosion is observed, as in the case of the partially immersed specimen.   Instead, uniform corrosion takes place over the entire surface of the metal sample. The existence of local anodes and cathodes under these circumstance is not as clear-cut as in the partially submerged metal plate.   As explained by Evans (43), the corrosion process begins at points on the metal surface where the material is less tightly bound to the bulk metal because of lattice dislocations and defects, corners, edges, or mechanical damage to the crystal lattice.   When at a given site this material, which is in a state of higher energy than that at surrounding sites, is entirely dissolved, the anodic attack shifts to another more active site, leaving the old site cathodic.   In this way, the anodic attack wanders from one active site to another during the metal dissolution process until finally all sites on the metal surface have had an anodic character at one time or another.   Such a mechanism would produce general corrosion of the entire metal surface.   According to Evans (43), Noordhof studied the local potential behavior of a completely submerged metal surface with two Ag/AgCl reference electrodes fitted with Luggin capillaries. In the early stages of corrosion, the visibly corroding sites were identical to those with an anodic potential.   He was able to determine the local cell currents which agreed reasonably well with the corrosion directly determined from weight-loss measurements.

The rate of corrosion of a totally submerged specimen is lower than that of a partially submerged one (43).   When the oxide precipitates at a point remote from the corroding anodic region, as in the partially immersed metal plate, very little protection is obtained from the fluffy membranous hydroxide layer.   However, when the anodic and cathodic areas are small and close together, as in the cases where the solution was stirred (42,136), an adherent, less porous film is formed directly at the point where the metal ions enter the solution.   As a result, the attack of the metal may be less, although the total amount of corrosion may be greater with stirring because the corrosion is spread over the entire surface.

If an electrolyte, such as an alkaline solution, is used in which a protective oxide film may be maintained, the rate of corrosion is limited by the rate at which the metal ions can diffuse through the protective film. Under these conditions, the process comes under anodic control (14), and the rate of corrosion becomes independent of the diffusion of O_2. Because of the reduction of O_2, the pH of the solution can rise, and if a point is reached where films form on the anodic sites, the corrosion rate may be lessened. Should the pH rise to a sufficiently high value, the corrosion of Zn can increase again because the formation of zincate ion can set in. In buffered solutions, however, the pH cannot build up during the course of the corrosion process, and a constant rate of corrosion can be obtained.

Hoar (79) has discussed the breakdown of protective films on iron. He considers the film to contain pores, and a local cell is set up between the dissolution of the metal at the bottom of the pore and the reduction of O_2 at the oxide covered sites. Because the precipitation of metal oxides or hydroxides in the pore consumes OH^- ions, the pH of solution in the pore decreases. Unless the OH^- ions can be replaced rapidly enough by diffusion from the cathodic sites, the acidity in the pore will increase to the point where the anodic products are soluble. In such cases, localized corrosion at the bottom of the pore may proceed unchecked, producing a pit, or the film over neighboring sites may be undermined leading to a breakdown of the protective film and general corrosion spreading over the metal surface. Where a good supply of OH^- ions or other film-producing ions, such as CO_3^{2-} ions, is available, the pore can be plugged with the prevention of further corrosion at that pore site. The behavior of porous oxide films has been explained by Hoar from the viewpoint of the formation and healing of pores in the protective film for the corrosion of iron (79), the pitting of tin (77), and the rest potential of the O_2 electrode on noble metals (76).

3. Corrosion Under Drops

When a drop of KCl solution is placed on a horizontally held Zn plate, a definite pattern of corrosion is set up (41,42,46). Corrosion of the metal takes place in the center of the drop around which lies an unattacked zone along the periphery of the drop. As in the case of the partially immersed Zn, those regions where the sites have easy access to oxygen become cathodic and are protected. At sites where the diffusion path for O_2 is long, anodic regions appear, and metal dissolution sets in. A ring of zinc hydroxide builds up between the protected and attacked zones. With decreasing salt concentrations, the protected band becomes narrower until in distilled water only a very narrow region along the edge of the drop is free from attack. The ring of zinc hydroxide is also very close to the edge of the drop.   In the case of distilled water, the system is under resistance control, and only a few sites along the periphery of the drop are protected.

Evans (46) reported that not all drops of electrolyte placed on an iron surface produced corrosion.   A possible explanation was offered in which those drops which covered a weak spot in the oxide film produced corrosion.   On the other hand, well-protected sites covered by the drop remained unattacked.

A striking demonstration of the existence of local cell currents in drops of electrolyte on a metal surface was reported by Blaha (16). He placed a drop of KCl solution on the face of one of the pole pieces of an electromagnet.   After the corrosion process had set in, the field was turned on, and the drop was observed to rotate.   If the polarity of the field was reversed by reversing the current through the coil, the drop rotated in the opposite direction.   The rotation became very slow when air was replaced with nitrogen.

4. Pitting

As mentioned by Evans (ref. 40, p. 99), it is reasonable to expect that the corrosion of metals in distilled water should be much less than in salt solutions.   Even if corrosion were present, a protective film should be formed by the anodic products when the water has become saturated with the metal oxide or hydroxide, and the corrosion process should be quickly halted.   In highly dilute solutions, such as the saturated solutions of the oxides or hydroxides of Fe, Zn, Al, etc., the problems of film nucleation play an important role.   For such a situation, the film forms slowly, and extended periods of time are required to spread over the entire surface of the metal.   As most of the surface becomes protected by the spreading film, the probability of the initiation of corrosion at a point on the unprotected area increases.   If corrosion begins at some unprotected site, the attack can become extremely intense because the large protected cathodic region can support a relatively large current which is concentrated at the small corroding anode.   Although the total amount of corrosion is small, it is concentrated at one small point, and serious damage to the specimen may take place very rapidly.   In a short time, thin sheets of metal may be perforated at the site of a pit.

When a Zn plate is placed in distilled water that is agitated (42), the corrosion is small, and after a protective film is formed, attack of the metal ceases.   In stagnant water, corrosion remains localized, and deep pits are produced (53).   A ring of solid corrosion products was built up around the pit, and the purer the distilled water used the more localized was the corrosion in the pit. When a Zn plate was immersed vertically in distilled water, Bengough and Hudson (8) observed that the pits were arranged along vertical lines. This phenomenon was also observed by Evans and Davies (53) who found that the pattern of pits was independent of the purity of the Zn, and of the direction of rolling or abrasion of the surface. Evans and Davies attributed the vertical line arrangement of pits to oxygen-screening of sites below an active pit by particles of corrosion product which happened to lodge on the site after sinking by the force of gravity from the edge of the pit. If a site which is not sufficiently active to produce a center of corrosion when exposed to the normal supply of oxygen is screened by a particle of corrosion product, it will become an active anode as the oxygen supply is reduced below normal by the shielding effect of the solid particle covering the site. In this way, a new pit may form below the first, and an extension of this process can produce vertical lines of pits.

To find experimental evidence to support these conclusions, Evans and Davies observed the corrosion patterns on a Zn plate inclined at an angle from the vertical in distilled water. On the front side, the usual arrangements of pits in vertical lines was observed; but on the back, such an arrangement was not found. This is to be expected because the particles of corrosion product fall directly to the bottom of the container on the back side without drifting over the metal surface as on the front.

When threads of polyethylene, linen, or glass were stretched around a sample of metal immersed in distilled water, two lines of pits or trenches were formed on either side of the region where the thread was in physical contact with the metal (53). At these points, the supply of oxygen to the sites was reduced by the screening effect of the thread, and localized, intense corrosion produced the pitting pattern.

When samples of Zn were whirled in distilled water, no pitting was observed (53), and the concentration of Zn^{2+} ion in solution corresponded to a saturated solution of zinc oxide or hydroxide (59). In this case, the same amount of corrosion, which if concentrated at a pit would cause severe damage, was spread over the entire surface, producing negligible attack of the sample. In distilled water, any regions of the metal surface which are shielded from the normal diffusion of O_2 to the surface (regions of lower O_2 concentration), such as the area around the head of a loose rivet, are a likely point of attack. Such types of corrosion are known as crevice corrosion (ref. 40, p. 208) and can be explained in terms of differential aeration (107).

The extremely destructive nature of pitting is traced to its auto-catalytic nature (1–3,22,38).   After a pit starts at a weak point on the metal surface, a relatively large number of metal ions enter the solution at this point because of the high anodic current density.   With the formation of oxides or hydroxides, the pH in the pit decreases, and since the cathodic reaction is spread over such a large area, hydroxyl ions cannot diffuse to the vicinity of the pit fast enough to prevent the continued decrease in pH.   As the solution in the pit becomes more acid, the anodic product becomes more soluble, and the metal dissolution at the bottom of the pit is accelerated which in turn lowers the pH further.   Within a relatively short time, a thin metal sheet is perforated.

Aziz (2) investigated the patterns of corrosion in the pitting of Al in distilled water with the radioactive tracers ^{60}Co and ^{210}Pb.   Pure Al samples were corroded in 150 cc of water until a well-defined pattern of corrosion was obtained.   In many cases, pitting had to be initiated artifically by damaging the oxide film with a small scratch.   Both natural and artificial initiation of pitting gave the same overall results. After the required time had elapsed, the plate was removed horizontally, and a solution of the chloride of the tracer was added to the film of water lying on the metal surface.   Since Co or Pb is more electropositive than Al, the tracer metal is plated out on the local cathodes.   After ten minutes the plate was washed and dried, and an autoradiograph of the surface taken.   He found no correlation between the distribution of the local cathodes and any observable structural property of the underlying metal, and he concluded that the pit formed by the breakdown of the film.   Film breakdown occurs at weak points, such as places where inclusions of impurities (53) exist.

5. Mutual Protection

Whenever intense corrosion occurs at a given site on the metal sample corroding in an aerated liquid medium, those sites in the neighborhood become cathodic, and hence, are protected.   This phenomenon is referred to as mutual protection by Evans (ref. 40, p. 113).   In strong solutions of KCl ($0.5M$), the intense corrosion at the edges of a steel plate protected the entire face of the plate.   To demonstrate the truth of this conclusion, Britton and Evans (21) covered the edges of the steel plate with baked varnish.   In this case, attack of the metal occurred at points on the face of the steel plate, producing arch-shaped areas of corrosion.

As the conductivity of the electrolyte is lowered, the distance over which protection extends becomes less and less.   In $0.001M$ KCl, the entire face is no longer protected, and in the middle, arch-shaped areas of corrosion appear. Finally, the protected area is reduced to a very narrow band next to the corroding metal edge in distilled water. On the face of the plate, each point of attack protects the immediate vicinity about it, and the pattern of corrosion on the metal face appears as a random distribution of bright circles of protected sites, each containing a dark center of corrosion.

If a point of intense corrosion develops for any given reason on a metal surface, a circular region about the center of corrosion will be protected from attack. However, a concentric band of sites around the periphery of the protected area in turn becomes subject to attack, which in turn produces a concentric, protected band of sites further out. In this way, alternate concentric bands of corrosion and bright metal are formed. Chaudron (25) gives an excellent example of concentric rings of corrosion on Al, and Bengough and Hudson (8) show circular patterns of corrosion on Cu. Such patterns have also been observed on Fe and Zn by Evans (ref. 40, p. 115).

6. Stress Corrosion

Metal sites under tensile stress are anodic to areas free of stress. Such anodic areas are susceptible to attack, and this type of attack is known as stress corrosion (ref. 40, p. 108). When a metal foil is bent, centers of corrosion are initiated along the bending axis (44) either on the concave or convex side. As reported by Simnad and Evans (138), cold-worked iron or steel is anodic to the annealed material. Scratching the surface of a metal sample not only removes most of the film, but also compresses some metal in the bottom of the groove and piles some metal on the sides (140). Brown and Mears (22) found that the sites along the scratch line on Al were anodic to the rest of the metal, and the current flowing was equal to the corrosion rate. Evans (ref. 40, p. 112) shows some of the patterns observed along the scratch line on a metal surface.

When the corrosive liquid attacks the intergranular regions of a metal more easily than the grains themselves, deep trenches between the relatively untouched grains are formed with very destructive results (26,93,131). If the process is under cathodic control, the dissolution of metal at the bottom of the trench may occur at a high rate because of the autocatalytic nature of the process (similar to a pit). Lascombe and Yannaquis (93) suggested that intergranular corrosion was produced by the relative orientation of adjoining grains. The region between the grains attracts impurities, and these may be a source of intergranular corrosion (43).

When the intergranular corrosion is spread over all the grain boundaries at the metal surface, the rate of destruction is low, but when concentrated on a small number of regions, the penetration and weakening of the metal may occur at a rapid pace.   Under stress, a metal exposed to a corrosive liquid may be attacked at a limited number of grain boundaries which run approximately perpendicular to the direction of the applied stress.   The combination of stress and corrosion can cause the rapid failure of a material which in the absence of corrosion would not fail by the application of the given stress alone (68,118).   In the absence of stress, the intergranular attack of the material would also not be initiated under ordinary conditions.

As explained by Evans (43), the disastrous effect of stress–corrosion cracking may be visualized in terms of mechanical and chemical barriers to the propagation of a crack through a given metal under stress and in contact with a corrosive liquid.   Consider a crack initiated by local cell attack of the intergranular region.   As the apex of the crack proceeds, it may reach a new grain which is not susceptible to attack, and the corrosion process halts.   However, the material has been weakened by this intergranular crack and the crack is further propagated mechanically until a region of strong chemical bonding is reached.   At this point, a local cell may be set up, and electrochemical corrosion can set in to weaken the material further.   Thus, the combination of stress and corrosion can propagate the crack indefinitely unless a structure is encountered which is both a mechanical and a chemical barrier combined.

When a specimen is broken by stress–corrosion cracking, the life is nearly the same whether the stress is applied at first or later.   In either case, the initial stage is general intergranular attack.   During later stages, corrosion is an important step in furthering the progress of a crack, because if a strong cathodic current is applied to the sample, cracking can be arrested (43,56).   It is concluded that the strong cathodic current causes the pH to rise at the bottom of the crack so that a protective film is formed which prevents further propagation of the crack.   Gilbert and Hadden (68) observed that the cracking of Al-Mg alloys could be stopped by excluding all contact of the sample with oxygen.   When O_2 was readmitted, the cracking process resumed. Apparently, those conditions which promote pitting also promote stress-corrosion cracking.   Stress–corrosion cracking occurs on light metal alloys only after certain heat treatments (43,68,118).

Although stress–corrosion cracking under a steady stress occurs only on certain metals and alloys under certain conditions, corrosion fatigue produced by an alternating stress can cause failure of nearly any material subjected to a corrosive medium.   In the absence of a corrosive liquid, a given material has a definite fatigue limit, and as long as the stress is below this value, failure will not occur.   This fatigue limit disappears in the presence of the corroding medium.   Under a steady stress, stress corrosion cracking only occurs if intergranular paths of susceptible material exist in certain materials after certain treatments.   With an alternating stress, susceptible material can be produced in almost any material.   Consequently, fatigue corrosion is the more generally occurring process.

According to Edeleanu and Evans (38), corrosion fatigue follows the direction along which gliding is expected in the stressed metal and is due to the attack of the disarrayed material formed during gliding along the slip bands.   As found by Simnad and Evans (137), this disarrayed material is most susceptible to corrosion while it is being formed.   If the system is allowed to settle down after a period of stress before contact with the corrosive medium is made, the intergranular material is much less sensitive to corrosion.   Whitwham and Evans (152) studied the effect of a period of dry fatigue on the life of a sample of iron or steel subjected to corrosion fatigue afterward.   They report that the life of a sample which has failed by corrosion fatigue is virtually unaffected whether or not it has been subjected to an alternating stress before contact with the corrosive medium.

7. Corrosion Inhibitor Addition Agents

The corrosion process may be retarded, if not halted completely, by adding to the corrosive medium a chemical agent which can interfere with the proper operation of the local cell.   Such a soluble inhibitor may interfere with the local cell process by increasing the ohmic resistance (formation of a thin, complete, impervious oxide film), by hindering the cathodic process (increase of the hydrogen overvoltage in acid media) by impeding the anodic process (increase of activation polarization for metal dissolution), or by interfering with both the anodic and cathodic processes (70,71).   In neutral and alkaline solutions, the passivity of a metal surface in a normally corrosive medium is due to the presence of a thin oxide layer on its surface (65).

In alkaline solutions, such as solutions of NaOH or KOH, oxygen must be present to prevent the corrosion of iron (103,104,127).   When a freshly abraded iron plate is placed in a solution of NaOH, iron dissolves at random anodic sites producing $Fe(OH)_2$ which is immediately oxidized to cubic iron oxide. Cubic iron oxide is anhydrous, i.e., γ-Fe_2O_3 or Fe_3O_4, or a solid solution of the two. As pointed out by Iitaka (85), the electron diffraction patterns for Fe_3O_4 and γ-Fe_2O_3 are very similar, but the protective film of iron oxide is generally believed to be γ-Fe_2O_3. If the iron plate is exposed to air or oxygen before immersion in a NaOH solution, the same thin film of γ-Fe_2O_3 is formed. Initially, this thin film is not completely protective, and Fe^{2+} ions leak through weak spots and film discontinuities. The emerging Fe^{2+} ions react with the electrolyte to form hydrated iron oxide, γ-$FeOOH$ (or γ-$Fe_2O_3 \cdot H_2O$), which plugs up the gaps or thickens the weak spots until complete protection results. Pryor and Cohen (127) estimate the thickness of the protective film to be about 200 A. From electron diffraction studies (103,104), both γ-Fe_2O_3 and γ-$FeOOH$ are detected in the protective film.

Disodium and trisodium phosphate are used as corrosion inhibitors in neutral and alkaline electrolytes (102,115–117,125–128), but O_2 must be present to prevent metal dissolution. When freshly abraded iron is placed in $0.1M$ Na_2HPO_4 solution saturated with air, a film of γ-Fe_2O_3 is formed by the oxidation of dissolved O_2. In the early stages of film formation, the film is not complete, and the process is accompanied by slow electrochemical attack. The emerging Fe^{2+} ions react with the electrolyte to form relatively thick plugs of $FePO_4 \cdot 2H_2O$, and after about 24 hr (126), the film is thick enough to prevent the outward diffusion of Fe^{2+} ions. At this point, corrosion ceases. By using radiophosphorus, ^{32}P, to tag the phosphate, it was concluded (128) that $FePO_4 \cdot 2H_2O$ was present in the film and was formed at discontinuities in the film.

In the absence of O_2, corrosion takes place on iron samples in $0.1M$ Na_2HPO_4, and the corrosion product consists of an adherent and a loose deposit. Analysis with x-rays (126) shows that the adherent material is $Fe_3(PO_4)_2 \cdot 8H_2O$ and the loose material is $FePO_4 \cdot nH_2O$. The ferrous salt forms as a membraneous layer and offers little protection against metal dissolution. A crystalline deposit, such as the ferric salt, is needed for protection (115).

If an insufficient amount of inhibitor is present to provide complete protection, localized attack which can be very intense may be concentrated at a small area (48). In the absence of the inhibitor, such an attack is spread over the surface with relatively little damage. Consequently, too little inhibitor is worse than none at all because the large portion of the surface covered acts as a large cathode so that the large corrosion current is confined to a small anode producing the intense attack.   Palmer (115) found that the pits obtained with insufficient phosphate inhibitor were covered with a membrane of ferrous phosphate in the form of a blister.   The presence of this membrane prevented the inhibitor from reaching the center of corrosion, and severe damage to the specimen continued unchecked.   By adding some $K_2Cr_2O_7$ to the phosphate solution to oxidize the membraneous ferrous phosphate to the crystalline ferric salt, the intense localized corrosion was prevented, and even with insufficient inhibitor present, only mild general attack prevailed.

Other inhibitors which require the presence of dissolved O_2 for protection of the sample are Na_2CO_3, sodium acetate, and sodium benzoate (74).   In these cases, the protective film is γ-Fe_2O_3 (127) produced by the O_2 dissolved in the corrosive medium, and the presence of the inhibitor keeps the protective film in repair.   Uhlig and co-workers (146) found that in the presence of O_2, polyphosphates inhibit corrosion, particularly if Ca^{2+} ion is present.   If O_2 is not present, corrosion is accelerated by the addition of the inhibitor.

When iron is partly immersed in a NaCl solution containing phosphate or carbonate, intense corrosion may occur at the water line under certain conditions (116,117).   Peers and Evans (117) found that the attack was affected by the shape of the meniscus at the water line.   The shape of the meniscus was changed by varying the degree of slant of a partially immersed plate in the solution.   They concluded that the meniscus acted like a protected niche where diffusion of the inhibitor to these surface sites was slower than to other sites.   As before, the combination of large cathode and small anode produces intense corrosion at the water line.   Such attack may be considered as a form of crevice corrosion and may be an important consideration for half-filled boilers which remain in disuse for long periods of time.

Two important corrosion inhibitors, chromate and nitrite, are very effective in neutral and alkaline solutions without depending on the presence of dissolved oxygen (127).   In the case of Na_2CrO_4, the protective film on iron is a thin layer of γ-Fe_2O_3 (20,29,80,85,89,105). Hoar and Evans (80) traced the progress of the protection of an iron plate in a KCl solution as the chromate ion was increased.   With little ($<0.001M$) or no CrO_4^{2-} present, general corrosion took place, producing flocculant rust, since the oxide was formed remote from the anodic dissolution sites.   For small concentrations of CrO_4^{2-} ($0.01M$), much of the surface is protected, but pitting occurs at the small anodic areas.   As the CrO_4^{2-} ion concentration is increased, interference tints are observed at first, but the film becomes thinner with increasing chromate concentration. Finally, at high concentrations ($1M$), the film is invisible, and complete protection of the iron is maintained. Chromic oxide was detected in the passivating film.

Since $CrO_4{}^{2-}$ ion does not require the presence of O_2 for the formation of a protective film, it is tempting to assume that the reason for the inhibiting properties of $CrO_4{}^{2-}$ may be related to its high oxidizing power. But this cannot be the complete answer, since $KMnO_4$, a very good oxidizing agent, is useless as a corrosion inhibitor (80). According to Hoar and Evans (80), the chromate oxidizes all of the ferrous ions produced by the corrosion process to form a film of cubic oxide at the anodic sites. Any discontinuities in the oxide film are plugged up by a mixture of iron and chromic oxides. A modification of this viewpoint proposed by Mayne and Pryor (105) suggests that the initial film is formed by the direct oxidation of the iron surface sites by adsorbed chromate ions. The film thickens by a crack-and-heal mechanism (49) until an impermeable barrier to the diffusion of iron ions is formed. Although they (105) could not detect any Cr in the γ-Fe_2O_3 film, Cohen and Beck (29) found the composition of the passive film on Fe in $CrO_4{}^{2-}$ solutions to be about 75% γ-Fe_2O_3 and 25% Cr_2O_3 from tracer studies using ^{51}Cr. Similar results were found by Brasher and Stove (20). Powers and Hackerman (123) reported that freshly abraded steel took up Cr from a solution containing ^{51}Cr as chromate, and this adsorbed chromate could not be removed by washing alone. According to Uhlig (144,145), the primary film responsible for the passivation of Fe in solutions containing chromate is an adsorbed layer of $CrO_4{}^{2-}$ ions, although with time an oxide film of γ-Fe_2O_3 is formed. The layer of adsorbed chromate satisfies the valence forces of the metal surface atoms by chemical bonding. In any event, the protective film finally obtained on an iron surface in chromate solutions is mainly γ-Fe_2O_3 containing some Cr_2O_3.

A protective film composed chiefly of γ-Fe_2O_3 is formed on an iron surface in solutions containing $NaNO_2$ (28,30,109,130). Although O_2 is not required for complete protection, the amount of $NaNO_2$ required for complete protection decreases as the partial pressure of O_2 increases (30). Hydrated ferric oxide, γ-$FeOOH$, has also been detected by Cohen and co-workers (28,109) by electron diffraction analysis in films stripped from iron which was passivated in solutions containing $NaNO_2$ (particularly if Cl^- ion was present). At weak places in the film, Fe^{2+} ions diffuse through the film to react homogeneously to form plugs of γ-$FeOOH$. In solutions with a low concentration of nitrite ($\sim$100 ppm), the film formed on the Fe surface is Fe_3O_4 (109).

Pryor and Cohen (127) studied the rate of corrosion by weight-loss determinations of iron in water solutions of the sodium salts of carbonate, hydroxide, dibasic and tribasic phosphate, acetate, benzoate, nitrite, chromate, tungstate, molybdate, and silicate as a function of the salt concentration. The results in the presence of air are shown in Fig. 9.6. Only Na_2CrO_4 and $NaNO_2$ form a protective film in the absence of O_2, and at no concentration is the corrosion rate greater than the rate in the absence of the inhibitor. However, for those salts such as Na_2HPO_4, NaAc, or Na_2CO_3, the rate of corrosion increases with increasing concentration of inhibitor until enough is present to produce a complete protective film after which the corrosion rate falls to zero. This points up the danger of adding an insufficient amount of inhibitor to a given system. Using sodium benzoate tagged with ^{14}C, Brasher and Stove (20) were able to detect the presence of small amounts of benzoate in the protective film formed on Fe in solutions of sodium benzoate. As reported by Cartledge (24), the pertechnetate ion (TcO_4^-) is an effective corrosion inhibitor for iron or steel requiring as little as three orders of magnitude less than the amount required for CrO_4^{2-} ions to give the same protection.

Evans (ref. 40, p. 154) suggests that those inhibitors which contain anionic oxygen in combination with a cationic radical, such as chromyl, benzoyl, acetyl, etc., may form an oxide film instead of a film of some insoluble compound because they interact with the surface by splitting off a CrO_2^{2+} ion from the adsorbed CrO_4^{2-} ion, leaving two O^{2-}ions on the metal surface to form the film of metal oxide. The CrO_2^{2+} ion can react with water to form the corresponding acid which is neutralized by hydroxyls from the reduction of oxygen. If O_2 is not present, Cr_2O_3 is formed. It has been observed (80,105) that as the O_2 content increases, the amount of Cr in the passive film decreases.

At an early date, Friend and co-workers (62,63) reported that the presence of colloids such as 0.1% solutions of agar or gelatin inhibited the corrosion of Fe, Zn, and Pb. The inhibiting effect of these addition agents is attributed (81,99) to the formation of an adsorbed layer on the metal surface which interferes with the operation of the local cell such as blocking of the passage of ions into solution or raising the hydrogen overvoltage to high values on metals in acid solutions.

Hackerman and co-workers (69,72,73) studied the effectiveness of polar organic molecules, such as stearate, as corrosion inhibitors in acid solutions. Long chain ($>C_{10}$) acids, alcohols, esters, and amines gave protection by forming adsorbed layers. It was suggested by Hackerman and Makrides (70,71) that the adsorbed organic molecules donated

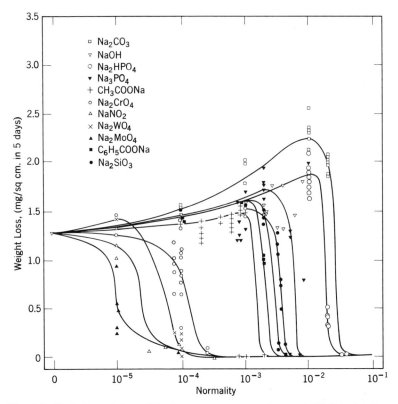

Fig. 9.6. Variation of the weight loss per square centimeter in 5 days for iron as a function of the concentration of 11 inhibitors in the presence of air (127).   (By permission of The Electrochemical Society.)

electrons to the metal d band and the effectiveness of the adsorbed layer was traced to the difficulty of transferring a positive ion from the metal to the solution through the layer of positively charged adsorbed organic ions.   If this mechanism is operative, a given polar organic should protect Fe with holes in the d band, but not Zn without d band holes. From data obtained by King and Hillner (90) for weight-loss determinations of the corrosion of Fe and Zn in $0.02M$ HCl containing sodium gluconate, ethylene diaminetetraacetic acid, or citric acid, they (70,71) calculated that the weight loss was over an order of magnitude greater for Zn than for Fe.

The use of chelating agents to form a more insoluble film on the metal surface was studied by King and co-workers (90,91).

B. BIMETALLIC CORROSION PROCESSES

Wet corrosion processes take place by local cell mechanisms in which certain sites on a metal surface act as cathodes, while others act as anodes where metal dissolution occurs. If a second metal, on which the cathodic reaction can occur more readily, is connected to the first electrically, the rate of corrosion of the first metal may be greatly increased. However, when the cathodic reaction is more inhibited on the second than on the first metal, the first metal becomes protected and the second metal is attacked. Such processes are known as bimetallic corrosion (51,129), where one metal in electrical contact with another is anodic to it. This type of corrosion is important in considerations of design and construction of marine (34,94), electrical (101), plumbing (87), and aircraft (139) equipment.

The greater the difference in potential between the anodic reaction on the one metal and the cathodic reaction on the other, the larger is the cell or corrosion current. In general, the greater the ratio of the electrode area of the cathodic reaction to the electrode area of the anodic reaction, the more intense is the corrosion of the metal anode. It is tempting to refer to the electromotive series or a table of the standard electrode potentials (33) to predict, by their relative positions in the standard table, how severe the bimetallic corrosion should be between two metals in electrical contact in a given electrolyte. The table of standard electrode potentials lists the equilibrium potentials for a given electrode reaction for which the reactants and products are in their standard states, but in a corrosion cell, the system is not at equilibrium, nor are the reactants or products usually in their standard states. The potential is related to the concentration of the reactants and products by the Nernst equation. By merely varying the concentrations of a given system, the positions of a pair of metals may be reversed with respect to potential in the standard table. Since current is drawn in an actual corrosion cell, the anodic and cathodic processes are polarized because they are irreversible. These polarized potentials may be very different from the equilibrium values quoted in the standard table (33). Consequently, predictions of corrosion behavior of a metallic couple obtained from tables of potentials are meaningless (ref. 40, p. 189). Polarization studies (see next section) provide the most reliable information.

It is known that Zn is anodic to iron or steel at ordinary temperatures, and when in electrical contact with steel in a corrosive medium, the Zn will dissolve as a sacrificial anode, giving protection to the iron. As a result, iron and steel were zinc plated so that the steel at exposed places would be protected by the local cell action.   Such material is called galvanized iron or steel (84).   The duration of protection depends on the thickness of the Zn plate.   Cadmium plating has also been used (139).

In hot water systems, incidents have been reported (84,88) where galvanized tanks have failed prematurely by severe pitting.   Kenworthy and Smith (88) found that galvanized samples that for months did not fail in cold water failed within days in hot water.   At high temperatures, Zn becomes cathodic to iron, and the phenomenon is known as temperature inversion of bimetallic corrosion (67,84,88). According to Gilbert (67) the anodic product of Zn in cold water is $Zn(OH)_2$ or a basic salt, but in hot water it is ZnO.   The enoblement of Zn in hot water is attributed to the small electronic conductivity of ZnO films.   In hot water, the large Zn cathode supports a large corrosion current concentrated at small exposed places on the steel.   Highly destructive pitting of the steel occurs.

Under certain circumstances, one metal is protected by being in contact with another, but the corrosion products interfere with the O_2 supply needed to maintain the film protecting the first metal (ref. 40, p. 193).   Although a metal may be protected at first, it later begins to corrode because of the interference of the corrosion products.

Bimetallic corrosion may take place even though the metals are not in electrical contact.   Galvanized tanks have failed because in the plumbing system copper pipes were used.   Minute amounts of Cu dissolved in the water and were deposited on the Zn plate.   The resultant local cell caused premature failure of the tank (87,122).

When bimetallic contacts must be made, either insulation must be provided or corrosion inhibiting paints must be used.   One of the most widely used primers is zinc yellow, which is a complex potassium zinc chromate (31,32,139).   Cole (31) studied a number of complex chromes as corrosion-inhibiting primers for Mg and Al alloys.   Apparently, strontium chromate is the best all around inhibitor, but alkali zinc chromates are good.

C. ANODIC CORROSION PROCESSES

Ordinarily, the rate of corrosion is increased on a given metal corroding by a differential aeration process by contacting it with another more noble metal so that it corrodes by a bimetallic process.   The corrosion rate can be increased further by applying an external source of potential to the system so that the given metal is anodized.   Differential aeration processes are oxygen-diffusion controlled, and bimetallic corrosion processes are limited by the difference in the electrode potentials of the two metals.   With anodic corrosion, these limits do not necessarily exist, since the driving force of the reaction is controlled by the external source of potential.   In many solutions, most metals become passive eventually, so unlimited corrosion rates are not possible.

If a sample of iron is polarized anodically against a suitable indifferent counterelectrode in a KCl solution, the iron goes into solution, while the cathodic reaction at least at low current densities is the reduction of O_2.   At high current densities, the O_2 reduction process may not be able to support the demanded current, and then, evolution of H_2 from water takes place.   In either case, the cathodic reaction adds OH^- ions to the system raising the pH.   As the pH increases, eventually a point is reached when iron ions in solution are precipitated out as hydrated ferric oxide (FeOOH, loose rust) at points remote from the anode. Such precipitates are not protective, and the iron dissolves continuously. If the oxygen supply to the system is restricted, the loose precipitate may be Fe_3O_4 or a complex green hydroxide (60).

When a dilute solution of H_2SO_4 is the corrosive medium, the pH is slow to shift because such a solution has a certain amount of buffering action.   Consequently, the pH does not reach a point where FeOOH or Fe_3O_4 separates out, and the solution remains clear.   As the process continues, however, the solution in the vicinity of the anode may become supersaturated with the metallic sulfate, and a hydrated salt layer is formed on the metal surface, as suggested by Turner (143) in the case of Ni in H_2SO_4 solutions.   Such a salt layer may interfere with the metal dissolution process causing passivation of the metal surface as proposed by Müller (110), or the layer can concentrate the current at uncovered sites so that these sites are polarized to potentials where oxide formation may occur at this high current density (143).   Piontelli and Serravalle (121) and Kolotyrkin (92) point out that such salt layers are not completely protective.   Passivity comes about by the formation of a film of a metal salt or a metal oxide (65), and the current goes into the evolution of oxygen.

Since film formation is a function of potential, much more information about the stability of a metal in a given electrolyte can be obtained from potentiostatic polarization studies than from galvanostatic ones.   An idealized polarization curve at controlled potential is shown in Fig. 9.7. A curve similar to this has been obtained (83) on Pd in H_2SO_4 solutions (see Fig. 3.9).

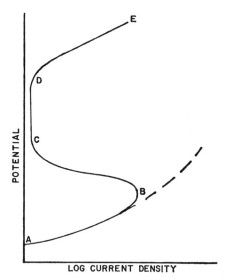

Fig. 9.7. A sketch of the anodic polarization curve obtained potentiostatically on an active metal in a given electrolyte.   See text for details.

At potentials below A in Fig. 9.7, the metal is inert in the given electrolyte.   Above A, the metal goes into solution, and between A and B, the rate of dissolution (current density) increases with increasing potential.   If a passivating film is not formed, the metal continues to dissolve with an ever increasing rate along the dashed line as the potential is raised.   In the usual case, a film of oxide begins to form at B where the rate of corrosion is maximum.   As the potential is shifted to more noble values, the fraction of the surface covered by the film increases, leaving less sites for anodic dissolution of the metal.   Consequently, the current density falls to smaller values until the point C is reached where the surface is completely covered and the metal becomes passive.   This point is known as the Flade potential (148).

In the region of potentials between C and B , the oxide film may be unstable under certain circumstances, and periodic transients or oscillations of the potential with time may occur.   Bartlett and Stephenson (6,7) have reported such transients on iron in sulfuric acid solutions.   Between C and D, a stable protective film exists on the metal surface, and the current density falls to very low values.   Finally, a potential is reached at D where another electrode reaction, the evolution of oxygen on the oxide film, can take place, and the current once again increases with potential between D and E.   The magnitude of the current flowing in the potential region CD will depend on how protective the film is. For a completely passive surface, the current is zero. Such a polarization curve as that in Fig. 9.7 predicts those experimental conditions where and to what extent a given metal is stable in contact with a given solution.

The polarization curve for Fe in $1N$ H_2SO_4 solution has been recorded by Prozak and co-workers (124). The point A is close to the Fe/Fe^{2+} potential,

$$Fe^{2+} + 2e \rightleftharpoons Fe \qquad E_0 = -0.440 \text{ V} \qquad (9.1)$$

Point C appears at about 0.58 V (vs. Pt/H_2 in same solution), and the corrosion current between C and D is virtually zero. Oxygen evolution does not take place until a potential of about 1.8 V is reached.

Steady-state O_2 overvoltage measurements have been carried out in alkaline (75,142,149) and acid (149) solutions. A Tafel slope of about 0.05 is obtained in alkaline solutions of about pH 14, and a mechanism in which the so-called electrochemical step, Eq. 9.3, is rate determining is favored according to the following scheme.

$$OH^- \rightarrow (OH)_{ads} + e \qquad (9.2)$$

$$(OH)_{ads} + OH^- \rightarrow (O)_{ads} + H_2O + e \qquad (9.3)$$

$$2(O)_{ads} \rightarrow O_2 \qquad (9.4)$$

These reactions take place on the thin electronically conducting (151) film of $\gamma\text{-}Fe_2O_3$, which Bonhoeffer (18) calculated to be about a monolayer thick from passivation studies of iron in nitric acid solutions. According to Weil (151) the thickness of the film is a function of the potential. Evans (47) has stripped these invisible $\gamma\text{-}Fe_2O_3$ films from passivated iron surfaces, and Kruger and Calver (92a) have studied the anodic formation of the oxide films on iron ellipsometrically. In acid solutions, the Tafel slope is 0.03, and the combination step of adsorbed OH radicals is preferred as the rate-determining step by Wade and Hackerman (149). For this mechanism, Eq. 9.2 is followed by

$$2(OH)_{ads} \rightarrow (O)_{ads} + H_2O \qquad (9.5)$$

followed in turn by Eq. 9.4.

Frumkin and co-workers (86) studied the constant current charging curves on iron in KOH solution. The curves exhibit two plateaus. The one at the lower potentials is associated with the formation of a layer of $Fe(OH)_2$. At higher potentials, the $Fe(OH)_2$ is oxidized to FeOOH which provides the passivating film.

When the polarizing circuit of an iron electrode, anodized for a given time in the passive region of potentials (above C in Fig. 9.7), is broken, the potential falls rapidly to a plateau.   This behavior was first reported by Flade (61), and the potential at which the plateau occurs is called (148) the Flade potential.   The Flade potential on iron is 0.58 V (151) and is rather reproducible since about the same value is obtained whether the point C is approached by increased anodization of an active sample or by following the potential decay of the open-circuit potential of a passivated sample.   After the plateau, the potential drops rapidly to the active region.   From this viewpoint, the Flade potential is that value of potential below which a passive metal becomes active.

The Flade potential on iron has questionable importance since it does not correspond to any known iron oxide equilibrium potential (148).

$$FeO + 2H^+ + 2e \rightleftharpoons Fe + H_2O \qquad E_0 = -0.060 \text{ V} \qquad (9.6)$$

$$Fe_3O_4 + 8H^+ + 8e \rightleftharpoons 3Fe + 4H_2O \qquad E_0 = -0.082 \text{ V} \qquad (9.7)$$

$$Fe_2O_3 + 6H^+ + 6e \rightleftharpoons 2Fe + 3H_2O \qquad E_0 = -0.040 \text{ V} \qquad (9.8)$$

$$Fe(OH)_2 + 2e \rightleftharpoons Fe + 2OH^- \qquad E_0 = -0.877 \text{ V} \qquad (9.9)$$

$$Fe(OH)_3 + e \rightleftharpoons Fe(OH)_2 + OH^- \quad E_0 = -0.56 \text{ V} \qquad (9.10)$$

Probably, a large overvoltage is present in the formation or dissolution of the passive film which could produce the 0.6 V shift from the Fe/Fe_2O_3 (Eq. 9.8) equilibrium potential (ref. 153, p. 238).

For more detailed discussions of the nature and theory of passivation mechanisms, the reader is referred to the International Colloquium on the Passivity of Metals of 1957 (58).

In recent years, the importance of microbial corrosion has been recognized, and a survey of the work done in this field has been given by Foley (61a).

D. ACCELERATED CORROSION TESTS

In commercial applications, such as the durability of metal finishes for jewelry or decorative trim for automobiles, it is important to know how well the particular finish will resist atmospheric corrosion.   If a given finish is required to be corrosion resistant for five years, the only way to determine this is to wait five years, but such a procedure is not economically practical.   As a result, the need for accelerated corrosion tests which faithfully reproduce the long-term-in-service attack in the laboratory in a matter of hours or minutes is essential.   A review of the history of plating in the automobile industry is given by Phillips (120).

The first acceptable accelerated corrosion test was introduced by Capp (23) as the neutral salt spray test, which is now identified as ASTM test B117. In this test, the sample was exposed to a spray of brine. Nixon (111), in 1945, proposed a modification of B117 known as the acidified salt spray (ASTM test B287) in which acetic acid was added to the spray. From these tests the so-called CASS (Copper chloride modified Acetic acid Salt Spray) test was developed (112) and is described as ASTM test B368 in which copper chloride was added to the acidified salt spray. At about the same time, the Corrodkote test (ASTM test B 380) was announced (15). With this test, the part to be tested was pasted with a slurry of copper nitrate, ferric chloride, ammonium chloride, and kaolin in water and placed in a humidity box for a given time. In England, a sulfur dioxide test (39) is used in which the part is exposed to a salt spray in a SO_2 atmosphere.

For these tests, an exposure of a part to the spray or the paste for 16–24 hr is equivalent to an exposure to the atmosphere in service for one year in Detroit. It is important to specify the locality where the part will be placed in service because the composition of the atmosphere changes radically from place to place (from the countryside to the seashore to an industrial center). With the advent of newer and more rapid methods of electrodepositing decorative metal finishes, an even more rapid corrosion test was needed to continue the improvement of the durability of exterior automotive trim.

The exterior automotive trim consists, in general, of consecutive deposits of copper, nickel, and finally chromium. Under ordinary circumstances, the chromium is passive with respect to the underlying nickel, and corrosion of the nickel occurs by the exposure of the Ni through pores and cracks in the Cr plate. The cathodic reaction (reduction of O_2) occurs on the passive Cr surface, and the corrosion process is under cathodic control, although Saur (132) has shown it to be under anodic control initially.

When only a few cracks or pores exist in the layer of Cr, the anodic current is concentrated at these points, producing intense attack. Within a relatively short time, the corrosion process can penetrate the Ni layer to the steel with the result that unsightly rust spots appear on the metal surface. If the anodic area is increased by producing more cracks in the Cr layer, the intensity of attack will be reduced. Although the amount of Ni corrosion is greater, it is spread over a greater area so that penetration to the basis metal is delayed, thus increasing the durability. Lovell and co-workers (96) have discussed the properties of such a Cr layer which is called microcracked chromium (37).

Since the corrosion rate is limited by the diffusion of O_2 to the cathodic local sites, the tests based on this process are slow. The rate of corrosion can be increased by the application of an external source of potential between the test part and a suitable counterelectrode. In this way, cathodic control is eliminated, and much higher corrosion rates are obtained. Regulation of the rate is carried out by controlling the anodic potential with respect to a suitable reference electrode. The use of the potentiostat has been reported (100,108,119) for studying the corrosion behavior of various metals.

A potentiostatically controlled anodic corrosion test known as the EC (electrolytic corrosion) test has been developed by Saur and Basco (134). In this test, the chromium plated part is immersed in a solution of NaCl, $NaNO_3$, and HNO_3 to keep the Ni and Cu corrosion products in solution and is anodized at 0.3 V vs. SCE using an inert cathode (platinized tantalum). An indicator, 1,10-phenanthroline hydrochloride, was added to give a red coloration at the pit sites when Fe^{2+} ions from the basis metal corrosion first appeared. Since the pH rises because of the cathodic reaction (H_2 evolution), the solution is discarded after the pH reaches a value of 2 to prevent the precipitation of Cu and Ni hydroxides.

To maintain the corrosion rate as a faithful reproduction of the in-service exposure, the current was not allowed (134) to exceed 3.3 mA/cm^2 (135). For several chromium plates, the EC test gave a faithful reproduction of the in-service atmospheric corrosion (133). Two minutes in the EC test correspond to one year exposure in Detroit. For example, to obtain the same bright Ni corrosion of a chromium plated part that required 23 hr in the CASS test, only 2.7 min was required in the EC test.

References

1. Aziz, P. M., *Corrosion*, **9**, 85 (1953).
2. Aziz, P. M., *J. Electrochem. Soc.*, **101**, 120 (1954).
3. Aziz, P. M., and H. P. Goddard, *Ind. Eng. Chem.*, **44**, 1791 (1952).
4. Barnard, K. N., *Corrosion*, **7**, 114 (1951).
5. Barnard, K. N., and G. L. Christie, *Corrosion*, **6**, 234 (1950).
6. Bartlett, J. H., *Trans. Electrochem. Soc.*, **987**, 521 (1945).
7. Bartlett, J. H., and L. Stephenson, *J. Electrochem. Soc.*, **99**, 504 (1952).
8. Bengough, G. D., and O. F. Hudson, *J. Inst. Metals*, **21**, 37 (1919).
9. Bengough, G. D., and A. E. Lee, *J. Iron Steel Inst.*, **125**, 285 (1932).
10. Bengough, G. D., A. R. Lee, and F. Wormwell, *Proc. Roy. Soc. (London)*, **A131**, 494 (1931); **A134**, 308 (1931).
11. Bengough, G. D., J. M. Stuart, and A. R. Lee, *Proc. Roy. Soc. (London)*, **A116**, 425 (1927); **A121**, 88 (1928); **A127**, 42 (1931).

12. Bengough, G. D., and F. Wormwell, *Proc. Roy. Soc.* (*London*), **A140**, 399 (1933).
13. Bianchi, G., *Corrosion Anti-Corrosion*, **5**, 146 (1957).
14. Bianchi, G., *Corrosion*, **14**, 245t (1958).
15. Bigge, D. M., *Ann. Tech. Proc. Am. Electroplaters' Soc.*, **46**, 149 (1959).
16. Blaha, F., *Nature*, **166**, 607 (1960).
17. Blaha, F., *Monatsh.*, **82**, 170 (1951).
18. Bonhoeffer, K. F., *Z. Elektrochem.*, **47**, 147 (1941).
19. Borgmann, C. W., and U. R. Evans, *Trans. Electrochem. Soc.*, **65**, 249 (1934).
20. Brasher, D. M., and E. R. Stove, *Chem. Ind.* (*London*), **1952**, 171.
21. Britton, S. C., and U. R. Evans, *Trans. Electrochem. Soc.*, **61**, 441 (1932).
22. Brown, R. H., and R. B. Mears, *Trans. Electrochem. Soc.*, **74**, 495 (1938).
23. Capp, J. A., *Proc. ASTM*, **14**, Part II, 474 (1914).
24. Cartledge, G. H., *J. Phys. Chem.*, **59**, 979 (1955); **60**, 28, 32, 1037 (1956); **61**, 973 (1957).
25. Chaudron, G., *Helv. Chim. Acta*, **31**, 1553 (1948).
26. Chaudron, G., P. Lascombe, and N. Yannaquis, *Rev. Met.*, **45**, 68 (1948).
27. Churchill, J. R., *Trans. Electrochem. Soc.*, **76**, 341 (1939).
28. Cohen, M., *Trans. Electrochem. Soc.*, **93**, 26 (1948); *J. Phys. Chem.*, **56**, 451 (1952).
29. Cohen, M., and A. F. Beck, *Z. Elektrochem.*, **62**, 696 (1958).
30. Cohen, M., R. Pyke, and P. Marier, *Trans. Electrochem. Soc.*, **96**, 254 (1949).
31. Cole, H. G., *J. Appl. Chem.* (*London*), **5**, 197 (1955).
32. Cole, H. G., and L. F. Le Brocq, *J. Appl. Chem.* (*London*), **5**, 149 (1955).
33. de Bethune, A. J., and N. A. Loud, *Standard Aqueous Electrode Potentials and Temperature Coefficients*, Clifford A. Hampel, Skokie, Ill., 1964.
34. Dorey, S. F., *J. Inst. Metals*, **82**, 497 (1953).
35. Downing, A. L., and G. A. Truesdale, *J. Appl. Chem.* (*London*), **5**, 570 (1955).
36. Downing, A. L., G. A. Truesdale, and G. F. Lowden, *J. Appl. Chem.* (*London*), **5**, 53 (1955).
37. Durham, R. P., U.S. Pat. 3,157,585, Nov. 1964.
38. Edeleanu, C., and U. R. Evans, *Trans. Faraday Soc.*, **47**, 1121 (1951).
39. Edwards, J., *Ann. Tech. Proc. Am. Electroplaters' Soc.*, **46**, 154 (1959).
40. Evans, U. R., *The Corrosion and Oxidation of Metals*, St. Martin's Press, New York, 1960.
41. Evans, U. R., *J. Inst. Metals*, **30**, 239 (1923).
42. Evans, U. R., *Ind. Eng. Chem.*, **17**, 363 (1925).
43. Evans, U. R., *Corrosion*, **7**, 238 (1951).
44. Evans, U. R., *J. Chem. Soc.*, **131**, 92 (1929).
45. Evans, U. R., *J. Chem. Soc.*, **131**, 92, 111 (1929); *Trans. Am. Electrochem. Soc.*, **57**, 407 (1930).
46. Evans, U. R., *J. Chem. Soc.*, **132**, 478 (1930).
47. Evans, U. R., *Nature*, **126**, 130 (1930).
48. Evans, U. R., *Trans. Electrochem. Soc.*, **69**, 213 (1936).
49. Evans, U. R., *Nature*, **157**, 732 (1946).
50. Evans, U. R., *J. Chem. Soc.*, **129**, 1020 (1927).
51. Evans, U. R., L. C. Bannister, and S. C. Britton, *Proc. Roy. Soc.* (*London*), **A131**, 355 (1931).
52. Evans, U. R., and C. W. Borgmann, *Z. Physik. Chem.* (*Leipzig*), **A160**, 194 (1932); *Trans. Faraday Soc.*, **28**, 813 (1932).

53. Evans, U. R., and D. E. Davies, *J. Chem. Soc.*, **1951**, 2607.
54. Evans, U. R., and T. P. Hoar, *Proc. Roy. Soc. (London)*, **A137**, 343 (1932).
55. Evans, U. R., and R. B. Mears, *Proc. Roy. Soc. (London)*, **A146**, 153 (1934).
56. Evans, U. R., and M. T. Simnad, *Proc. Roy. Soc. (London)*, **A188**, 372 (1947).
57. Evans, U. R., and R. Tomlinson, *J. Appl. Chem. (London)*, **2**, 105 (1952).
58. Evans, U. R., et al., *Z. Elektrochem.*, **62**, 619 (1958).
59. Feitknecht, W., *Helv. Chim. Acta*, **32**, 2294 (1949).
60. Feitknecht, W., and G. Keller, *Z. Anorg. Allgem. Chem.*, **262**, 61 (1950).
61. Flade, F., *Z. Physik. Chem. (Leipzig)*, **76**, 513 (1911).
61a. Foley, R. T., *Electrochem. Tech.*, **5**, 72 (1967).
62. Friend, J. A. N., and J. S. Tidmus, *J. Inst. Metals*, **33**, 19 (1925).
63. Friend, J. A. N., and R. H. Vallance, *J. Chem. Soc.*, **121**, 466 (1922).
64. Gameson, A. L. H., and K. G. Robertson, *J. Appl. Chem. (London)*, **5**, 502 (1955).
65. Gerischer, H., *Angew. Chem.*, **70**, 285 (1958).
66. Gessow, I. D., *Corrosion*, **12**, 100t (1956).
67. Gilbert, P. T., *J. Electrochem. Soc.*, **99**, 16 (1952).
68. Gilbert, P. T., and S. E. Hadden, *J. Inst. Metals*, **77**, 237 (1950).
69. Hackerman, N., and E. L. Cook, *J. Electrochem. Soc.*, **97**, 1 (1950).
70. Hackerman, N., and A. C. Makrides, *Ind. Eng. Chem.*, **46**, 523 (1954).
71. Hackerman, N., and A. C. Makrides, *J. Phys. Chem. (Leipzig)*, **59**, 707 (1955).
72. Hackerman, N., and H. R. Schmidt, *Corrosion*, **5**, 237 (1949).
73. Hackerman, N., and J. D. Sudbury, *J. Electrochem. Soc.*, **97**, 109 (1950).
74. Hancock, P., and J. E. O. Mayne, *J. Appl. Chem. (London)*, **9**, 345 (1949).
75. Hicking, A., and S. Hill, *Discussions Faraday Soc.*, **1**, 236 (1947).
76. Hoar, T. P., *Proc. Roy. Soc. (London)*, **A142**, 628 (1933).
77. Hoar, T. P., *Trans. Faraday Soc.*, **33**, 1152 (1937).
78. Hoar, T. P., *J. Electrodepositors Tech. Soc.*, **14**, 33 (1938).
79. Hoar, T. P., *Trans. Faraday Soc.*, **45**, 683 (1949).
80. Hoar, T. P., and R. U. Evans, *J. Chem. Soc.*, **134**, 2476 (1932).
81. Hoar, T. P., and R. D. Holliday, *J. Appl. Chem. (London)*, **3**, 502 (1953).
82. Hoare, J. P., *J. Electrochem. Soc.*, **112**, 602, 608, 849 (1965).
83. Hoare, J. P., *J. Electrochem. Soc.*, **112**, 1129 (1965).
84. Hoxeng, R. B., and C. F. Prutton, *Corrosion*, **5**, 330 (1949).
85. Iitaka, I., S. Miyake, and T. Iimori, *Nature*, **139**, 156 (1937).
86. Kabanov, B., R. Burshtein, and A. N. Frumkin, *Discussions Faraday Soc.*, **1**, 259 (1947).
87. Kenworth, L., *J. Inst. Metals*, **69**, 67 (1943).
88. Kenworthy, L., and M. D. Smith, *J. Inst. Metals*, **70**, 463 (1944).
89. King, C. V., E. Goldschmidt, and N. Mayer, *J. Electrochem. Soc.*, **99**, 423 (1952).
90. King, C. V., and E. Hillner, *J. Electrochem. Soc.*, **101**, 79 (1954).
91. King, C. V., and E. Rau, *J. Electrochem. Soc.*, **103**, 331 (1956).
92. Kolotyrkin, Ya. M., *Z. Elektrochem.*, **62**, 664 (1958).
92a. Kruger, J., and J. P. Calvert, *J. Electrochem. Soc.*, **114**, 43 (1967); **110**, 654 (1963).
93. Lascombe, P., and N. Yannaquis, *Compt. Rend.*, **224**, 921 (1947); **226**, 498 (1948).
94. Lees, D. C. G., *Chem. Ind. (London)*, **1954**, 949.

95. Livingston, J., R. Morgan, and A. H. Richardson, *J. Phys. Chem.*, **34**, 2356 (1930).
96. Lovell, W. E., E. H. Shotwell, and J. Boyd, *Ann. Tech. Proc. Am. Electroplaters' Soc.*, **47**, 215 (1960).
97. Lynes, W., *J. Electrochem. Soc.*, **103**, 467 (1956); **98**, 3C (1951).
98. McArthur, C. G., *J. Phys. Chem.*, **20**, 495 (1916).
99. Machu, W., *Trans. Electrochem. Soc.*, **72**, 333 (1937).
100. Makrides, A. C., *Corrosion*, **18**, 338t (1962).
101. March, E. C. J., *Electroplating*, **7**, 88 (1954).
102. Mayne, J. O. E., and J. W. Menter, *J. Chem. Soc.*, **1954**, 103.
103. Mayne, J. E. O., and J. W. Menter, *J. Chem. Soc.*, **1954**, 99.
104. Mayne, J. E. O., J. W. Menter, and M. J. Pryor, *J. Chem. Soc.*, **1950**, 3229.
105. Mayne, J. E. O., and M. J. Pryor, *J. Chem. Soc.*, **1949**, 1831.
106. Mears, R. B., and R. H. Brown, *Trans. Electrochem. Soc.*, **74**, 519 (1938); **81**, 455 (1942).
107. Mears, R. B., and U. R. Evans, *Trans. Faraday Soc.*, **30**, 417 (1934).
108. Melborne, S. H., and G. N. Flint, *Trans. Inst. Metal Finishing*, **39**, 85 (1962).
109. Mellors, G. W., M. Cohen, and A. F. Beck, *J. Electrochem. Soc.*, **105**, 332 (1958).
110. Müller, W. J., *Z. Elektrochem.*, **33**, 401 (1927); **36**, 365 (1930); *Monatsh. Chem.*, **48**, 61 (1927).
111. Nixon, C. F., *Monthly Review Am. Electroplaters' Soc.*, Nov. 1945.
112. Nixon, C. F., J. D. Thomas, and D. W. Hardesty, *Ann. Tech. Proc. Am. Electroplaters' Soc.* **46**, 159 (1959); **47**, 90 (1960).
113. Osborn, O., *Corrosion*, **7**, 2 (1951).
114. Osborn, O., and H. A. Robinson, *Corrosion*, **8**, 114 (1952).
115. Palmer, W. G., *J. Iron Steel Inst.*, **163**, 421 (1949); *Corrosion*, **7**, 10 (1951).
116. Peers, A. M., *Trans. Faraday Soc.*, **51**, 1748 (1955).
117. Peers, A. M., and U. R. Evans, *J. Chem. Soc.*, **1953**, 1093.
118. Perryman, E. C. W., and S. E. Hadden, *J. Inst. Metal*, **77**, 207 (1950).
119. Petrocelli, J. V., V. Hospadaruk, and G. A. DiBari, *Plating*, **49**, 50 (1962).
120. Phillips, W. M., *Metal Finishing*, **52**, 60 [7], 73 [8], 76 [9] (1954).
121. Piontelli, R., and G. Serravalle, *Z. Elektrochem.*, **62**, 759 (1958).
122. Porter, F. C., and S. E. Hadden, *J. Appl. Chem. (London)*, **3**, 385 (1953).
123. Powers, R., and N. Hackerman, *J. Electrochem. Soc.*, **100**, 314 (1953).
124. Prozak, M., V. Prozak, and Vl. Cihal, *Z. Elektrochem.*, **62**, 739 (1958).
125. Pryor, M. J., *J. Electrochem. Soc.*, **102**, 163 (1955).
126. Pryor, M. J., and M. Cohen, *J. Electrochem. Soc.*, **98**, 263 (1951).
127. Pryor, M. J., and M. Cohen, *J. Electrochem. Soc.*, **100**, 203 (1953).
128. Pryor, M. J., M. Cohen, and F. Brown, *J. Electrochem. Soc.*, **99**, 542 (1952).
129. Pryor, M. J., and D. S. Keir, *J. Electrochem. Soc.*, **104**, 269 (1957).
130. Pyke, R., and M. Cohen, *Trans. Electrochem. Soc.*, **93**, 63 (1948).
131. Roald, B., and M. A. Streicher, *J. Electrochem. Soc.*, **97**, 283 (1950).
132. Saur, R. L., *Plating*, **48**, 1310 (1961).
133. Saur, R. L., and R. P. Basco, *Plating*, **53**, 320, 981 (1966).
134. Saur, R. L., and R. P. Basco, *Plating*, **53**, 35 (1966).
135. Saur, R. L., *Plating*, **53**, 1124 (1966).
136. Schaschl, E., and G. A. Marsh, *Corrosion*, **13**, 243t (1957).
137. Simnad, M. T., and U. R. Evans, *J. Iron Steel Inst.*, **156**, 531 (1947).

138. Simnad, M. T., and U. R. Evans, *Trans. Faraday Soc.*, **46**, 175 (1950).

139. Simpson, N. H., *Corrosion*, **6**, 51 (1950).

140. Thornhill, R. S., and U. R. Evans, *J. Chem. Soc.*, **1938**, 614.

141. Trouard, S. E., *Corrosion*, **13**, 151t (1957).

142. Tsinman, A. I., *Zh. Fiz. Khim.*, **37**, 1598 (1963).

143. Turner, D. R., *J. Electrochem. Soc.*, **98**, 434 (1951).

144. Uhlig, H. H. *Chem. Eng. News*, **24**, 3154 (1946); *Metaux Corrosion*, **22**, 204 (1947).

145. Uhlig, H. H., and A. Geary, *J. Electrochem. Soc.*, **101**, 215 (1954).

146. Uhlig, H. H., D. N. Triadis, and M. Stern, *J. Electrochem. Soc.*, **102**, 59 (1955).

147. Vernon, W. H. J., *J. Chem. Soc.*, **128**, 2273 (1926); *Trans. Faraday Soc.*, **23**, 113 (1927).

148. Vetter, K. J., *Z. Elektrochem.*, **62**, 642 (1958).

149. Wade, W. H., and N. Hackerman, *Trans. Faraday Soc.*, **53**, 1636 (1957).

150. Wagner, C., and W. Traud, *Z. Elektrochem.*, **44**, 391 (1938).

151. Weil, K. G., *Z. Elektrochem.*, **62**, 638 (1958).

152. Whitwham, D., and U. R. Evans, *J. Iron Steel Inst.*, **165**, 72 (1950); *Corrosion*, **7**, 28 (1951).

153. Young, L., *Anodic Oxide Films*, Academic Press, New York, 1961, p. 228.

Author Index

Numbers in parentheses are references and indicate that the author's work is referred to although his name is not mentioned in the text. Numbers in *italics* show the pages on which complete references are listed.

A

Acheson, E. G., 314(1), *348*
Ackermann, P., 159(75), *206*
Adams, R., 17(248,1), 53, *73, 78, 79*
Adams, R. N., 21(157), 24(157), 42 (177), 48(177), *77*, 323, *348, 353*
Adlhart, O. J., 332(3), *348*
Afanas'ev, A. S., 236(1), *293*
Agar, J. N., 10(1), *10*, 310(31), *348*
Aggarwal, P. S., 274(2), *293*
Akamatu, H., 318(237), *354*
Akopyan, A. U., 117(1), 123, 129(1), 131(1), *140*, 161(1), 170(1), *204*, 236 (3), *293*
Aladjalova, N., 54(84), *75*, 173(78), *206*
Aladzhalova, N. A., 117(49), *142*, 312 (187a), *353*
Albery, W. J., 124(1a), *140*
Alder, B. J., 314, *348*
Allessandrini, E. I., 274(4), *293*
Alexanian, C., 320, *348*
Alford, H. R., 336, *353*
Allen, T. H., 192(136), *207*
Allmand, A. J., 159(2), 160(2,65) *204*, *205*
Altmann, S., 18(2), *73*, 173(3), *204*
Amlie, R. F., 219(90), 221–225(90), 229 (90), 233, 259(5,433), 260(5), 262(5, 433), 263(5,433), 271(433), *293, 295, 304*
Ammar, I. A., 284, 287, *293*
Anbar, M., 233(7), 234, 235, *293*
Anderson, F. C., 271(9), 289(9), *293*
Anderson, J. R., 242, *293*
Anderson, J. S., 245, 246, *293*
Andre, H. G., 212, *293*
Andreeva, E. P., 161(130), 162(130), *207*
Andreeva, V. A., *300*
Andrews, K. F., 274(221), *298*

Angerstein, H., 30(46), *74*
Angstadt, R. T., 244(12,434), 245(12), 246, 250(430,434), 251, 253(429), 255 (12,429), 256, 258, 259(430,434), 262 (430), 263(430,434), 264(430), 266 (12), 267(429), 268, 269(429), 271 (428,429,434), *293, 304*
Anson, F. C., 21(3), 39, 40, 43, *73*, 171, 172, 174(6,8), *204*
Antonova, L. G., 277(502), 284(502), *306*
Antony, A. P., 333(267), *355*
Antropov, L. I., 270(13), *293*
Appelt, K., 241(15), *293*
Appleby, M. P., 240(14,14a), 241(14a), 242, *293*
Archer, D. H., 342, *348*
Archibald, E. H., 173(249), *210*
Argersinger, W. J., 42(177), 48(177), *77*
Armstrong, G., 21(41), 23(41), 24(41), 41(41), 48(4), *73, 74*
Arnold, K., 284(497), 285, *305*
Aronowitz, G., 171(49b), *205*
Asbury, W. C., 273(106), *296*
Askill, S., 197(36), 198(36), *204*
Astakhov, I. I., 244, 258, 263, 267, 270 (17), *293*
Aston, J. G., 41(76), *75*
Audubert, R., 96, *112*
Austin, P. C., 215, *293*
Avseevich, G. P., *80*
Awad, S. A., 72(137), *76*
Aziz, P. M., 371, *387*
Azzam, A. M., 34(5), *73*

B

Baborovski, G., 215(19), *293*
Bacon, F. T., 308, 320(8), 332, *348*
Bacon, G. E., 314, *348*
Bagg, J., 212, 213, *293*

Bigelow, J. E., 271(40), *294*
Bigge, D. M., 386(15), *388*
Binnie, W. P., 248(41), *294*
Bircumshaw, L. L., 240, 241(410), *294*, *303*
Biscoe, J., 314, *348*
Bitton, H. T. S., 216, *294*
Blackburn, T. R., 53(17), *73*, 102(8), *112*
Blackman, J. G., 319, *348*
Blackman, L. C. F., 315, *348*
Blaedel, W. J., 193, *204*
Blaha, F., 363(17), 369, *388*
Blinks, L. R., 201, *204*
Bloch, O., 117(38), 119(38), *141*
Blocher, J. M., 214, *294*
Blondel, M., 18(18), *73*
Bloom, M., 243(423), *303*
Blum, S. L., 315, 324, *353*
Blumberg, P., 317(72), *350*
Bockris, J. O'M., 9(24), 10(3), *10, 11*, 31, 32, 34, 35(196), 37, 38, 41, 44, 45, 50, 51(196,197), 58(50), 65, *73, 74, 78*, 82, 83, 84(11), 85(11), 86, 87, 90, 91, 95(11), 96–98, 105(30b), 110(30b), 111, *112, 113, 115*, 117(10,14b), 119 (10), 121(9), 125, 128(9,14b), 130(59), 132(14d), 133(14a, 14d), *141,142*, 143 (3), 144(3,7,7a), 145, 147(2), 148–150, 151(3), *151*, 203(18), *209*, 281(41b), 282(41b), *303*
Bockros, J. C., 315(28), *348*
Bode, H., 244(46), 248(47), 250, 272 (45), 275, *294*
Boehm, H. P., 319(57), 320(57), *349*
Böld, W., 24(21), 28(21), 29, 30, 35(21), 41(21), 56, 57(22), 58(21), 59, 60(22), 61, *73*, 89(12), 109(13), 110, *112*, 132 (11), *141*, 172(24), 173(24), *204*
Bogatskii, D. P., 273, *294*
Boies, D. B., 338(78), *350*
Bond, G. C., 42(23), *73*
Bone, S. J., 244(51), 250, 253(51), *294*
Bonhoeffer, K. F., 384, *388*
Bonilla, C. F., 251(526), 255(526), 267, 276, *295, 306*
Bonk, J. F., 218, *294*
Bonnemay, M., 171(25), *204*, 227(53), 228(53), *294*, 310(29), 312(30), *348*

Bonoël, G., 227(53), 228(53), *294*
Bordonali, C., 190, *204*
Boreskov, G. K., 212(55), 213, *294*
Borgmann, C. W., 362, 363(52), *388*
Borisova, T. I., 171(26), *204*, 223, 252, 284(157), 287–289(157), *297, 306*
Borkowski, J., 191(137), *207*
Borneman, K., 64(24), *73*
Bortlisz, H., 196, *208*
Borucka, A., 310(31), *348*
Bose, E., 14, *73*
Bose, I., 199(62), *205*
Boswell, M. C., 228(58), 255(58), 273 (58), *294*
Bottomley, R. A., 185(143), *207*
Bottomley, G. A., 319, *348*
Bourgault, P. L., 84(29), *113*, 277(59), 281, 284(115), 289, *294, 296*
Boussingault, J., 240(60), *294*
Bowden, F. P., 21(26), 23, 34, 37, 41 (26), *73*, 82, 90, 96(16), 97, *112*, 145 (4), *151*, 172, *204*
Bowen, R. J., 65, *73*, 120(11a), *141*
Bowers, R. C., 183, *204*
Bowers, W. G., 330(129), *351*
Boyd, J., 386(96), *390*
Bozzan, T., 190, *204*
Brackett, F. S., 182(195), 185(195), 195 (195), *209*
Bradley, A. F., 197(231), 198(231), *210*
Bradshaw, B. C., 281, *294*
Braekken, H., 226(62), *294*
Bragg, W. H., 313, *349*
Bragg, W. L., 313, *349*
Brasher, D. M., 376(20), 377, 378, *388*
Bratzler, K., 318(34), 331(34), *349*
Braun, T., 292(355), *302*
Brauner, B., 213(63), *294*
Brcec, B. C., 271(64), *294*
Brdicka, R., 167(29), *204*
Bredig, G., 68(28), *73*
Bregman, J. I., 338(78), *350*
Breiter, M. W., 9(4), *10*, 21(32,79), 24 (21), 28(21), 29, 30, 35(21), 40, 41 (21), 43, 49(30), 51(30), 56, 57(22), 58 (21), 59, 60(22), 61, *73, 75*, 89, 92, 109(13), 110, 112, *112*, 117(13), 120, 123, 124(13), 125(13), 127, 128(12,14, 19), 132(11), *141*, 148(5), 149(5), *151,*

Subject Index

A

Adam's catalyst, 17
Adsorption of oxygen, on carbon, 318
  on gold, 49
  on nickel, 274
  on platinum, 23
  on silver, 212
Alpha PbO_2, 244
Anode, carbon, definition of, 1
  properties of, 321
  sacrificial, 361

B

Basic lead sulfates, 247
Battery, Edison, 271
  lead–acid, 238, 265
  nickel–cadmium, 272, 291
   active material on, 291
   effect of lithium on, 291
  paste, 265
  silver–zinc, 211, 232
Beta PbO_2, 244

C

Carbon, modifications of, 313
  porous, formation of, 315
  turbostratic, 314
Catalytic decomposition of peroxide, 68
Cathode, definition of, 1
Cathodic protection, 361
Cell, electrochemical, 3
  Grove, 14, 15
  Hersch, 190
  local, 8, 358
   on platinum, 33
  Mancy, 190
  oxygen, concentration, 341
  zinc–air, 330
Charging curves, on gold, 48
  on iridium, 59
  on iron, 384
  on lead, 251
  on mercury, 160
  on nickel, 277
  on osmium, 64
  on palladium, 53
  on platinum, 21

on rhodium, 56
on ruthenium, 64
on silver, 219
Chronopotentiometry, 155
Coefficient, transfer, 8
Conduction, electrolytic, 3
  electronic, 3
Corrosion, accelerated, CASS test for, 386
  Corrodkote test for, 386
  EC test for, 387
  anodic, 382
  under anodic control, 359
  bimetallic, 380
  under cathodic control, 359
  concentric rings of, 372
  crevice, 370
  under drops, 368
  fatigue, 374
  grid, 267
  inhibition of, by chromate, 376
   by nitrite, 377
   by organics, 378
   by phosphate, 375
  insufficient inhibitor, 375
  intergranular, 372
  of partially immersed plates, 362
  pitting, 369
  under resistance control, 359
  stress, 372
  temperature inversion, 381
  of totally submerged plates, 366
  water line, 376
Coulometric analytical techniques, 154
Current, local cell, 8, 364
  maximum, in polarograms, 162
  suppressor, 163

D

Differential aeration, 361

E

Electrical double layer, 2
Electrode, boron carbide, 348
  Clark, 183
   temperature compensation of, 184

419